Lecture Notes in Physics

Monographs

Springer
Berlin
Heidelberg
New York
Barcelona
Hong Kong
London
Milan
Paris
Singapore
Tokyo

The Editorial Policy for Monographs

The series Lecture Notes in Physics reports new developments in physical research and teaching - quickly, informally, and at a high level. The type of material considered for publication in the monograph Series includes monographs presenting original research or new angles in a classical field. The timeliness of a manuscript is more important than its form, which may be preliminary or tentative. Manuscripts should be reasonably self-contained. They will often present not only results of the author(s) but also related work by other people and will provide sufficient motivation, examples, and applications.
The manuscripts or a detailed description thereof should be submitted either to one of the series editors or to the managing editor. The proposal is then carefully refereed. A final decision concerning publication can often only be made on the basis of the complete manuscript, but otherwise the editors will try to make a preliminary decision as definite as they can on the basis of the available information.
Manuscripts should be no less than 100 and preferably no more than 400 pages in length. Final manuscripts should be in English. They should include a table of contents and an informative introduction accessible also to readers not particularly familiar with the topic treated. Authors are free to use the material in other publications. However, if extensive use is made elsewhere, the publisher should be informed. Authors receive jointly 30 complimentary copies of their book. They are entitled to purchase further copies of their book at a reduced rate. No reprints of individual contributions can be supplied. No royalty is paid on Lecture Notes in Physics volumes. Commitment to publish is made by letter of interest rather than by signing a formal contract. Springer-Verlag secures the copyright for each volume.

The Production Process

The books are hardbound, and quality paper appropriate to the needs of the author(s) is used. Publication time is about ten weeks. More than twenty years of experience guarantee authors the best possible service. To reach the goal of rapid publication at a low price the technique of photographic reproduction from a camera-ready manuscript was chosen. This process shifts the main responsibility for the technical quality considerably from the publisher to the author. We therefore urge all authors to observe very carefully our guidelines for the preparation of camera-ready manuscripts, which we will supply on request. This applies especially to the quality of figures and halftones submitted for publication. Figures should be submitted as originals or glossy prints, as very often Xerox copies are not suitable for reproduction. For the same reason, any writing within figures should not be smaller than 2.5 mm. It might be useful to look at some of the volumes already published or, especially if some atypical text is planned, to write to the Physics Editorial Department of Springer-Verlag direct. This avoids mistakes and time-consuming correspondence during the production period.
As a special service, we offer free of charge LaTeX and TeX macro packages to format the text according to Springer-Verlag's quality requirements. We strongly recommend authors to make use of this offer, as the result will be a book of considerably improved technical quality.
For further information please contact Springer-Verlag, Physics Editorial Department II, Tiergartenstrasse 17, D-69121 Heidelberg, Germany.

Series homepage – http://www.springer.de/phys/books/lnpm

Michel Hénon

Generating Families in the Restricted Three-Body Problem

II. Quantitative Study of Bifurcations

Springer

Author

Michel Hénon
CNRS
Observatoire de la Côte d'Azur
B.P. 4229
06304 Nice Cédex 4, France

Library of Congress Cataloging-in-Publication Data applied for.

Die Deutsche Bibliothek - CIP-Einheitsaufnahme
Hénon, Michel: Generating families in the restricted three-body problem / Michel Hénon. - Berlin ; Heidelberg ; New York ; Barcelona ; Hong Kong ; London ; Milan ; Paris ; Singapore ; Tokyo : Springer
(Physics and astronomy online library)
2. Quantitative study of bifurcations. - 2001
(Lecture notes in physics : N.s. M, Monographs ; 65)
ISBN 3-540-41733-8

ISSN 0940-7677 (Lecture Notes in Physics. Monographs)
ISBN 3-540-41733-8 Springer-Verlag Berlin Heidelberg New York

Springer-Verlag Berlin Heidelberg New York
a member of BertelsmannSpringer Science+Business Media GmbH

http://www.springer.de

Typesetting: Camera-ready copy from the author
Cover design: *design & production*, Heidelberg

Printed on acid-free paper
SPIN: 10644571 55/3141/du - 5 4 3 2 1 0

Preface

In a previous volume (Hénon 1997, hereafter called volume I), the study of generating families in the restricted three-body problem was initiated. (We recall that generating families are defined as the limits of families of periodic orbits for $\mu \to 0$.) The main problem was found to lie in the determination of the junctions between the branches at a bifurcation orbit, where two or more families of generating orbits intersect. A partial solution to this problem was given by the use of invariants: symmetries and sides of passage. Many simple bifurcations can be solved in this way. In particular, the evolution of the nine natural families of periodic orbits can be described almost completely. However, as the bifurcations become more complex, i.e. when the number of families passing through the bifurcation orbit increases, the method fails.

This volume describes another approach to the problem, consisting of a detailed, quantitative analysis of the families in the vicinity of a bifurcation orbit. This requires more work than the qualitative approach used in Vol. I. However, it has the advantage of allowing us, in principle at least, to determine in all cases how the branches are joined. In fact it gives more than that: we will see that, in almost all cases, the first-order asymptotic approximation of the families in the neighbourhood of the bifurcation can be derived. This allows, in particular, a quantitative comparison with numerically found families.

Chapter 11 deals with the relevant definitions and general equations. The quantitative study of bifurcations of type 1 is described in Chaps. 12–16. The analysis of type 2 is more involved; it is described in Chaps. 17–23. Type 3 is even more complex; its analysis had not yet been completed at the time of writing.

As was the case for the previous volume, this work is sometimes lacking in mathematical rigor; there is certainly much room for improvement. However, a number of factors lead me to believe that the results are correct: agreement with the results of the qualitative analysis of Vol. I; agreement with numerical computations; internal consistency; and, simple intuition.

My thanks go to Larry Perko, who read a draft version of this volume and made many helpful comments and suggestions. I also thank Alexander Bruno for many discussions by e-mail, and for sending an english translation of parts of his new book in advance of publication.

Nice, March 2001 *Michel Hénon*

Contents

11. Definitions and General Equations

11.1 Introduction

The present volume II begins with Chapter 11, as it is the direct continuation of Chapters 1 to 10 of volume I (Hénon 1997). This avoids any ambiguity about cross-references. All references to Chapters 1 to 10 (for instance, the reference to Sect. 2.1 below) are to be found in volume I, while all references to Chapters 11 to 23 are to be found in the present volume II.

In the present chapter, we describe the general method for the quantitative study of bifurcations, and we derive equations common to the three types of bifurcations. Later chapters will be devoted to detailed treatments of individual types 1 and 2.

Some bifurcations have already been studied quantitatively by Guillaume (1971), in the case of symmetric orbits and total bifurcation (ibid., pages 9 and 98), and by Perko (1977a, 1981a, 1981b), again in the case of symmetric orbits. Here we consider more generally symmetric or asymmetric orbits, and partial or total bifurcations.

11.2 The O Notation

The notation $O()$ was briefly described in Sect. 2.1, and some simple uses of that notation took place in Chapters 4 and 5. In the present volume, we will make a much more intensive use of that notation. Therefore we recall here in some detail its definition and properties.

11.2.1 Definitions

We follow Graham et al. 1989, Sect. 9.2, with adaptations to our problem. We consider expressions involving small quantities x_1, ..., x_n; we are interested in the limit $x_1 \to 0$, ..., $x_n \to 0$. We define the O notation as follows:

Definition 11.2.1. *The expression $O[g(x_1, \ldots, x_n)]$ represents the set of all functions $f(x_1, \ldots, x_n)$ such that*

$$|f| \leqq C|g| \qquad \text{for} \qquad |x_1| \leqq \epsilon_1, \ldots, |x_n| \leqq \epsilon_n \,, \tag{11.1}$$

for some set of positive constants C, ϵ_1, ..., ϵ_n.

f and g themselves do not have to tend to zero in the limit, or even to remain finite. For example, the notation $O(1/x_1^2)$ is perfectly valid: it represents the set of all functions which grow no faster than $1/x_1^2$ for $x_1 \to 0$.

In the present work, the function g will generally be a product of powers: $g = x_1^{q_1} \dots x_n^{q_n}$. The exponents q_i do not have to be positive; nor do they have to be integers.

More generally, we consider expressions involving also other variables y_1, $\dots$, y_m, which remain finite in the limit $\epsilon_i \to 0$. The above definition applies then for every choice of the y_i. The constants C, ϵ_1, $\dots$, ϵ_n are functions of $y_1, \dots, y_m$.

The O notation can be used alone, as for instance in: $y_0 = O(\mu)$. In many cases, however, it will appear inside an expression. The rule is then

Definition 11.2.2. *An expression containing one or more instances of the O notation represents the set of the expressions obtained by letting each $O()$ range over its domain (given by Definition 11.2.1).*

For instance, the expression

$$O[g_1(x)] + O[g_2(x)]O[g_3(x)] \tag{11.2}$$

represents the set of all functions of the form $f_1(x) + f_2(x)f_3(x)$, where f_1, f_2, f_3 belong respectively to the sets defined by $O[g_1(x)]$, $O[g_2(x)]$, $O[g_3(x)]$.

This will be called a *O-expression.* For completeness, we include also the case of an expression containing no O notation; in that case the set reduces to a single element.

The O notation cannot be used recursively: in $O(g)$, the expression g cannot itself contain instances of the $O()$ notation, because Definition 11.2.1 would then be meaningless. Apart from that restriction, the O notation can be freely used in an expression.

We consider next equations involving the O notation, i.e. equations of the form $f_1 = f_2$, where f_1 and f_2 are O-expressions. We call this a *O-relation.* The fundamental rule is then

Definition 11.2.3. *In an equality containing instances of the notation O, the '$=$' sign should be taken to mean '$\subseteq$'.*

Thus, the O-relation $f_1 = f_2$ means that the set of functions defined by f_1 is a subset of the set of functions defined by f_2. Note the dissymetry (which is commented upon in Graham et al. 1989).

11.2.2 Computation Rules

Many rules can be derived for the manipulation of O-expressions and O-relations (see Graham et al. 1989, Sect. 9.3). We quote only some of the most important rules which will be needed later. To simplify the equations, we adopt the following conventions: f represents a O-expression; g represents a function of x, i.e. of the small quantities $x_1, \ldots, x_n$; h represents a function of y, i.e. of the variables $y_1, \ldots, y_m$.

1. First we have, trivially:

$$g = O(g) \ . \tag{11.3}$$

2. Substitution: If the two O-relations $f_1 = f_2$ and $f_1' = f_2'$ hold, and if the O-expression f_1' appears inside f_2, then it can be replaced by f_2'.
   This follows from the transitive property of the subset relation. Note, however, that only the substitution of a left-hand side f_1' into a right-hand side f_2 is allowed.
3. Simplifications: Elimination of identical terms:

$$O(g) + O(g) = O(g) \ . \tag{11.4}$$

4. Elimination of a multiplicative factor:

$$hO(g) = O(g) \ . \tag{11.5}$$

5. Using (11.3), we obtain a very frequently used relation:

$$hg = O(g) \ . \tag{11.6}$$

6. Sum:

$$O(g_1 + g_2) = O(g_1) + O(g_2) \ . \tag{11.7}$$

   The reverse relation is *not* true.
7. Product:

$$O(g_1)O(g_2) = O(g_1 g_2) \ . \tag{11.8}$$

   Here the reverse relation is also true:

$$O(g_1 g_2) = O(g_1)O(g_2) \ . \tag{11.9}$$

8. Powers: if $q_1' \geqq q_1, \ldots, q_n' \geqq q_n$, then

$$O(x_1^{q_1'} \ldots x_n^{q_n'}) = O(x_1^{q_1} \ldots x_n^{q_n}) \ . \tag{11.10}$$

9. Elimination of a composite term:

$$O(g_1)+O(g_1^{\lambda} g_2^{1-\lambda})+O(g_2) = O(g_1)+O(g_2) \quad \text{for} \quad 0 \leqq \lambda \leqq 1 \ . \tag{11.11}$$

   As an example, we have

$$O(x_1^2) + O(x_1 x_2) + O(x_2^2) = O(x_1^2) + O(x_2^2). \tag{11.12}$$

10. Truncation: Consider a function involving only one small parameter x_1, which can be expanded as a converging series:

$$f = h_0 + h_1 x_1 + h_2 x_1^2 + \ldots \ . \tag{11.13}$$

   Using the previous rules, we can truncate the expansion at some order $q \leqq 0$:

$$f = \sum_{j=0}^{q} h_j x_1^j + O(x_1^{q+1}) . \tag{11.14}$$

Our equations will frequently involve sums of O terms, of the form $O(g_1)+O(g_2)+\ldots$. To abbreviate, we will write this as: $O(g_1, g_2, \ldots)$.

11.3 General Equations

11.3.1 Definitions

We consider a second species bifurcation orbit Ω (Sect. 6.5). We introduce a fundamental numbering, which will be used throughout the present volume:

- For a partial bifurcation of order n, we number the collisions of the bifurcating arc (including the two end collisions) from 0 to n, and the basic arcs from 1 to n.
- For a total bifurcation of order n, we number the collisions from 0 to $n-1$, with collision 0 at the origin, and we number the basic arcs from 1 to n, starting from the origin.

Thus in both cases, the basic arc i joins the collisions $i-1$ and i (except for the arc n in a total bifurcation, which joins the collisions $n-1$ and 0).

We define $\mathcal{C}$, the *set of internal collisions*, as

$$\mathcal{C} = \begin{cases} \{1, \ldots, n-1\} & \text{for a partial bifurcation,} \\ \{0, \ldots, n-1\} & \text{for a total bifurcation.} \end{cases} \tag{11.15}$$

In the case of a partial bifurcation, the two end collisions are excluded from $\mathcal{C}$. We define also $\mathcal{A}$, the *set of basic arcs*, as

$$\mathcal{A} = \{1, \ldots, n\} . \tag{11.16}$$

Here the definition is the same for partial and total bifurcations.

We consider now the restricted problem, with μ small but not zero, and a periodic orbit Ω_μ close to the bifurcation orbit Ω. (Such orbits exist by virtue of the definition 2.9.1 of generating orbits, of which bifurcation orbits are a special case.) We use a system of coordinates $(\hat{X}, \hat{Y})$ with the origin in M_2 and with fixed directions, related to the coordinates (x, y) and (X, Y) defined in Sect. 2.2 by

$$\begin{aligned} \hat{X} &= (x-1)\cos t - y \sin t = X - \cos t , \\ \hat{Y} &= (x-1)\sin t + y \cos t = Y - \sin t . \end{aligned} \tag{11.17}$$

The equations of motion are in the (X, Y) system

$$\begin{aligned} \ddot{X} &= -\mu \cos t - (1-\mu) X r^{-3} - \mu (X - \cos t) \rho^{-3} , \\ \ddot{Y} &= -\mu \sin t - (1-\mu) Y r^{-3} - \mu (Y - \sin t) \rho^{-3} , \end{aligned} \tag{11.18}$$

and in the $(\hat{X}, \hat{Y})$ system

$$\ddot{\hat{X}} = (1-\mu)\cos t - (1-\mu)(\hat{X} + \cos t)r^{-3} - \mu\hat{X}\rho^{-3} ,$$
$$\ddot{\hat{Y}} = (1-\mu)\sin t - (1-\mu)(\hat{Y} + \sin t)r^{-3} - \mu\hat{Y}\rho^{-3} . \tag{11.19}$$

We call $\boldsymbol{\rho}$ the vector from M_2 to M_3 with coordinates $(\hat{X}, \hat{Y})$. The orbit is defined by $\boldsymbol{\rho}(t)$.

Each collision is now generally replaced by a close approach of M_2 by M_3. Following the tradition of stellar dynamics, we will call *encounter* such a close approach. We number the encounters in the same way as the corresponding collisions.

We consider an encounter $i \in \mathcal{C}$. The motion of M_3 in the vicinity of M_2 is quasi-hyperbolic. We call ϵ_i the minimal distance of approach.

Consider a point P_i of the orbit Ω_μ during encounter i, or more precisely a point corresponding to a time t_i such that $\rho(t_i) = O(\epsilon_i)$. (The exact placement of this point is deliberately left unspecified.) Ω_μ is divided by the P_i into *pieces*. For $\mu \to 0$, each piece tends towards a basic arc of the bifurcation orbit Ω. We number the pieces in the same way as the corresponding basic arcs.

The relative velocity of M_3 with respect to M_2 has coordinates $\dot{\hat{X}}$, $\dot{\hat{Y}}$. At a collision, there is $x = 1$, $y = 0$; from (11.17) we have then

$$\dot{\hat{X}} = \dot{x}\cos t - \dot{y}\sin t , \qquad \dot{\hat{Y}} = \dot{x}\sin t + \dot{y}\cos t , \tag{11.20}$$

and

$$\sqrt{\dot{\hat{X}}^2 + \dot{\hat{Y}}^2} = \sqrt{\dot{x}^2 + \dot{y}^2} = v , \tag{11.21}$$

where v is defined in Sect. 8.2. Thus, the relative velocity modulus at a collision has the same value v in the $(\hat{X}, \hat{Y})$ system and in the (x, y) system.

For bifurcation orbits of types 1, 2, or 3, v is finite, as shown by Figs. 4.1 and 4.3, or by relations (8.3) and (4.8):

$$v = \Theta(1) . \tag{11.22}$$

v has the same value at all collisions by virtue of (8.3); thus it is not necessary to give it a subscript.

For μ small but non-zero, the orbit of M_3 passes in the vicinity of M_2 with essentially the same velocity v.

The attraction of M_2 produces a deflection by an angle ψ_i, given by (Mihalas and Routly, 1968, p. 174)

$$\psi_i = \frac{2\mu}{v^2\epsilon_i} . \tag{11.23}$$

We find from (11.20) that ψ_i also has the same value in the $(\hat{X}, \hat{Y})$ and (x, y) systems.

Next we consider a range of values of μ :

$$0 < \mu \leqq \mu_0 \tag{11.24}$$

and a set of orbits Ω_μ, defined by a function $\rho(t, \mu)$, with the condition that, as $\mu \to 0$, Ω_μ tends to Ω. In what follows, we will generally omit μ for simplicity and write $\rho(t)$.

We have

$$\epsilon_i \to 0 \qquad \text{for} \qquad \mu \to 0 \, . \tag{11.25}$$

The deflection angle ψ must also vanish as we approach the bifurcation orbit:

$$\psi_i \to 0 \qquad \text{for} \qquad \mu \to 0 \, . \tag{11.26}$$

We make now the fundamental assumption that, for $\mu \to 0$, ϵ_i varies as a power of μ:

$$\epsilon_i = \Theta(\mu^{\nu_i}) \, . \tag{11.27}$$

From (11.23) and 11.27) we have then

$$\psi_i = \Theta(\mu^{1-\nu_i}) \, . \tag{11.28}$$

(11.25) and (11.26) show that the exponent ν_i, thus defined, must satisfy (cf. Guillaume 1971, p. 98)

$$0 < \nu_i < 1 \, . \tag{11.29}$$

11.3.2 Intermediate Arcs

We will approximate each piece of Ω_μ by a *keplerian arc*, i.e. an arc from a *keplerian orbit*, defined as a solution of the problem $\mu = 0$ (Sects. 2.9, 2.10). Since we are in the vicinity of a bifurcation orbit, the arc must be elliptic (Sect. 6.2.2). It is thus a part of a keplerian elliptical motion around a body M_1 of mass 1.

Consider the piece i of Ω_μ, delimited by P_{i-1} and P_i. Consider on this piece a point which is far from M_2; or more precisely, a time t_{0i} such that $|\rho(t_{0i})| = \Theta(1)$. (We could for instance take $t_{0i} = (t_{i-1} + t_i)/2$; but here again the precise placement is not important.) We consider now the keplerian orbit which is osculating to Ω_μ at time t_{0i}. We call it the *intermediate orbit* i. The corresponding vector description (with origin in M_2) will be called $\boldsymbol{\rho}_i(t)$. The part of this orbit between times t_{i-1} and t_i will be called the *intermediate arc* i. We write

$$\boldsymbol{\rho}(t) = \boldsymbol{\rho}_i(t) + \Delta\boldsymbol{\rho}_i(t) \, . \tag{11.30}$$

The introduction of intermediate orbits will play a fundamental role in the analysis of the bifurcations. The real orbit Ω_μ will not be compared directly to the bifurcation orbit. Instead, we proceed in two steps: the real orbit is first compared to the intermediate orbit, and the intermediate orbit is then compared to the bifurcation orbit. (See Guillaume 1971, p. 72; 1975b, p. 452;

Perko 1977a, p. 277.) It will be shown in the next section that the real orbit and the intermediate orbit differ in position by $O(\mu)$ only inside the time interval $t_{i-1} < t < t_i$. Fortunately, it turns out that this $O(\mu)$ approximation is sufficient for our needs in nearly all cases (the only exception being the bifurcation 2P1; see Sect. 23.2). In other words, we will be able to replace the study of the true orbit by the study of the sequence of intermediate arcs. Since an intermediate arc is keplerian, it can be described by its four orbital elements. Thus we end up having to consider a finite system of ordinary equations, instead of a set of differential equations.

11.3.3 Orders of Magnitude of $\Delta\boldsymbol{\rho}_i$ and $\Delta\dot{\boldsymbol{\rho}}_i$

The equations of motion for the true orbit are given by (11.19), or in vector notation

$$\ddot{\boldsymbol{\rho}} = -(1-\mu)\left[\frac{\boldsymbol{\rho}-\boldsymbol{\rho}_m}{|\boldsymbol{\rho}-\boldsymbol{\rho}_m|^3} + \boldsymbol{\rho}_m\right] - \frac{\mu\boldsymbol{\rho}}{|\boldsymbol{\rho}|^3}\,, \tag{11.31}$$

where $\boldsymbol{\rho}_m$ is the vector from M_2 to M_1:

$$\boldsymbol{\rho}_m = \begin{pmatrix} -\cos t \\ -\sin t \end{pmatrix}\,. \tag{11.32}$$

The equations of motion for the intermediate orbit are given by the same equations with $\mu = 0$:

$$\ddot{\boldsymbol{\rho}}_i = -\left[\frac{\boldsymbol{\rho}_i-\boldsymbol{\rho}_m}{|\boldsymbol{\rho}_i-\boldsymbol{\rho}_m|^3} + \boldsymbol{\rho}_m\right]\,. \tag{11.33}$$

$\Delta\boldsymbol{\rho}_i(t)$ is completely defined by the initial values

$$\Delta\boldsymbol{\rho}_i(t_{0i}) = 0\,, \qquad \Delta\dot{\boldsymbol{\rho}}_i(t_{0i}) = 0\,, \tag{11.34}$$

and the differential equation

$$\Delta\ddot{\boldsymbol{\rho}}_i = -(1-\mu)\frac{\boldsymbol{\rho}-\boldsymbol{\rho}_m}{|\boldsymbol{\rho}-\boldsymbol{\rho}_m|^3} + \frac{\boldsymbol{\rho}_i-\boldsymbol{\rho}_m}{|\boldsymbol{\rho}_i-\boldsymbol{\rho}_m|^3} + \mu\boldsymbol{\rho}_m - \frac{\mu\boldsymbol{\rho}}{|\boldsymbol{\rho}|^3} \tag{11.35}$$

or

$$\Delta\ddot{\boldsymbol{\rho}}_i = O(\mu) + O(\Delta\boldsymbol{\rho}_i) - \frac{\mu\boldsymbol{\rho}}{|\boldsymbol{\rho}|^3}\,. \tag{11.36}$$

We estimate now the order of magnitude of $\Delta\boldsymbol{\rho}_i$ and $\Delta\dot{\boldsymbol{\rho}}_i$ in the interval $t_{0i} < t < t_i$. We integrate (11.36) twice, starting from $t = t_{0i}$. We consider first the interval $t_{0i} < t < t_i - \epsilon_i$. By definition, there is no other encounter in that interval, and the relative velocity v at encounter i is finite. It follows that

$$|\boldsymbol{\rho}(t)| = \Theta(t_i - t)\,. \tag{11.37}$$

The last term in (11.36) becomes large in the vicinity of an encounter, and we will assume that it is dominant, i.e. that the first two terms can be ignored for the purpose of estimating orders of magnitude. We have then

$$\Delta\ddot{\rho}_i(t) = O[\mu(t_i - t)^{-2}] . \quad (11.38)$$

Integrating, we find

$$\Delta\dot{\rho}_i(t) = \int_{t_{0i}}^{t} \Delta\ddot{\rho}_i(t')\mathrm{d}t' = O[\mu(t_i - t)^{-1}] ,$$

$$\Delta\rho_i(t) = \int_{t_{0i}}^{t} \Delta\dot{\rho}_i(t')\mathrm{d}t' = O[\mu \ln(t_i - t)] . \quad (11.39)$$

The last equation gives in particular

$$\Delta\rho_i(t) = O(\mu \ln \epsilon_i) . \quad (11.40)$$

We will neglect logarithmic factors in orders of magnitude, here and in what follows (cf. Perko 1976b, p. 399). Thus

$$\Delta\rho_i(t) = O(\mu) , \quad (11.41)$$

For $t = t_i - \epsilon_i$, we have

$$\Delta\dot{\rho}_i = O(\mu/\epsilon_i) , \qquad \Delta\rho_i = O(\mu) . \quad (11.42)$$

Finally, in the interval $t_i - \epsilon_i < t < t_i$, we have

$$|\rho(t)| = \Theta(\epsilon_i) , \quad (11.43)$$

and therefore, from (11.36), ignoring again the first two terms,

$$\Delta\ddot{\rho}_i(t) = O(\mu/\epsilon_i^2) . \quad (11.44)$$

Integrating, we find

$$\Delta\dot{\rho}_i(t) = \Delta\dot{\rho}_i(t_i - \epsilon_i) + \int_{t_i-\epsilon_i}^{t} \Delta\ddot{\rho}_i(t')\mathrm{d}t' = O(\mu/\epsilon_i) ,$$

$$\Delta\rho_i(t) = \Delta\rho_i(t_i - \epsilon_i) + \int_{t_i-\epsilon_i}^{t} \Delta\dot{\rho}_i(t')\mathrm{d}t' = O(\mu) . \quad (11.45)$$

Thus, we have

$$\Delta\dot{\rho}_i(t) = O(\mu/\epsilon_i) , \qquad \Delta\rho_i(t) = O(\mu) \qquad \text{for} \qquad t_{0i} < t < t_i . \quad (11.46)$$

A similar argument can be made by integrating backward in time, in the interval $t_{i-1} < t < t_{0i}$. Therefore we have $\Delta\rho_i(t) = O(\mu)$ for the whole piece $t_{i-1} < t < t_i$.

The time t_i has been defined with some arbitrariness, as a time during encounter i. Therefore

Proposition 11.3.1. *The intermediate orbit i gives an $O(\mu)$ approximation of the true orbit for the whole piece between the encounters $i-1$ and i, and also during these close encounters.*

Within a $O(\mu)$ approximation, therefore, the true orbit can be replaced by a sequence of keplerian arcs. This superficially resembles a second species orbit. However, the points at which the intermediate arcs are joined do not coincide with M_2 any more. The point corresponding to encounter i lies at a distance $\Theta(\epsilon_i) = \Theta(\mu^{\nu_i})$, where ν_i can take values in the range $0 < \nu_i < 1$.

These points can be chosen with some latitude, because the two successive intermediate arcs i and $i+1$ coincide to order $O(\mu)$ in the ϵ_i vicinity of encounter i. This fact can be understood intuitively as follows: the two intermediate arcs make locally an angle of the order of ψ, or $\mu^{1-\nu_i}$ as shown by (11.28); on the other hand, the span of the ϵ_i vicinity is of the order of μ^{ν_i} according to (11.27); multiplying, we find that the two arcs deviate from each other by a distance of order $O(\mu)$ inside the ϵ_i vicinity.

11.3.4 More Accurate Estimate of $\Delta\dot{\boldsymbol{\rho}}_i$

We compute now $\Delta\dot{\boldsymbol{\rho}}_i$ more accurately; this will be needed below to obtain the second *matching relation* at an encounter.

As in Sect. 8.2, we use subscripts b and a to designate quantities before and after an encounter. Consider the tangent to the intermediate arc i at time $t = t_i$ (Fig. 11.1, where the left and right figures correspond to the cases where M_3 passes to the left or to the right of M_2, respectively). We define the unit vector $\boldsymbol{i}_b$ on this tangent and the velocity modulus v_b by:

$$\dot{\boldsymbol{\rho}}_i(t_i) = v_b \boldsymbol{i}_b \,. \tag{11.47}$$

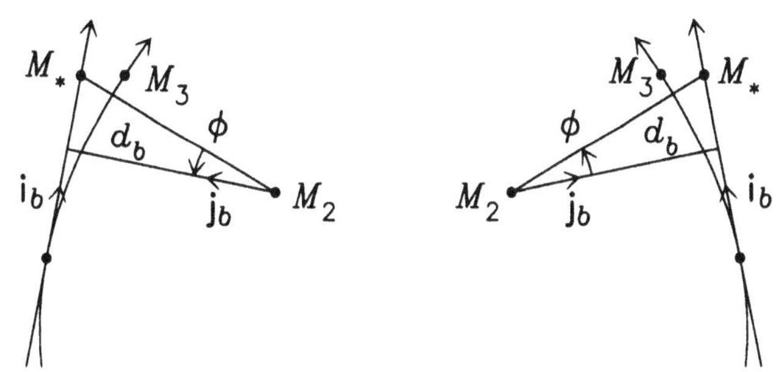

Fig. 11.1. Osculating straight-line motion.

We call $\boldsymbol{j}_b$ the unit vector orthogonal to $\boldsymbol{i}_b$ and directed from M_2 towards the tangent, and d_b the distance from M_2 to the tangent (see figure). There is

$$d_b = \Theta(\epsilon_i) \,. \tag{11.48}$$

Finally we consider a point M_* with a uniform motion on the tangent, osculating to the intermediate arc. This motion is given by

$$\boldsymbol{\rho}_* = d_b\,\boldsymbol{j}_b + (d_b \tan\phi)\boldsymbol{i}_b \,, \tag{11.49}$$

where $\boldsymbol{\rho}_*$ is the vector from M_2 to M_* and the angle ϕ is defined on the figure (in such a way that it always increases with time). We have

$$\dot{\boldsymbol{\rho}}_* = v_b\boldsymbol{i}_b = \frac{d_b}{\cos^2\phi}\frac{\mathrm{d}\phi}{\mathrm{d}t}\boldsymbol{i}_b \,. \tag{11.50}$$

We have for the intermediate arc

$$\boldsymbol{\rho}_i(t) = \boldsymbol{\rho}_*(t) + O(t_i - t)^2 \,. \tag{11.51}$$

The moduli are given by

$$|\boldsymbol{\rho}_*(t)| = \frac{d_b}{\cos\phi}\,, \qquad |\boldsymbol{\rho}_i(t)| = \frac{d_b}{\cos\phi} + O(t_i - t)^2 \,. \tag{11.52}$$

We rewrite (11.36), substituting (11.30) and using (11.46):

$$\Delta\ddot{\boldsymbol{\rho}}_i = -\frac{\mu\boldsymbol{\rho}_i}{|\boldsymbol{\rho}_i|^3} + O(\mu) + O(\mu^2/\rho_i^3) \,. \tag{11.53}$$

Substituting (11.51) and integrating, we obtain

$$\Delta\dot{\boldsymbol{\rho}}_i(t_i) = \int_{t_{0i}}^{t_i} \left\{ -\frac{\mu\boldsymbol{\rho}_*}{|\boldsymbol{\rho}_*|^3} + O[(t_i - t)^2\mu/\rho_*^3] + O(\mu) + O(\mu^2/\rho_*^3) \right\} \mathrm{d}t \,. \tag{11.54}$$

Using (11.49), (11.52) and changing the variable with (11.50), we find for the first term

$$\begin{aligned} -\int_{t_{0i}}^{t_i} \mu\frac{\boldsymbol{\rho}_*}{|\boldsymbol{\rho}_*|^3}\,\mathrm{d}t &= -\int \frac{\mu\cos^2\phi}{d_b^2}(\boldsymbol{j}_b\cos\phi + \boldsymbol{i}_b\sin\phi)\,\mathrm{d}t \\ &= -\frac{\mu}{v_b d_b}\int (\boldsymbol{j}_b\cos\phi + \boldsymbol{i}_b\sin\phi)\,\mathrm{d}\phi = -\frac{\mu}{v_b d_b}\Big[\boldsymbol{j}_b\sin\phi - \boldsymbol{i}_b\cos\phi\Big]_{\phi_{0i}}^{\phi_b} \\ &= -\frac{\mu}{v_b d_b}\Big[\boldsymbol{j}_b(1+\sin\phi_b) - \boldsymbol{i}_b\cos\phi_b\Big] \,. \end{aligned} \tag{11.55}$$

Here ϕ_b and ϕ_{0i} are the values of ϕ corresponding to $t = t_i$ and $t = t_{0i}$. In the last expression we have used the fact that for $t = t_{0i}$, there is $\phi_{0i} \simeq -\pi/2$.

For the second term of (11.54), we substitute (11.52), we change the variable with (11.50) and we use (11.22); we obtain

$$\int_{t_{0i}}^{t_i} O\left[\frac{\mu}{d_b^2}(t_i - t)^2\cos\phi\right]\mathrm{d}\phi \,. \tag{11.56}$$

There is

$$v_b(t_i - t) = d_b(\tan\phi_b - \tan\phi) \,, \tag{11.57}$$

and therefore

$$t_i - t = O(d_b) + O(d_b \tan\phi)\,, \quad (t_i - t)^2 = O(d_b^2) + O(d_b^2 \tan^2\phi)\,. \tag{11.58}$$

After substitution in (11.56), that expression becomes

$$O\left(\mu \int_{\phi_{0i}}^{\phi_b} \cos\phi \,\mathrm{d}\phi\right) + O\left(\mu \int_{\phi_{0i}}^{\phi_b} \frac{\sin^2\phi}{\cos\phi}\,\mathrm{d}\phi\right)\,. \tag{11.59}$$

The first term reduces to $O(\mu)$. In the second term, we neglect the logarithmic singularity which exists for $\phi \to \phi_{0i} \simeq -\pi/2$ and we obtain again $O(\mu)$.

The third term of (11.54) is $O(\mu)$. The fourth term becomes, using (11.52) and (11.50),

$$\int_{\phi_{0i}}^{\phi_b} O\left(\frac{\mu^2}{d_b^2}\cos\phi\right)\,\mathrm{d}\phi = O(\mu^2/d_b^2) = O(\mu^2/\epsilon_i^2)\,. \tag{11.60}$$

Finally we obtain

$$\Delta\dot{\boldsymbol{\rho}}_i(t_i) = \frac{\mu}{v_b d_b}\Big[\boldsymbol{i}_b \cos\phi_b - \boldsymbol{j}_b(1 + \sin\phi_b)\Big] + O(\mu) + O(\mu^2/\epsilon_i^2)\,. \tag{11.61}$$

Similar computations can be made for the intermediate arc $i+1$, near its beginning at encounter i. We define in the same way as above the quantities $\boldsymbol{i}_a$, v_a, $\boldsymbol{j}_a$, d_a, ϕ_a. We integrate from $t_{0,i+1}$ to t_i (in the direction of decreasing t). For $t = t_{0,i+1}$, there is $\phi_{0,i+1} \simeq +\pi/2$. We obtain

$$\Delta\dot{\boldsymbol{\rho}}_{i+1}(t_i) = \frac{\mu}{v_a d_a}\Big[\boldsymbol{i}_a \cos\phi_a + \boldsymbol{j}_a(1 - \sin\phi_a)\Big] + O(\mu) + O(\mu^2/\epsilon_i^2)\,. \tag{11.62}$$

11.3.5 Matching Relations

We consider now an internal encounter $i \in C$, and we effect the *matching* between the intermediate arcs i and $i+1$. First, (11.46b) gives

$$\boldsymbol{\rho}(t_i) - \boldsymbol{\rho}_i(t_i) = O(\mu)\,, \tag{11.63}$$

and, for the intermediate arc $i+1$

$$\boldsymbol{\rho}(t_i) - \boldsymbol{\rho}_{i+1}(t_i) = O(\mu)\,, \tag{11.64}$$

so that

$$\boldsymbol{\rho}_{i+1}(t_i) = \boldsymbol{\rho}_i(t_i) + O(\mu)\,. \tag{11.65}$$

This constitutes the *first matching relation.*

We consider now the relations (11.61) and (11.62). Equation (11.46a) shows that

$$\dot{\boldsymbol{\rho}}_{i+1}(t_i) = \dot{\boldsymbol{\rho}}_i(t_i) + O(\mu/\epsilon_i)\,; \tag{11.66}$$

it follows that

$$v_a = v_b + O(\mu/\epsilon_i)\,, \qquad i_a = i_b + O(\mu/\epsilon_i)\,, \qquad j_a = j_b + O(\mu/\epsilon_i)\,,$$
$$d_a = d_b + O(\mu)\,, \qquad \phi_a = \phi_b + O(\mu/\epsilon_i)\,. \tag{11.67}$$

Therefore we can replace a-subscripted by b-subscripted quantities in (11.62), incurring a relative error μ/ϵ_i. Since the first term of (11.62) is of order μ/ϵ_i, the absolute error is $O(\mu^2/\epsilon_i^2)$, and is included in the last term. So we have

$$\Delta\dot{\rho}_{i+1}(t_i) = \frac{\mu}{v_b d_b}\Big[i_b \cos\phi_b + j_b(1-\sin\phi_b)\Big] + O(\mu) + O(\mu^2/\epsilon_i^2)\,. \tag{11.68}$$

There is

$$\dot{\rho}_{i+1}(t_i) = \dot{\rho}(t_i) - \Delta\dot{\rho}_{i+1}(t_i) = \dot{\rho}_i(t_i) + \Delta\dot{\rho}_i(t_i) - \Delta\dot{\rho}_{i+1}(t_i)\,, \tag{11.69}$$

and, using (11.61) and (11.68),

$$\dot{\rho}_{i+1}(t_i) = \dot{\rho}_i(t_i) - \frac{2\mu j_b}{v_b d_b} + O(\mu) + O(\mu^2/d_b^2)\,. \tag{11.70}$$

This is the *second matching relation* (see Guillaume, 1971). Note that we might just as well have used j_a, v_a, d_a. We note also that j_b, v_b, d_b can be taken in an arbitrary point in an ϵ_i vicinity of M_2: the variations of j_b and v_b inside that vicinity are of order ϵ_i, and the variation of d_b is of order ϵ_i^2, so that we always have a relative error $O(\epsilon_i)$, which when applied to the term of order μ/ϵ_i gives an absolute error $O(\mu)$.

The relation (11.70), which holds in the $(\hat{X},\hat{Y})$ system of axes with their origin in M_2 and with fixed directions, is also true in a system of *rotating* axes with their origin in M_2, with v_b, j_b, d_b redefined in that system. Indeed the difference $\dot{\rho}_{i+1} - \dot{\rho}_i$ is invariant in this change of axes; within a distance ϵ_i of M_2, the relative speed of the two systems is of order ϵ_i, so that the variations of v_b and j_b are of order ϵ_i and the variation of d_b is of order ϵ_i^2.

The second term in the right-hand side of (11.70) is significant if $\mu/d_b \gg \mu$ and $\mu/d_b \gg \mu^2/d_b^2$; we recover precisely the conditions (11.25) and (11.26)! This is a confirmation that (11.70) is indeed the correct relation for bifurcations. When we move away from the bifurcation at a node, the terms 2 and 4 both become of order 1 and the relation becomes meaningless; this means that we must now compute more exactly the deflection angle, which is not small any more. Conversely, in an antinode, the terms 2 and 3 become both of order μ and (11.70), coupled with (11.65), shows that the intermediate arcs i and $i+1$ are identical within $O(\mu)$: the effect of M_2 has become negligible.

In a partial bifurcation, the boundary conditions at collisions 0 and n must also be considered. In these collisions, the deflection angle does not vanish at the bifurcation; therefore, for $\mu \neq 0$, the distance of approach is $O(\mu)$ and we have

$$\rho_1(t_0) = O(\mu)\,; \qquad \rho_n(t_n) = O(\mu)\,. \tag{11.71}$$

11.3.6 The Case $\mu = 0$

Until now in this chapter we have assumed that $\mu \neq 0$. We consider now the case $\mu = 0$ and we show that the matching equations are also valid for that case (i.e. for first and second species orbits), provided that (11.70) is written in a slightly different way. This extension will be useful in what follows.

ϵ_i is zero at a node, non-zero at an antinode. Conversely, ψ_i is non-zero at a node, zero at an antinode. Thus

$$\psi_i \epsilon_i = 0 \tag{11.72}$$

which takes the place of (11.23).

We define a time t_i as before; in the case of a node, t_i coincides with the time of the collision. The intermediate arcs coincide now with the true orbit: $\Delta\boldsymbol{\rho}_i = 0$. In particular, we have

$$\boldsymbol{\rho}_{i+1}(t_i) = \boldsymbol{\rho}_i(t_i) = \boldsymbol{\rho}(t_i) \ , \tag{11.73}$$

and therefore the first matching relation (11.65) is still true.

For a partial bifurcation, there is a collision at each end, and

$$\boldsymbol{\rho}_1(t_0) = \boldsymbol{\rho}(t_0) = 0 \ , \qquad \boldsymbol{\rho}_n(t_n) = \boldsymbol{\rho}(t_n) = 0 \ , \tag{11.74}$$

and therefore the equations (11.71) also are still true.

Finally we consider the second matching relation (11.70). For an antinode, the intermediate orbit is the same before and after t_i, and

$$\dot{\boldsymbol{\rho}}_{i+1}(t_i) = \dot{\boldsymbol{\rho}}_i(t_i) = \dot{\boldsymbol{\rho}}(t_i) \ . \tag{11.75}$$

On the other hand, d_b is non-zero. Therefore (11.70) is still verified. For a node, (11.70) becomes meaningless because μ/d_b has the indeterminate value $0/0$. However, we can rewrite the equation by multiplying both sides by d_b:

$$d_b \left[\dot{\boldsymbol{\rho}}_{i+1}(t_i) - \dot{\boldsymbol{\rho}}_i(t_i)\right] = -\frac{2\mu \boldsymbol{j}_b}{v_b} + O(\mu d_b) + O(\mu^2/d_b) \ . \tag{11.76}$$

We have from (11.70)

$$\frac{\mu}{d_b} = -\frac{v_b \left[\dot{\boldsymbol{\rho}}_{i+1}(t_i) - \dot{\boldsymbol{\rho}}_i(t_i)\right]}{2\boldsymbol{j}_b} + O(\mu) + O(\mu^2/d_b^2) \tag{11.77}$$

from which

$$O\left(\frac{\mu}{d_b}\right) = O\left[\dot{\boldsymbol{\rho}}_{i+1}(t_i) - \dot{\boldsymbol{\rho}}_i(t_i)\right] \ . \tag{11.78}$$

The last two terms of (11.77) have been dropped in view of (11.48), (11.27), and (11.29). Thus (11.76) can be written

$$d_b \left[\dot{\boldsymbol{\rho}}_{i+1}(t_i) - \dot{\boldsymbol{\rho}}_i(t_i)\right] = -\frac{2\mu \boldsymbol{j}_b}{v_b} + O(\mu d_b) + O\left\{\mu \left[\dot{\boldsymbol{\rho}}_{i+1}(t_i) - \dot{\boldsymbol{\rho}}_i(t_i)\right]\right\} \ . \tag{11.79}$$

This new equation is still true for $\mu = 0$: all terms vanish.

11.4 General Method

Since we are considering now small but non-zero values of μ, orbits in the vicinity of a bifurcation are characterized by two small parameters: μ and ΔC. The parameter μ specifies how far we are from the two-body problem, while ΔC specifies how far we are from the bifurcation orbit.

Fig. 1.1 essentially represents this two-parameter space; we will consider particularly the cases c and d which represent the generic situation in the vicinity of a bifurcation orbit Ω. The value $\mu = 0$ is represented by the full lines. A value $\mu > 0$ is represented by the dashed lines.

In the vicinity of Ω, the curves C = Const. are approximately parallel straight lines (not represented on Fig. 1.1). Thus, $|\Delta C|$ generally increases when we move away from Ω on a curve with fixed μ.

For $\mu = 0$, the solutions are known: they are second species orbits, formed by a sequence of T- and S-arcs. (Occasionally they can also be first species orbits.) The branches emanating from bifurcation orbits of various types have been studied in detail in Sect. 6.2.

For $\mu > 0$, a reasonable strategy seems to be to begin by studying the asymptotic branches of the dashed line characteristics in Figs. 1.1, i.e. the parts of these characteristics which extend away from Ω and approach the full lines of the $\mu = 0$ case. Then we progressively make our way toward the vicinity of Ω, until at last the branches are joined two by two.

We cannot simply do this for the characteristic corresponding to a small but fixed value of μ, because fundamentally our interest lies not in small values of μ, but in the limit $\mu \to 0$. Therefore a slightly subtler approach is required. We define a fundamental quantity ν by

$$\Delta C = \Theta(\mu^{\nu}) . \tag{11.80}$$

We consider then a fixed value of ν and we let $\mu \to 0$. We will determine the asymptotic form of the equations and of their solutions in that limit.

The relation (11.80) can also be written

$$\ln |\Delta C| = \nu \Theta(\ln \mu) . \tag{11.81}$$

Thus, the value $\nu = 0$ corresponds to $\mu = 0$, $\Delta C \neq 0$, i.e. the full lines in Fig. 1.1. A small value of ν corresponds to points close to the full lines, i.e. the distant parts of the dashed characteristics in Fig. 1.1. Increasing values of ν correspond to increasingly close parts of the characteristics. Thus, the strategy defined above will be implemented by starting from $\nu = 0$ and studying progressively increasing values of ν.

This strategy will be used for bifurcations of types 1 and 2, in Chaps. 12 and 17, respectively. The details differ from one type to another, but the general method is the same. We will find in each case that there exists a sequence of cases for ν, such that each case requires a separate treatment and results in specific asymptotic equations. Each case is either an open interval or an isolated value of ν. For bifurcations of type 1, we will find in Chap. 12 the following sequence of cases:

$$\nu = 0\,, \quad 0 < \nu < 1/2\,, \quad \nu = 1/2\,, \quad \nu > 1/2\,. \tag{11.82}$$

For bifurcations of type 2, we will find in Chap. 17 the following sequence:

$$\nu = 0\,, \quad 0 < \nu < 1/3\,, \quad \nu = 1/3\,, \quad 1/3 < \nu < 1/2\,, \quad \nu = 1/2. \tag{11.83}$$

The cases will be studied one by one in the order of the sequence, i.e. for increasing ν. In the first case, $\nu = 0$, the asymptotic expressions of the variables can be deduced from the study of bifurcation orbits made in Chap. 6 (see below Sects. 12.3 and 17.3). In each of the following cases with $\nu > 0$, the following procedure is applied:

1. We estimate the orders of magnitude of the variables by extrapolating from the previous case.
2. These orders of magnitude suggest an appropriate change of variables. Intuition plays a role here: the "good" change of variables is not always obvious. In most cases, the new variables are $\Theta(1)$ for $\mu \to 0$; but it can happen that some of them are $o(1)$.
3. At this stage it is generally convenient to make another change of variables which consists simply in a change of scale for each variable, so as to simplify the equations as much as possible.
4. In each equation, the dominant terms are collected in the left-hand member, and the equation is divided by an appropriate factor to make these terms $O(1)$. The right-hand member then should contain only terms which are $o(1)$.
5. We determine under which conditions the right-hand member is indeed $o(1)$. This may determine an upper limit for ν, which is the limit of the presently considered interval for ν.
   The *asymptotic equations* are then obtained by equating the right-hand member to zero.
6. We solve the asymptotic equations. If there are several solutions, the relevant solution is determined by continuity with the previous interval.
7. We compute the Jacobian $|J|$ of the system of equations, for the asymptotic solution. (Note: usage varies concerning the terms used to designate the *Jacobian matrix*, J, and the *Jacobian determinant*, $|J|$. In the present work, only the determinant will be needed, and we will call it simply *Jacobian*.) If $|J|$ is non-zero, the procedure is successful: the change of variables selected at step 2 above was indeed the good one. The *implicit function theorem* guarantees then the existence of a solution in the vicinity of the asymptotic solution, for $\mu \neq 0$. Moreover, the difference between this true solution and the asymptotic solution is given in order of magnitude by the maximum of the right-hand members.
8. Sometimes the error estimates can be refined for some variables by considering separately some equations.

9. We can go back to the initial variables by inverting the changes of variables.

12. Quantitative Study of Type 1

12.1 Fundamental Equations

12.1.1 Arc Relations

The intermediate orbit i, defined in Sect. 11.3.2, is a keplerian orbit. Therefore a definite relation exists between the positions and velocities at the two ends, corresponding to the times t_{i-1} and t_i. We compute now these relations explicitly.

Until now some arbitrariness has been left in the choice of t_i. Now we define it precisely as the time of the intersection of the true orbit Ω_μ with the orbit of M_2. This was found to give the simplest calculations in the case of type 1.

We define y_i as the oriented length of the arc from M_2 to M_3 at time t_i (Fig. 12.1). y_i is also the lead of M_3 over M_2 for the passage at the intersection point.

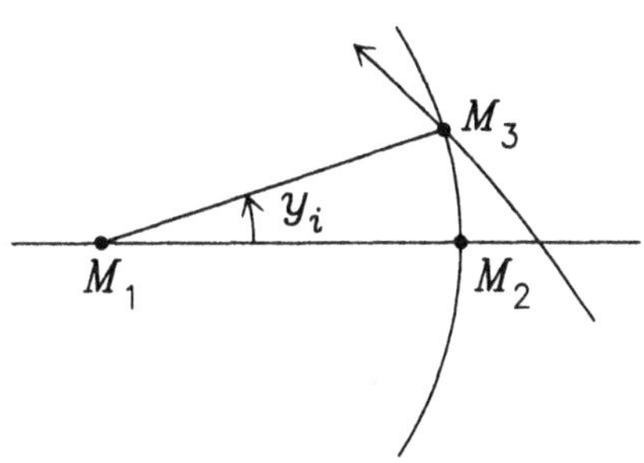

Fig. 12.1. Definition of y_i.

We consider now the intermediate orbit i ($i \in \mathcal{A}$). It intersects the orbit of M_2 at a time t_i''. Since the radial velocity is non-zero for bifurcation orbits of type 1, we have

$$t_i'' = t_i + O(\mu) . \tag{12.1}$$

We call y_i'' the corresponding lead. There is also

$$y_i'' = y_i + O(\mu) \,. \tag{12.2}$$

Similarly, at the previous encounter, the intermediate orbit i intersects the orbit of M_2 at a time t_i', with a lead y_i', and there is

$$t_i' = t_{i-1} + O(\mu) \,, \tag{12.3}$$

$$y_i' = y_{i-1} + O(\mu) \,. \tag{12.4}$$

The intermediate orbit is a keplerian orbit and therefore can be represented by a point in the (A, Z) plane. (We recall that the (A, Z) coordinates are related to the orbital elements (a, e) by a simple change of variables; see Chap. 4.) We call that point (A_i, Z_i). We call C_i the Jacobi constant of the intermediate orbit. We call (A, Z) the point representing the bifurcation orbit, and C its Jacobi constant. We define

$$A_i = A + \Delta A_i \,, \qquad Z_i = Z + \Delta Z_i \,, \qquad C_i = C + \Delta C_i \,. \tag{12.5}$$

The bifurcation orbit consists of a sequence of alternating basic arcs PQ and QP (Sect. 6.2.1.2). We consider first the case where the basic arc i is a first basic arc PQ. In the particular case where the intermediate orbit i has a collision in P, i.e. the lead y_i' is zero, we can use the analysis in Sect. 4.3.1, and the lead at the end is, from (4.20), (4.23), (4.28)),

$$y_i'' = t_2 - t_4 = t_{20} - t_{40} + 2\pi(\alpha - \beta A_i) = 2\pi(Z_i + \alpha - \beta A_i) \,. \tag{12.6}$$

In the general case where the initial lead y_i' is non-zero, we have then

$$y_i'' = y_i' + 2\pi(Z_i + \alpha - \beta A_i) \,. \tag{12.7}$$

We substitute (12.5) and we use the relation (6.8) for the bifurcation orbits; we note also that $\beta = \beta_0$ for the first basic arc according to (6.17). We obtain

$$y_i'' - y_i' = 2\pi(\Delta Z_i - \beta_0 \Delta A_i) \,. \tag{12.8}$$

A similar computation can be made when the basic arc i is a second basic arc QP. In the particular case where the intermediate orbit i has a collision in P, i.e. the final lead y_i'' is zero, we obtain from (4.20), (4.23), (4.28)),

$$y_i' = 2\pi(Z_i + \alpha - \beta A_i) \,. \tag{12.9}$$

In the general case where the final lead y_i'' is non-zero, we have then

$$y_i' = y_i'' + 2\pi(Z_i + \alpha - \beta A_i) \,. \tag{12.10}$$

There is $\beta = \beta_0 - J$ for the second basic arc according to (6.18), and we obtain

$$y_i' - y_i'' = 2\pi[\Delta Z_i - (\beta_0 - J)\Delta A_i] \,. \tag{12.11}$$

We define

$$s_i = \begin{cases} -1 & \text{if the arc } i \text{ is a first basic arc,} \\ +1 & \text{if the arc } i \text{ is a second basic arc.} \end{cases} \tag{12.12}$$

In other words, $s_i = +1$ if the encounter i is at P, -1 if it is at Q. The two equations (12.8) and (12.11) can then be merged into

$$y_i'' - y_i' = 2\pi s_i \left[-\Delta Z_i + \left(\beta_0 - \frac{J}{2} - s_i \frac{J}{2} \right) \Delta A_i \right] . \tag{12.13}$$

There is

$$\Delta Z_i = \left(\frac{\partial Z}{\partial C} \right)_A \Delta C_i + \left(\frac{\partial Z}{\partial A} \right)_C \Delta A_i + O(\Delta C_i^2) + O(\Delta A_i^2) . \tag{12.14}$$

We call ΔC the displacement of the Jacobi constant for the true orbit with respect to the bifurcation orbit. Since the true orbit and the intermediate orbit agree within $O(\mu)$ (Sect. 11.3.3), we have

$$C_i = C + O(\mu) , \qquad \Delta C_i = \Delta C + O(\mu) . \tag{12.15}$$

From (4.10) we have

$$\Delta A_i = \frac{3}{2} \sqrt{a} \Delta a_i + O(\Delta a_i^2) \tag{12.16}$$

where Δa_i is the variation of the semi-major axis of the intermediate orbit i with respect to the bifurcation orbit. Substituting (12.14) and (12.16) into (12.13), and using also (12.2), (12.4), and (12.15b), we obtain

$$\begin{aligned} y_i - y_{i-1} &= -2\pi s_i \left(\frac{\partial Z}{\partial C} \right)_A \Delta C \\ &+3\pi s_i \sqrt{a} \left[- \left(\frac{\partial Z}{\partial A} \right)_C + \beta_0 - \frac{J}{2} - s_i \frac{J}{2} \right] \Delta a_i \\ &+O(\mu) + O(\Delta C^2) + O(\Delta a_i^2) , \qquad i \in \mathcal{A} . \end{aligned} \tag{12.17}$$

This equation, linking the value a_i of an intermediate arc and the values y_{i-1} and y_i at its ends, will be called an *arc relation*.

12.1.2 Additional Relation for Two Arcs

We prove now a relation involving two consecutive basic arcs, which will be useful later on.

If we add (12.13) for the arcs i and $i+1$, and take into account (12.2) and (12.4), as well as $s_{i+1} = -s_i$, we obtain

$$\begin{aligned} y_{i+1} - y_{i-1} &= 2\pi s_i (\Delta Z_{i+1} - \Delta Z_i) - 2\pi s_i \left(\beta_0 - \frac{J}{2} \right) (\Delta A_{i+1} - \Delta A_i) \\ &-\pi J (\Delta A_i + \Delta A_{i+1}) + O(\mu) . \end{aligned} \tag{12.18}$$

The values of C_i and C_{i+1} are the same within $O(\mu)$ (see (12.15)). Therefore

$$\begin{aligned} \Delta Z_{i+1} - \Delta Z_i &= \left(\frac{\partial Z}{\partial A} \right)_C (\Delta A_{i+1} - \Delta A_i) \\ &+O(\mu) + O[(\Delta A_{i+1} - \Delta A_i)^2] . \end{aligned} \tag{12.19}$$

Substituting this into (12.18) and using (12.16), we obtain

$$y_{i+1} - y_{i-1} = 3\pi s_i \left[\left(\frac{\partial Z}{\partial A} \right)_C - \beta_0 + \frac{J}{2} \right] \sqrt{a}(\Delta a_{i+1} - \Delta a_i)$$
$$- \frac{3\pi J}{2} \sqrt{a}(\Delta a_i + \Delta a_{i+1}) + O(\mu) + O(\Delta a_i^2) + O(\Delta a_{i+1}^2) \,, \qquad i \in \mathcal{C} \,. \tag{12.20}$$

This relation is stronger than the relation which would have been obtained by simply adding (12.17) for the arcs i and $i+1$: there is no error term $O(\Delta C^2)$.

12.1.3 Encounter Relations

We establish now a relation for encounter i ($i \in \mathcal{C}$). We consider first the case where this encounter is near a point P ($s_i = +1$). The intermediate orbit i intersects the orbit of M_2 at time t_i'', in the vicinity of collision i. We call v_i'' and θ_i'' the polar coordinates of the velocity at that time, in rotating coordinates (see Fig. 8.2). There is (see (8.3))

$$v_i''^2 = 3 - C_i \tag{12.21}$$

and therefore from (12.15)

$$v_i'' = v + O(\mu) \tag{12.22}$$

where v is the velocity modulus for the true orbit.

$\cos\theta_i''$ can be expressed as a function of v_i'' and V_i'', the velocity modulus in fixed axes (Fig. 8.2):

$$\cos\theta_i'' = \frac{V_i''^2 - v_i''^2 - 1}{2v_i''} \,, \tag{12.23}$$

and we have (see (4.12a))

$$V_i''^2 = 2 - \frac{1}{a_i} \,, \tag{12.24}$$

so that

$$\cos\theta_i'' = \frac{1 - v^2 - 1/a_i}{2v} + O(\mu) \,. \tag{12.25}$$

Similarly, the intermediate orbit $i+1$ intersects the orbit of M_2 at time t_{i+1}', in the vicinity of collision i, and we define v_{i+1}' and θ_{i+1}' as the polar coordinates of the velocity at that time. A similar relation is obtained:

$$\cos\theta_{i+1}' = \frac{1 - v^2 - 1/a_{i+1}}{2v} + O(\mu) \tag{12.26}$$

and we have

$$\cos\theta'_{i+1} - \cos\theta''_i = \frac{1}{2v}\left(\frac{1}{a_i} - \frac{1}{a_{i+1}}\right) + O(\mu)$$
$$= \frac{\Delta a_{i+1} - \Delta a_i}{2va^2}\,[1 + O(\Delta a_i) + O(\Delta a_{i+1})] + O(\mu)\ . \tag{12.27}$$

We use now the modified second matching equation (11.79), which we project on the tangent to the orbit of M_2. (This direction is not exactly the same at t''_i and at t'_{i+1}; but the difference is $O(\mu)$ and therefore negligible.) In the term $\dot{\rho}_{i+1}(t_i)$, we can replace t_i by t'_{i+1} with an error $O(\mu)$; the projection is then $v'_{i+1}\cos\theta'_{i+1} + O(\mu)$, or $v\cos\theta'_{i+1} + O(\mu)$. In the same way, the term $\dot{\rho}_i(t_i)$ gives in projection $v\cos\theta''_i + O(\mu)$. On the other hand, the distance d from M_2 to the tangent to the orbit can be computed with an error $O(\epsilon_i^2)$ with M_2 replaced by a point M_2^* situated on the tangent to the circle, at the same distance y_i from the intersection point. We have thus: $d = |y_i|\sin\theta''_i + O(\epsilon_i^2)$. Finally the projection of $\jmath$ on the tangent to the circle is $\sin\theta''_i \times \operatorname{sign}(y_i)$. Therefore

$$y_i v(\cos\theta'_{i+1} - \cos\theta''_i) = -\frac{2\mu}{v} + O(\mu y_i) + O[\mu(\cos\theta'_{i+1} - \cos\theta''_i)]\ . \tag{12.28}$$

We can use for v the value of the bifurcation orbit, with an error $O(\epsilon_i)$. Substituting (12.27), we obtain

$$y_i(\Delta a_{i+1} - \Delta a_i) = -\frac{4\mu a^2}{v} + O(\mu y_i) + O(\mu\Delta a_i) + O(\mu\Delta a_{i+1})\ ,\quad i \in \mathcal{C}\ . \tag{12.29}$$

This equation, linking the value of y_i at an encounter and the values Δa_i and Δa_{i+1} of the adjacent arcs, will be called an *encounter relation*. It is identical with the equation obtained by Guillaume (1971, p.83, equation for Δa_1).

For an encounter near a point Q, exactly the same equation is obtained.

For a partial bifurcation, the equations (11.71) give

$$y_0 = O(\mu)\ ,\qquad y_n = O(\mu)\ . \tag{12.30}$$

12.1.4 Recapitulation

(12.17), (12.29), (12.30) constitute a set of equations for the y_i and the Δa_i. We simplify them by writing

$$\frac{4a^2}{v} = G_1\ ,\qquad 2\pi\left(\frac{\partial Z}{\partial C}\right)_A = G_2\ ,\qquad \frac{3\pi J\sqrt{a}}{2} = G_3\ ,$$
$$1 + \frac{2}{J}\left[\left(\frac{\partial Z}{\partial A}\right)_C - \beta_0\right] = K \tag{12.31}$$

(the last definition reproduces (8.14)). The quantities G_1, G_2, G_3, K are constants for a given bifurcation. G_1 and G_3 are always positive. G_2 and K are functions of the partial derivatives of the function $Z(A, C)$; in view of the complicated expression of Z (see (4.28); also 4.79)), and the appearance of the surface (Fig. 4.15), we expect them to be arbitrary, ordinary real numbers. Therefore we make the following fundamental conjectures:

Conjecture 12.1.1. *K never takes integer values.*

This conjecture is supported by the numerical computation of K for many bifurcations (see Sect. 8.2.1). It will be frequently used in what follows.

Conjecture 12.1.2. *G_2 never takes the value 0.*

This will also be needed in what follows.

We obtain the fundamental equations for bifurcations of type 1:

$$\begin{aligned} &y_i - y_{i-1} = -G_2 s_i \Delta C - G_3(1+Ks_i)\Delta a_i + O(\mu) + O(\Delta C^2) + O(\Delta a_i^2)\,, \\ &\qquad i \in \mathcal{A}\,, \\ &y_i(\Delta a_{i+1} - \Delta a_i) = -G_1\mu + O(\mu y_i) + O(\mu \Delta a_i) + O(\mu\Delta a_{i+1})\,, \quad i \in \mathcal{C}\,, \\ &y_0 = O(\mu)\,, \qquad y_n = O(\mu) \qquad \text{for a partial bifurcation.} \end{aligned} \tag{12.32}$$

The additional relation (12.20) becomes

$$\begin{aligned} y_{i+1} - y_{i-1} = &-G_3(1+Ks_i)\Delta a_i - G_3(1-Ks_i)\Delta a_{i+1} \\ &+O(\mu) + O(\Delta a_i^2) + O(\Delta a_{i+1}^2)\,, \qquad i \in \mathcal{C}\,. \end{aligned} \tag{12.33}$$

(12.32a) is a relation between ΔC, Δa_i, y_{i-1}, and y_i. (12.32b) is a relation between Δa_i, y_i, Δa_{i+1}. For ΔC given, successive elements can thus be computed step by step, in the order

$$\ldots \Delta a_i, y_i, \Delta a_{i+1}, y_{i+1}, \cdots\,, \tag{12.34}$$

as shown schematically by Fig. 12.2.

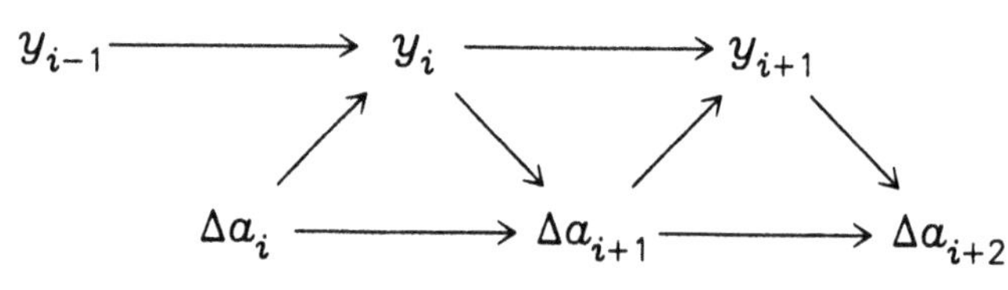

Fig. 12.2. Step by step computation.

We remark also that for a partial bifurcation, we have a system of $2n-1$ equations for the $2n$ variables ΔC, y_1 to y_{n-1}, Δa_1 to Δa_n (y_0 and y_n are given by (12.32c) and (12.32d)), and therefore a one-parameter family of solutions, which is the usual family of periodic orbits; these equations are therefore sufficient. For a total bifurcation, we have $2n$ relations for the $2n+1$ variables ΔC, y_1 to y_n, Δa_1 to Δa_n, and again the equations are sufficient.

The values of the s_i in (12.32) and (12.33) can be specified as follows. In a partial bifurcation, for the s_i to be determined according to (12.12), we must specify whether the starting point is in P or in Q. But the equations (12.32) and (12.33) show that for any solution, there exists another solution obtained by changing the signs of K and of the s_i. Therefore the study of a value K with the starting point in P is identical with the study of the value $-K$ with the starting point in Q (see Sect. 8.4.1), and we will consider only the case where the starting point is in P. In exchange for this simplification, we will have to consider all values of K, positive or negative.

In a total bifurcation, by convention, the origin is in P; see Sect. 6.2.1.2. It follows from these conventions that, in all cases:

$$s_i = (-1)^i \, . \tag{12.35}$$

12.2 Exclusion of Successive Identical T-Arcs

In this section we make a short excursion away from bifurcation orbits. It turns out that the formalism which has been developed in the previous chapter and in this one can be applied not only to a bifurcation orbit, but also to an ordinary generating orbit. We will use it to prove the fundamental Proposition 4.3.2, which has been already used many times in previous chapters: *An ordinary generating orbit of the second species cannot contain two identical T-arcs of type 1 in succession.*

We recall that an ordinary generating orbit is defined as a generating orbit which is not a bifurcation orbit, i.e. which belongs to only one family.

Assume that an ordinary generating orbit contains a sequence of two or more identical T-arcs of type 1. Each T-arc has collisions only in P or only in Q; otherwise the orbit would be a bifurcation orbit of type 1 (Sect. 6.2.1.2).

If the T-arcs have intermediate collisions, they can be decomposed into smaller T-arcs. We assume that this decomposition has been carried out. (This is not a restriction, because we still have a sequence of two or more identical T-arcs, to which the proposition applies.) Each T-arc is then a basic arc, with no intermediate collisions.

All T-arcs of the sequence begin and end in the same point (P or Q). The deflection angles between them vanish (all arcs have the same supporting ellipse). We distinguish two cases, by analogy with total and partial bifurcations (Sect. 6.2):

1) These T-arcs make up the whole orbit. We call this a *total T-sequence.*

2) The T-arcs make up only part of the generating orbit. We call this a *partial T-sequence.* The deflection angles at the two ends of the sequence do not vanish.

We call $\tilde{n}$ the number of T-arcs in the T-sequence. We number the collisions and the T-arcs as in Sect. 11.3.1. We define intermediate arcs as in Sect. 11.3.2. We obtain the same matching relations as in the case of bifurcations: (11.65), (11.70) or (11.79), (11.71). We define t_i, y_i, t_i'', y_i'', t_i', y_i', as in Sect. 12.1.1. The computations of the arc relation, however, must be redone since a basic arc is now from P to P or from Q to Q. In the particular case where the intermediate orbit i has a collision in its initial point, i.e. the lead y_i' is zero, the lead at the end is

$$y_i'' = -2\pi J \Delta A_i \; . \tag{12.36}$$

In the general case where the initial lead y_i' is non-zero, we have then

$$y_i'' - y_i' = -2\pi J \Delta A_i \; . \tag{12.37}$$

Using (12.2), (12.4), and (12.16), we obtain the arc relation

$$y_i - y_{i-1} = -3\pi J \sqrt{a} \Delta a_i + O(\mu) + O(\Delta a_i^2) \; , \qquad i \in \mathcal{A} \; . \tag{12.38}$$

The collision relation (12.29) is unchanged. So are the boundary conditions (12.30) for a partial T-sequence.

We multiply (12.38) by Δa_i:

$$y_i \Delta a_i - y_{i-1} \Delta a_i = -3\pi J \sqrt{a} (\Delta a_i)^2 + O(\mu \Delta a_i) + O(\Delta a_i^3) \; , \qquad i \in \mathcal{A} \; . \tag{12.39}$$

We consider first the case of a total T-sequence. Adding the equations (12.39) for $i = 1$ to $\tilde{n}$ and the equations (12.29) for $i = 1$ to $\tilde{n}$, we obtain

$$0 = -3\pi J \sqrt{a} \sum_{i=1}^{\tilde{n}} (\Delta a_i)^2 - \frac{4\mu a^2}{v} \tilde{n} + O(\mu \Delta a) + O(\Delta a^3) + O(\mu y) \tag{12.40}$$

where we have written

$$\Delta a = \max(|\Delta a_i|) \; , \qquad y = \max(|y_i|) \; . \tag{12.41}$$

For $\mu \to 0$, the last three terms become negligible in comparison to the first two, which are of order Δa^2 and μ, respectively. On the other hand, the first term is negative or zero, and the second term is negative. We have reached an impossibility, and therefore an ordinary generating orbit cannot be made of a total T-sequence.

Next we consider the case of a partial T-sequence. We add the equations (12.39) for $i = 1$ to $\tilde{n}$ and the equations (12.29) for $i = 1$ to $\tilde{n} - 1$. Using also the boundary conditions (12.30), we obtain

$$0 = -3\pi J \sqrt{a} \sum_{i=1}^{\tilde{n}} (\Delta a_i)^2 - \frac{4\mu a^2}{v} (\tilde{n} - 1) + O(\mu \Delta a) + O(\Delta a^3) + O(\mu y) \; . \tag{12.42}$$

For $\tilde{n} > 1$, this is again impossible. Thus, an ordinary generating orbit cannot contain a partial T-sequence of two or more T-arcs.

The case $\tilde{n} = 1$ is possible, because the second term vanishes. This corresponds to a single T-arc, not adjacent to identical T-arcs.

The proposition 4.3.2 concerns only *identical* T-arcs. It is perfectly possible for a T-arc to be followed by another T-arc which differs in the values of I and J, or in the superscript i or e (Sect. 4.3.4). In that case the deflection angle does not vanish.

12.3 The Case $\nu = 0$

We return now to the study of the vicinity of bifurcations. We consider first the case $\mu = 0$, $\Delta C \neq 0$, corresponding to $\nu = 0$. This case is simple and it is not necessary to use the machinery described in Sect. 11.4. The solutions are known: they correspond to a displacement from the bifurcation orbit along one of the branches which emanate from it. For a total bifurcation, this can be either a first species elliptical orbit (particular case) or a second species orbit, formed of S- and T-arcs (general case) (Sects 6.2.1.2). For a partial bifurcation, we have a bifurcating arc, again formed of S- and T-arcs (Sect. 6.2.2.1).

We determine first the orders of magnitude in ΔC of the quantities Δa_i and y_i. The fundamental equations (12.32) reduce to

$$\begin{aligned} & y_i - y_{i-1} = -G_2 s_i \Delta C - G_3(1 + K s_i)\Delta a_i + O(\Delta C^2) + O(\Delta a_i^2) , \\ & \qquad i \in \mathcal{A} , \\ & y_i(\Delta a_{i+1} - \Delta a_i) = 0 , \qquad i \in \mathcal{C} , \\ & y_0 = 0 , \qquad y_n = 0 \qquad \text{for a partial bifurcation.} \end{aligned} \tag{12.43}$$

The additional relation (12.33) becomes

$$\begin{aligned} & y_{i+1} - y_{i-1} = -G_3(1 + K s_i)\Delta a_i - G_3(1 - K s_i)\Delta a_{i+1} \\ & \qquad + O(\Delta a_i^2) + O(\Delta a_{i+1}^2) , \qquad i \in \mathcal{C} . \end{aligned} \tag{12.44}$$

12.3.1 First Species Orbit

This orbit consists of the two basic arcs. Using (12.35), we find that the equations (12.43) give 4 equations for the 4 unknowns y_0, y_1, Δa_1, Δa_2:

$$\begin{aligned} & y_1 - y_0 = G_2 \Delta C - G_3(1 - K)\Delta a_1 + O(\Delta C^2) + O(\Delta a_1^2) , \\ & y_0 - y_1 = -G_2 \Delta C - G_3(1 + K)\Delta a_2 + O(\Delta C^2) + O(\Delta a_2^2) , \\ & y_0(\Delta a_1 - \Delta a_2) = 0 , \qquad y_1(\Delta a_2 - \Delta a_1) = 0 , \end{aligned} \tag{12.45}$$

and the additional relation (12.44) becomes

$$0 = -G_3(1 + K)\Delta a_2 - G_3(1 - K)\Delta a_1 + O(\Delta a_1^2) + O(\Delta a_2^2) . \tag{12.46}$$

y_0 and y_1 are non-zero for a first species orbit outside of the bifurcation; therefore, from (12.45c,d)

$$\Delta a_1 = \Delta a_2 . \tag{12.47}$$

From (12.46) we have then

$$\Delta a_1 = \Delta a_2 = 0 . \tag{12.48}$$

We recover the invariance of the period along a first species elliptical family. (This is an exact invariance, not simply an approximation for small ΔC.) (12.45a,b) reduce to a single equation for two unknowns:

$$y_1 - y_0 = G_2 \Delta C + O(\Delta C^2) . \tag{12.49}$$

The system is degenerate. This degeneracy can be removed by using the fact that the first species elliptical orbits of interest here are symmetric. Therefore $y_1 = -y_0$, and

$$y_1 = -y_0 = \frac{G_2}{2} \Delta C + O(\Delta C^2) . \tag{12.50}$$

12.3.2 Second Species Orbit

The orbit or bifurcating arc, in a family emanating from the bifurcation, consists in a sequence of S- and/or T-arcs. We have, for every collision i which is a node, and also for the two ends of a bifurcating arc

$$y_i = 0 . \tag{12.51}$$

As a consequence, we obtain for each arc an independent set of equations, which can be solved separately.

12.3.2.1 *T*-Arc. We consider first the case of a T-arc, running from i to $i+2$. The equations (12.43) become

$$\begin{aligned} y_{i+1} - y_i &= G_2 s_i \Delta C - G_3(1 - K s_i)\Delta a_{i+1} + O(\Delta C^2) + O(\Delta a_{i+1}^2) , \\ y_{i+2} - y_{i+1} &= -G_2 s_i \Delta C - G_3(1 + K s_i)\Delta a_{i+2} + O(\Delta C^2) + O(\Delta a_{i+2}^2) , \\ y_{i+1}(\Delta a_{i+2} &- \Delta a_{i+1}) = 0 ; \end{aligned} \tag{12.52}$$

the additional relation (12.44) becomes

$$\begin{aligned} y_{i+2} - y_i = &-G_3(1 - K s_i)\Delta a_{i+1} - G_3(1 + K s_i)\Delta a_{i+2} \\ &+ O(\Delta a_{i+1}^2) + O(\Delta a_{i+2}^2) \end{aligned} \tag{12.53}$$

and (12.51) gives

$$y_i = 0 , \qquad y_{i+2} = 0 . \tag{12.54}$$

y_{i+1} is non-zero for the T-arc outside of the bifurcation (it is an antinode); therefore, from (12.52c)

$$\Delta a_{i+1} = \Delta a_{i+2} . \tag{12.55}$$

From (12.53) and (12.54), we have then

$$\Delta a_{i+1} = \Delta a_{i+2} = 0 . \tag{12.56}$$

We recover the invariance of the period along a T family. (Here again this is an exact invariance.) Finally, from (12.52a,b), we have

$$y_{i+1} = G_2 s_i \Delta C + O(\Delta C^2) . \tag{12.57}$$

12.3.2.2 *S*-Arc. We consider now the case of a S-arc, running from i to $i+m$. We have $y_{i+j} \neq 0$ for the intermediate antinodes ($j = 1, 2, \ldots, m-1$); therefore, from (12.43b),

$$\Delta a_{i+1} = \Delta a_{i+2} = \ldots = \Delta a_{i+m} \ . \tag{12.58}$$

We call Δa this common value. (12.51) gives

$$y_i = 0 \ , \qquad y_{i+m} = 0 \ . \tag{12.59}$$

Summing (12.43a) over the m basic arcs and using (12.59), we find

$$0 = G_2 s_i \Delta C - m G_3 \Delta a + G_3 K s_i \Delta a + O(\Delta C^2) + O(\Delta a^2) \tag{12.60}$$

from which we extract

$$\Delta a = \frac{G_2}{(m s_i - K) G_3} \Delta C + O(\Delta C^2) \ . \tag{12.61}$$

From (12.59) and (12.43a), we can then compute the successive values of the y_i:

$$y_{i+j} = \begin{cases} \dfrac{m-j}{m s_i - K} G_2 \Delta C + O(\Delta C^2) & \text{if } j \text{ is odd,} \\ \dfrac{-j}{m s_i - K} G_2 \Delta C + O(\Delta C^2) & \text{if } j \text{ is even.} \end{cases} \tag{12.62}$$

12.4 The Case $0 < \nu < 1/2$

We consider now the case $\mu > 0$, $\Delta C \neq 0$, corresponding to $\nu > 0$. We recall the fundamental relation (11.80): $\Delta C = \Theta(\mu^\nu)$. We will try to extend each solution obtained in the previous section for $\nu = 0$ to the present case $\nu > 0$. This corresponds to the asymptotic branches of the families of orbits for $\mu \neq 0$, which are close to the branches for $\mu = 0$ (see Fig. 1.1).

Here we use for the first time the general method described in Sect. 11.4. Reference is made below to the successive steps.

Step 1. We estimate orders of magnitude by extrapolating from the previous case $\nu = 0$. The equations (12.50), (12.57), (12.62) suggest that y_i is of the order of ΔC inside a first species orbit, a T-arc, or a S-arc; (12.61) suggests that Δa_i is of the order of ΔC in a S-arc.

On the other hand, some quantities vanish for $\nu = 0$: y_i at a node (see (12.52)), Δa_i in a first species orbit or a T-arc (see (12.48), (12.56)). We cannot at the moment estimate their order of magnitude for $\nu > 0$ (this will be remedied below; see Step 8 in Sects. 12.4.1 and 12.4.2). These quantities will become generally non-zero for $\nu > 0$; it seems reasonable to assume that they will be small compared to ΔC if ν is sufficiently small.

Step 2. This suggests the following change of variables:

$$y_i = \Delta C \, \bar{Y}_i^* \ , \qquad \Delta a_i = \Delta C \, \bar{X}_i^* \ . \tag{12.63}$$

Step 3. It will be convenient to make a second change of variables in order to simplify the equations:

$$\bar{Y}_i^* = -G_2 s_i \bar{Y}_i \,, \qquad \bar{X}_i^* = \frac{G_2}{G_3} s_i \bar{X}_i \,. \tag{12.64}$$

Step 4. Substituting in (12.32), collecting the dominant terms in the left-hand members, and dividing by ΔC or ΔC^2 as appropriate, we obtain

$$\begin{aligned}
&\bar{Y}_i + \bar{Y}_{i-1} - 1 - (1 + K s_i)\bar{X}_i = O(\mu \Delta C^{-1}) + O(\Delta C) \,, \qquad i \in \mathcal{A} \,, \\
&\bar{Y}_i(\bar{X}_{i+1} + \bar{X}_i) = -\frac{G_1 G_3}{G_2^2} \mu \Delta C^{-2} + O(\mu \Delta C^{-1}) \,, \qquad i \in \mathcal{C} \,, \\
&\bar{Y}_0 = O(\mu \Delta C^{-1}) \,, \qquad \bar{Y}_n = O(\mu \Delta C^{-1}) \qquad \text{for a partial bifurcation.}
\end{aligned} \tag{12.65}$$

The additional relation (12.33) becomes

$$\begin{aligned}
&\bar{Y}_{i+1} - \bar{Y}_{i-1} + (1 + K s_i)\bar{X}_i - (1 - K s_i)\bar{X}_{i+1} = \\
&\qquad O(\mu \Delta C^{-1}) + O(\bar{X}_i^2 \Delta C) + O(\bar{X}_{i+1}^2 \Delta C) \,, \qquad i \in \mathcal{C} \,.
\end{aligned} \tag{12.66}$$

Step 5. We want all terms in the right-hand members to vanish for $\mu \to 0$. This is the case if

$$0 < \nu < \frac{1}{2} \,. \tag{12.67}$$

The asymptotic equations are then obtained by equating the right-hand members to zero:

$$\begin{aligned}
&\bar{Y}_i + \bar{Y}_{i-1} - 1 - (1 + K s_i)\bar{X}_i = 0 \,, \qquad i \in \mathcal{A} \,, \\
&\bar{Y}_i(\bar{X}_{i+1} + \bar{X}_i) = 0 \,, \qquad i \in \mathcal{C} \,, \\
&\bar{Y}_0 = 0 \,, \qquad \bar{Y}_n = 0 \qquad \text{for a partial bifurcation,}
\end{aligned} \tag{12.68}$$

and

$$\bar{Y}_{i+1} - \bar{Y}_{i-1} + (1 + K s_i)\bar{X}_i - (1 - K s_i)\bar{X}_{i+1} = 0 \,, \qquad i \in \mathcal{C} \,. \tag{12.69}$$

We solve now these equations. We observe that (12.68b) has two distinct solutions. The solution $\bar{Y}_i = 0$ corresponds to the case where i becomes a node in the asymptotic limit; the case $\bar{Y}_i \neq 0$, $\bar{X}_i + \bar{X}_{i+1} = 0$ corresponds to the case where i becomes an antinode. Accordingly, the form of the solution depends on whether the asymptotic orbit belongs to the first species or the second species. Steps 6 to 9 will be executed separately for each species.

12.4.1 First Species Orbit

Step 6. The two collisions become antinodes; therefore the asymptotic equations (12.68) become

$$\begin{aligned}
&\bar{Y}_1 + \bar{Y}_0 - 1 - (1 - K)\bar{X}_1 = 0 \,, \qquad \bar{Y}_0 + \bar{Y}_1 - 1 - (1 + K)\bar{X}_2 = 0 \,, \\
&\bar{X}_1 + \bar{X}_2 = 0 \,.
\end{aligned} \tag{12.70}$$

These equations reduce to

$$\bar{X}_1 = \bar{X}_2 = 0 \,, \qquad \bar{Y}_0 + \bar{Y}_1 = 1 \,. \tag{12.71}$$

As before, there is a degeneracy. It can be removed by using the symmetry property, which gives $\bar{Y}_0 = \bar{Y}_1$, so that

$$\bar{Y}_0 = \bar{Y}_1 = \frac{1}{2} \,. \tag{12.72}$$

Step 7. For μ positive and small, the solution of the system of equations (12.65) continues to exist, provided that

1. the error terms in the right-hand members are small,
2. the Jacobian of the asymptotic equations (12.68) is non-zero (implicit function theorem).

The first condition is satisfied if $\nu < 1/2$. If we compute the Jacobian of the equations (12.70), we find that it vanishes; this is a consequence of the degeneracy. We eliminate this degeneracy by using the symmetry relation $\bar{Y}_0 = \bar{Y}_1$. The system reduces to 3 equations for the 3 variables $(\bar{Y}_1, \bar{X}_1, \bar{X}_2)$:

$$\begin{aligned} &2\bar{Y}_0 - 1 - (1-K)\bar{X}_1 = 0 \,, \qquad 2\bar{Y}_1 - 1 - (1+K)\bar{X}_2 = 0 \,, \\ &\bar{Y}_0(\bar{X}_1 + \bar{X}_2) = 0 \,. \end{aligned} \tag{12.73}$$

The Jacobian is

$$|J| = \begin{vmatrix} 2 & -(1-K) & 0 \\ 2 & 0 & -(1+K) \\ 0 & 1/2 & 1/2 \end{vmatrix} = 2 \,. \tag{12.74}$$

Therefore: *For $0 < \nu < 1/2$, there exists for $\mu > 0$ a solution of the fundamental equations (12.65) close to the solution (12.71a), (12.72) of the asymptotic equations*. The error is of the order of the largest term in the right-hand members of (12.65):

$$\begin{aligned} &\bar{X}_1, \bar{X}_2 = O(\Delta C) + O(\mu\Delta C^{-2}) \,, \\ &\bar{Y}_0, \bar{Y}_1 = \frac{1}{2} + O(\Delta C) + O(\mu\,\Delta C^{-2}) \,. \end{aligned} \tag{12.75}$$

Step 8. We can (although it is not necessary) obtain now finer estimates for $\bar{X}_1$ and $\bar{X}_2$. From (12.65b) and (12.75b) we have

$$\bar{X}_1 + \bar{X}_2 = -2\frac{G_1 G_3}{G_2^2}\mu\Delta C^{-2}\left[1 + O(\Delta C) + O(\mu\Delta C^{-2})\right] \,. \tag{12.76}$$

Combining with (12.66) (with $\bar{Y}_{i+1} = \bar{Y}_{i-1}$), we obtain

$$\begin{aligned} &\bar{X}_1 = -(1+K)\frac{G_1 G_3}{G_2^2}\mu\Delta C^{-2}\left[1 + O(\Delta C) + O(\mu\Delta C^{-2})\right] \,, \\ &\bar{X}_2 = -(1-K)\frac{G_1 G_3}{G_2^2}\mu\Delta C^{-2}\left[1 + O(\Delta C) + O(\mu\Delta C^{-2})\right] \,. \end{aligned} \tag{12.77}$$

Step 9. Finally, we can go back to the physical variables, using (12.64) and (12.63):

$$\begin{aligned} y_0 &= -\frac{G_2}{2}\Delta C\left[1+O(\Delta C)+O(\mu\Delta C^{-2})\right] , \\ y_1 &= \frac{G_2}{2}\Delta C\left[1+O(\Delta C)+O(\mu\,\Delta C^{-2})\right] , \\ \Delta a_1 &= (K+1)\frac{G_1}{G_2}\mu\Delta C^{-1}\left[1+O(\Delta C)+O(\mu\Delta C^{-2})\right] , \\ \Delta a_2 &= (K-1)\frac{G_1}{G_2}\mu\Delta C^{-1}\left[1+O(\Delta C)+O(\mu\Delta C^{-2})\right] . \end{aligned} \tag{12.78}$$

12.4.2 Second Species Orbit

Step 6. For $\nu > 0$, we continue to call *node* a value of i for which

$$\bar{Y}_i = 0 \tag{12.79}$$

(excluding the two ends of a bifurcating arc) and *antinode* a value of i for which $\bar{Y}_i \neq 0$. We call also T-*arc* or S-*arc* the part of the orbit between two nodes, or between one node and an end of the bifurcating arc. It reduces to a T- or S-arc as defined in Sect. 4.3 for $\nu \to 0$.

We again obtain for each arc an independent set of equations. We consider first the case of a T-arc. The equations are

$$\begin{aligned} &\bar{Y}_i = \bar{Y}_{i+2} = 0 , \\ &\bar{Y}_{i+1} + \bar{Y}_i - 1 - (1 - Ks_i)\bar{X}_{i+1} = 0 , \\ &\bar{Y}_{i+2} + \bar{Y}_{i+1} - 1 - (1 + Ks_i)\bar{X}_{i+2} = 0 , \\ &\bar{X}_{i+1} + \bar{X}_{i+2} = 0 . \end{aligned} \tag{12.80}$$

This linear system of equations is easily solved:

$$\bar{X}_{i+1} = \bar{X}_{i+2} = 0 , \qquad \bar{Y}_{i+1} = 1 . \tag{12.81}$$

We consider next the case of a S-arc. There is for the intermediate antinodes

$$\bar{X}_{i+j} + \bar{X}_{i+j+1} = 0 \qquad (j = 1, \ldots, m-1) . \tag{12.82}$$

We write $\bar{X}_{i+1} = -\bar{X}$; we have then

$$\bar{X}_{i+j} = (-1)^j \bar{X} . \tag{12.83}$$

(12.68a) gives for the successive arcs

$$\begin{aligned} &\bar{Y}_{i+1} + \bar{Y}_i - 1 + (1 - Ks_i)\bar{X} = 0 , \\ &\bar{Y}_{i+2} + \bar{Y}_{i+1} - 1 - (1 + Ks_i)\bar{X} = 0 , \\ &\ldots \\ &\bar{Y}_{i+m} + \bar{Y}_{i+m-1} - 1 + (1 - Ks_i)\bar{X} = 0 . \end{aligned} \tag{12.84}$$

Adding the odd equations and subtracting the even equations, and using (12.68c,d), we obtain

$$-1 + (m - Ks_i)\bar{X} = 0 , \tag{12.85}$$

from which

$$\bar{X} = \frac{1}{m - Ks_i} \tag{12.86}$$

and

$$\bar{X}_{i+j} = \frac{(-1)^j}{m - Ks_i} , \qquad \bar{Y}_{i+j} = \begin{cases} \dfrac{m-j}{m - Ks_i} & \text{if } j \text{ is odd,} \\ \dfrac{j}{m - Ks_i} & \text{if } j \text{ is even.} \end{cases} \tag{12.87}$$

Step 7. We have thus obtained the asymptotic solution for an orbit or a bifurcating arc of the second species. For $0 < \mu \ll 1$, the solution of the system (12.65) continues to exist if $\nu < 1/2$ and if the Jacobian $|J|$ of the asymptotic equations (12.68) is not zero. We must now compute the latter.

There are $2n$ variables and $2n$ equations for a total bifurcation; $2n + 1$ variables and $2n + 1$ equations for a partial bifurcation. We consider a node in i. The partial derivatives of the expression f appearing on the left-hand side of (12.68b) are

$$\frac{\partial f}{\partial \bar{Y}_i} = \bar{X}_{i+1} + \bar{X}_i , \qquad \frac{\partial f}{\partial \bar{X}_{i+1}} = \bar{Y}_i , \qquad \frac{\partial f}{\partial \bar{X}_i} = \bar{Y}_i . \tag{12.88}$$

Only the first of these derivatives is non-zero. Therefore the corresponding line in $|J|$ has a single non-zero element. We can then factor out this element and eliminate the corresponding line and column.

After this has been done for all nodes, $|J|$ is decomposed into a product of smaller determinants, each of which corresponds to a T- or S-arc. For a T-arc, there are 3 variables $\bar{X}_{i+1}$, $\bar{Y}_{i+1}$, $\bar{X}_{i+2}$, and 3 equations of the form (12.80b,c,d); the determinant is

$$\begin{vmatrix} -1 + Ks_i & 1 & 0 \\ 0 & 1 & -1 - Ks_i \\ \bar{Y}_{i+1} & 0 & \bar{Y}_{i+1} \end{vmatrix} = -2\bar{Y}_{i+1} . \tag{12.89}$$

This is non-zero in view of (12.81).

For a S-arc, formed of m basic arcs, there are $2m - 1$ variables $\bar{X}_{i+1}$, $\bar{Y}_{i+1}$, $\bar{X}_{i+2}$, $\ldots$, $\bar{X}_{i+m}$, and $2m - 1$ equations (12.68a) and (12.68b). The determinant is easily computed by successive eliminations; we obtain

$$-m \prod_{j=1}^{m-1} \bar{Y}_{i+j} . \tag{12.90}$$

This is non-zero in view of (12.87b).

Thus $|J|$ is non-zero. *For $\nu < 1/2$, there exists for $\mu \neq 0$ a solution of (12.65) close to (12.79), (12.81), (12.87).* The error is of the order of the largest term in the right-hand members of (12.65), so that the solution has the following form:

- For a node:

$$\bar{Y}_i = O(\Delta C) + O(\mu \Delta C^{-2}) . \tag{12.91}$$

- For a T-arc:

$$\begin{aligned} \bar{X}_{i+1} &= O(\Delta C) + O(\mu \Delta C^{-2}) , \\ \bar{X}_{i+2} &= O(\Delta C) + O(\mu \Delta C^{-2}) , \\ \bar{Y}_{i+1} &= 1 + O(\Delta C) + O(\mu \Delta C^{-2}) . \end{aligned} \tag{12.92}$$

- For a S-arc:

$$\begin{aligned} \bar{X}_{i+j} &= \frac{(-1)^j}{m - Ks_i} + O(\Delta C) + O(\mu \Delta C^{-2}) , \\ \bar{Y}_{i+j} &= \begin{cases} \dfrac{m-j}{m - Ks_i} + O(\Delta C) + O(\mu \Delta C^{-2}) & \text{if } j \text{ is odd,} \\ \dfrac{j}{m - Ks_i} + O(\Delta C) + O(\mu \Delta C^{-2}) & \text{if } j \text{ is even.} \end{cases} \end{aligned} \tag{12.93}$$

Step 8. We can obtain now finer estimates for some variables, giving the dominant term explicitly. We consider first the value $\bar{Y}$ at a node i. As in Sect. 8.2, we designate by the subscripts b and a (before and after) the quantities relating to the two arcs which are joined in i; in particular we call m_b and m_a the values of m for these arcs.

- At a TS node (joining a T-arc to a S-arc), we have for the S-arc, according to (12.93):

$$\bar{X}_{i+1} = \frac{-1}{m_a - Ks_i} + O(\Delta C) + O(\mu \Delta C^{-2}) , \tag{12.94}$$

and for the T-arc, according to (12.92)

$$\bar{X}_i = O(\Delta C) + O(\mu \Delta C^{-2}) . \tag{12.95}$$

(12.65b) gives then

$$\bar{Y}_i = (m_a - Ks_i) \frac{G_1 G_3}{G_2^2} \mu \Delta C^{-2} \left[1 + O(\Delta C) + O(\mu \Delta C^{-2}) \right] . \tag{12.96}$$

- For a ST node, we have for the S arc

$$\bar{X}_i = \frac{-1}{m_b + Ks_i} + O(\Delta C) + O(\mu \Delta C^{-2}) , \tag{12.97}$$

and for the T-arc

$$\bar{X}_{i+1} = O(\Delta C) + O(\mu \Delta C^{-2}) , \tag{12.98}$$

so that

$$\bar{Y}_i = (m_b + Ks_i) \frac{G_1 G_3}{G_2^2} \mu \Delta C^{-2} \left[1 + O(\Delta C) + O(\mu \Delta C^{-2}) \right] . \tag{12.99}$$

– Finally, for a SS node, the equations (12.94) and (12.97) give

$$\bar{Y}_i = \frac{(m_b + Ks_i)(m_a - Ks_i)}{m_b + m_a} \frac{G_1 G_3}{G_2^2} \mu \Delta C^{-2} \left[1 + O(\Delta C) + O(\mu \Delta C^{-2})\right] . \qquad (12.100)$$

Next we consider the values $\bar{X}$ for a T-arc. (12.65b) gives, using (12.92),

$$\bar{X}_{i+2} + \bar{X}_{i+1} = -\frac{G_1 G_3}{G_2^2} \mu \Delta C^{-2} \left[1 + O(\Delta C) + O(\mu \Delta C^{-2})\right] . \qquad (12.101)$$

Combining with the additional relation (12.66), we obtain

$$\begin{aligned} \bar{X}_{i+1} &= \frac{\bar{Y}_i - \bar{Y}_{i+2}}{2} - (1 + Ks_i) \frac{G_1 G_3}{2G_2^2} \mu \Delta C^{-2} \\ &\quad + O(\mu \Delta C^{-1}) + O(\bar{X}_{i+1}^2 \Delta C) + O(\bar{X}_{i+2}^2 \Delta C) + O(\mu^2 \Delta C^{-4}) , \\ \bar{X}_{i+2} &= \frac{\bar{Y}_{i+2} - \bar{Y}_i}{2} - (1 - Ks_i) \frac{G_1 G_3}{2G_2^2} \mu \Delta C^{-2} \\ &\quad + O(\mu \Delta C^{-1}) + O(\bar{X}_{i+1}^2 \Delta C) + O(\bar{X}_{i+2}^2 \Delta C) + O(\mu^2 \Delta C^{-4}) . \end{aligned} \qquad (12.102)$$

$\bar{Y}_i$ is given by (12.65c) if the node i is the end of a bifurcating arc, or by (12.99) if it is a junction of the T-arc with a preceding S-arc. We define: $g' = 0$ in the first case, $g' = 1$ in the second case, so that both cases can be written as a single formula

$$\bar{Y}_i = g'(m_b + Ks_i) \frac{G_1 G_3}{G_2^2} \mu \Delta C^{-2} + O(\mu \Delta C^{-1}) + O(\mu^2 \Delta C^{-4}) . \qquad (12.103)$$

Similarly, we define $g'' = 0$ if the node $i + 2$ is the end of a bifurcating arc, $g'' = 1$ if it is a junction of the T-arc with a following S-arc. Using (12.96), we obtain

$$\bar{Y}_{i+2} = g''(m_a - Ks_i) \frac{G_1 G_3}{G_2^2} \mu \Delta C^{-2} + O(\mu \Delta C^{-1}) + O(\mu^2 \Delta C^{-4}) . \qquad (12.104)$$

Substituting into (12.102), we obtain

$$\begin{aligned} \bar{X}_{i+1} &= \left[-(1 + Ks_i) + g'(m_b + Ks_i) - g''(m_a - Ks_i)\right] \frac{G_1 G_3}{2G_2^2} \mu \Delta C^{-2} \\ &\quad + O(\mu \Delta C^{-1}) + O(\mu^2 \Delta C^{-4}) , \\ \bar{X}_{i+2} &= \left[-(1 - Ks_i) - g'(m_b + Ks_i) + g''(m_a - Ks_i)\right] \frac{G_1 G_3}{2G_2^2} \mu \Delta C^{-2} \\ &\quad + O(\mu \Delta C^{-1}) + O(\mu^2 \Delta C^{-4}) . \end{aligned} \qquad (12.105)$$

Step 9. Finally, we can go back to the physical variables, using (12.64) and (12.63); we obtain a generalization of the equations (12.51), (12.56), (12.57), (12.61), (12.62):

– For a TS node:

$$y_i = -(m_a s_i - K)\frac{G_1 G_3}{G_2}\mu\Delta C^{-1}\left[1 + O(\Delta C) + O(\mu\Delta C^{-2})\right] . \quad (12.106)$$

– For a ST node:

$$y_i = -(m_b s_i + K)\frac{G_1 G_3}{G_2}\mu\Delta C^{-1}\left[1 + O(\Delta C) + O(\mu\Delta C^{-2})\right] . \quad (12.107)$$

– For a SS node:

$$y_i = -\frac{(m_b s_i + K)(m_a s_i - K)}{(m_b + m_a)s_i}\frac{G_1 G_3}{G_2}\mu\Delta C^{-1} \left[1 + O(\Delta C) + O(\mu\Delta C^{-2})\right] . \quad (12.108)$$

– Antinode: For a T-arc:

$$\Delta a_{i+1} = \left[(K + s_i) - g'(m_b s_i + K) + g''(m_a s_i - K)\right]\frac{G_1}{2G_2}\mu\Delta C^{-1} \left[1 + O(\Delta C) + O(\mu\Delta C^{-2})\right] ,$$

$$\Delta a_{i+2} = \left[(K - s_i) - g'(m_b s_i + K) + g''(m_a s_i - K)\right]\frac{G_1}{2G_2}\mu\Delta C^{-1} \left[1 + O(\Delta C) + O(\mu\Delta C^{-2})\right] ,$$

$$y_{i+1} = G_2 s_i \Delta C\left[1 + O(\Delta C) + O(\mu\Delta C^{-2})\right] . \quad (12.109)$$

– For a S-arc:

$$\Delta a_{i+j} = \frac{1}{m s_i - K}\frac{G_2}{G_3}\Delta C\left[1 + O(\Delta C) + O(\mu\Delta C^{-2})\right] ,$$

$$y_{i+j} = \begin{cases} \dfrac{m-j}{m s_i - K} G_2 \Delta C\left[1 + O(\Delta C) + O(\mu\Delta C^{-2})\right] & \text{for } j \text{ odd,} \\ -\dfrac{j}{m s_i - K} G_2 \Delta C\left[1 + O(\Delta C) + O(\mu\Delta C^{-2})\right] & \text{for } j \text{ even.} \end{cases} \quad (12.110)$$

Fig. 12.3 shows in log-log plots the variations in order of magnitude of the quantities y_i and Δa_i as functions of ΔC, given by (12.106) to (12.110) in the interval $\mu^{1/2} < \Delta C < 1$.

12.4.3 Sides of Passage

Incidentally, we recover the rules for the side of passage established in chapter 8. As is easily seen from Fig. 12.1, y_i has the sign of $s_i \sigma \cos\theta$, with σ defined in Sect. 8.1. Therefore, from the definition (8.22) of s and the definition (12.31) of G_2, σ has the sign of $-s s_i G_2 \Delta C\, y_i$. On the other hand, m_a equals k_a if the point i is P, $-k_a$ if it is Q, or $s_i k_a$ in general. Similarly, m_b equals $-s_i k_b$. The above equations show then that for a TS node, σ has the sign of $s s_i(k_a - K)$; for a ST node, the sign of $s s_i(K - k_b)$; and for a SS node, the sign of $s(K - k_b)(k_a - K)$. We recover the results of Sect. 8.2.1.

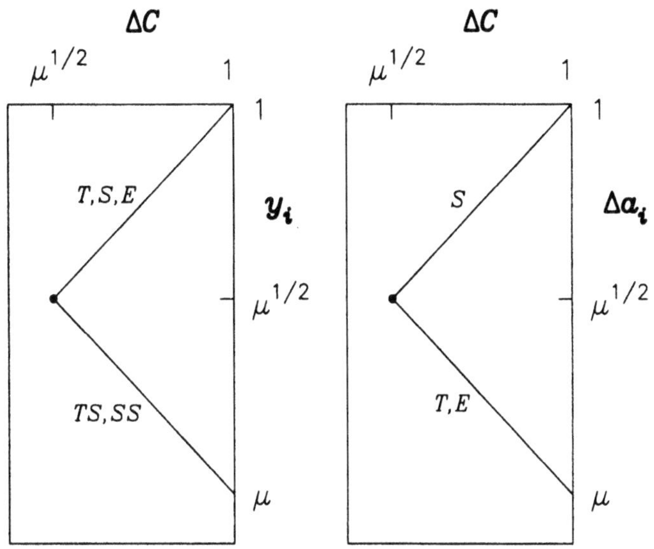

Fig. 12.3. Variations of y_i and Δa_i as functions of ΔC. Both scales are logarithmic. In the left panel, TS and SS represent values of y_i at a node, while T and S represent values at an antinode, inside a T-arc or S-arc respectively.

Next we consider antinodes. For a first species orbit E or a T-arc, we obtain: $\sigma = s$; for a S-arc, we have: $\sigma = \text{sign}[ss_i(ms_i - K)] = \text{sign}[sk(k-K)]$. We recover the results of Sect. 8.3.1.

12.5 The Case $\nu = 1/2$

We study now what happens when the value $\nu = 1/2$ is reached. (Remember that this means $\Delta C = \Theta(\mu^{1/2})$.)

Step 1. The results (12.78), (12.106) to (12.110) of the previous section show that when the value $\nu = 1/2$ is reached, the variables y_i and Δa_i are always of order $\mu^{1/2}$. The distinction between nodes and antinodes disappears; there is a *fusion* of the consecutive arcs. There is also a fusion between the first and second species. The S- and T-arcs lose their identity. This is also apparent on Fig. 12.3.

Step 2. This suggests the following change of variables:

$$y_i = \mu^{1/2} Y_i^* \,, \qquad \Delta a_i = \mu^{1/2} X_i^* \,, \qquad \Delta C = \mu^{1/2} W^* \,. \tag{12.111}$$

Step 3. It will be convenient to make a second change of variables:

$$\begin{aligned} Y_i^* &= s_i \,\text{sign}(\cos\theta) \sqrt{G_1 G_3}\, Y_i \,, \qquad X_i^* = -s_i \,\text{sign}(\cos\theta) \sqrt{\frac{G_1}{G_3}}\, X_i \,, \\ W^* &= -\text{sign}(\cos\theta) \frac{1}{G_2} \sqrt{G_1 G_3}\, W \,. \end{aligned} \tag{12.112}$$

(G_1 and G_3 are positive; see (12.31).)

Step 4. Substituting in (12.32), collecting the dominant terms in the left-hand member, and dividing by appropriate factors, we obtain

$$\begin{aligned} &Y_i + Y_{i-1} - W - (1+Ks_i)X_i = O(\mu^{1/2})\,, \qquad i \in \mathcal{A}\,,\\ &Y_i(X_{i+1}+X_i) + 1 = O(\mu^{1/2})\,, \qquad i \in \mathcal{C}\,,\\ &Y_0 = O(\mu^{1/2})\,, \qquad Y_n = O(\mu^{1/2}) \qquad \text{for a partial bifurcation.} \end{aligned} \tag{12.113}$$

Step 5. The right-hand members do tend to zero for $\mu \to 0$. The asymptotic equations are obtained by equating them to zero:

$$\begin{aligned} &Y_i + Y_{i-1} - W - (1+Ks_i)X_i = 0\,, \qquad i \in \mathcal{A}\,,\\ &Y_i(X_{i+1}+X_i) + 1 = 0\,, \qquad i \in \mathcal{C}\,,\\ &Y_0 = 0\,, \qquad Y_n = 0 \qquad \text{for a partial bifurcation.} \end{aligned} \tag{12.114}$$

Step 6. The available evidence (see below) suggests that this system cannot be solved explicitly in general. Its study will be the object of chapters 13 and 14.

Step 7. Also it will only be possible to compute the Jacobian after the system has been solved (see below). We will see that it is non-zero in general (but it can vanish in isolated points on the characteristics).

Therefore there exists in general, for $\mu > 0$, a solution close to the asymptotic solution, the distance being of the order of $\mu^{1/2}$.

Step 8. does not apply here.

Step 9. After an asymptotic solution (Y_i, X_i, W) is found, we will be able to come back to the physical variables with (12.112) and (12.111), obtaining

$$\begin{aligned} &y_i = s_i\sqrt{G_1G_3}\,\mu^{1/2}Y_i + O(\mu)\,, \qquad \Delta a_i = -s_i\sqrt{\frac{G_1}{G_3}}\,\mu^{1/2}X_i + O(\mu)\,,\\ &\Delta C = -\frac{1}{G_2}\sqrt{G_1G_3}\,\mu^{1/2}W\,. \end{aligned} \tag{12.115}$$

The dots in Fig. 12.3 represent the orders of magnitude of ΔC, y_i, Δa_i.

12.6 The Case $\nu > 1/2$

The detailed study of the case $\nu = 1/2$ which will be made in Chaps. 13 and 14 shows that all branches are joined two by two. Thus the program outlined in Sect. 11.4 is completed.

One might ask whether there exist any solutions of the fundamental equations corresponding to values $\nu > 1/2$. In the variables Y_i, X_i, W defined in Sect. 12.5, this range corresponds to the limit $W \to 0$. Indeed an examination of the characteristics representing the families of solutions of (12.114) in a plane (W, Y_1) or (W, Y_0) reveals that they intersect sometimes the axis $W = 0$ (see for instance Figs. 13.1 to 13.7, 14.1 to 14.4). Any such point with

$W = 0$ must formally be excluded in the study of the case $\nu = 1/2$, because it violates 11.80). In order to study those points and their vicinity, we apply once more the method described in Sect. 11.4.

Step 1. The previous considerations suggest that the y_i and Δa_i should remain of the same order of magnitude as in the case $\nu = 1/2$. On the other hand, we assume $\Delta C \ll \mu^{1/2}$.

Steps 2 and 3. This suggests that we use for the y_i and Δa_i the same changes of variables as in (12.111) and (12.112):

$$y_i = \mu^{1/2} Y_i^* , \qquad \Delta a_i = \mu^{1/2} X_i^* , \tag{12.116}$$

$$Y_i^* = s_i \operatorname{sign}(\cos\theta) \sqrt{G_1 G_3}\, Y_i , \quad X_i^* = -s_i \operatorname{sign}(\cos\theta) \sqrt{\frac{G_1}{G_3}}\, X_i . \tag{12.117}$$

Step 4. Substituting in (12.32), collecting the dominant terms in the left-hand member, and dividing by appropriate factors, we obtain

$$\begin{aligned} &Y_i + Y_{i-1} - (1 + K s_i) X_i = O(\mu^{1/2}) + O(\mu^{-1/2} \Delta C) , \qquad i \in \mathcal{A} , \\ &Y_i (X_{i+1} + X_i) + 1 = O(\mu^{1/2}) , \qquad i \in \mathcal{C} , \\ &Y_0 = O(\mu^{1/2}) , \qquad Y_n = O(\mu^{1/2}) \qquad \text{for a partial bifurcation.} \end{aligned} \tag{12.118}$$

Step 5. The right-hand members do tend to zero for $\mu \to 0$. The asymptotic equations are obtained by equating them to zero:

$$\begin{aligned} &Y_i + Y_{i-1} - (1 + K s_i) X_i = 0 , \qquad i \in \mathcal{A} , \\ &Y_i (X_{i+1} + X_i) + 1 = 0 , \qquad i \in \mathcal{C} , \\ &Y_0 = 0 , \qquad Y_n = 0 \qquad \text{for a partial bifurcation.} \end{aligned} \tag{12.119}$$

Step 6. Even though the system (12.119) is slightly simpler than (12.114), it still apparently cannot be solved explicitly in general. For a partial bifurcation, (12.119) is a system of $2n + 1$ equations for $2n + 1$ variables; for a total bifurcation, it is a system of $2n$ equations for $2n$ variables. Thus we expect isolated solutions.

Step 7. The results found in the next two chapters suggest that the Jacobian does not vanish in general.

Step 8. does not apply here.

Step 9. After an asymptotic solution (Y_i, X_i) is found, we come back to the physical variables with (12.117) and (12.116), obtaining

$$y_i = s_i \sqrt{G_1 G_3}\, \mu^{1/2} Y_i + O(\mu) , \qquad \Delta a_i = -s_i \sqrt{\frac{G_1}{G_3}}\, \mu^{1/2} X_i + O(\mu) . \tag{12.120}$$

It would seem that we need to study the solutions of (12.119). Actually this is not necessary. The solutions of that system are obviously a subset of the solutions of the system (12.114), which will be studied in great detail in Chaps. 13 and 14. All that will be necessary is to select among the solutions of (12.114) those for which $W = 0$.

In practice it is not even necessary to consider these solutions separately. They are not fundamentally different from the other solutions of (12.114). The need to treat them separately is an artifact, resulting from the assumption (11.80), namely that ΔC should be strictly of the order of μ^ν. This assumption is appropriate in general: it allows us to define ν unambiguously. Here, however, it is in a sense too strong. If we replace it by

$$\Delta C = O(\mu^{1/2}) , \tag{12.121}$$

then the points with $W = 0$ can be included in the previous Section 12.5, and the present section becomes unnecessary.

13. Partial Bifurcation of Type 1

We begin now the study of the system of equations (12.114), in the case of a partial bifurcation. The equations are then

$$\begin{aligned}
&Y_i + Y_{i-1} - W - (1 + Ks_i)X_i = 0\,, \qquad i = 1, \ldots, n\,,\\
&Y_i(X_{i+1} + X_i) + 1 = 0\,, \qquad i = 1, \ldots, n-1\,,\\
&Y_0 = 0\,, \qquad Y_n = 0. \qquad\qquad (13.1)
\end{aligned}$$

These equations form a system of $2n+1$ equations for the $2n+2$ variables W, Y_0 to Y_n, X_1 to X_n. Therefore we expect one-parameter families of solutions. For a given value of μ, there is a one-to-one correspondence, given by (12.111) and (12.112), between the present variables W, Y_i, X_i and the physical variables ΔC, y_i, Δa_i. Thus, the one-parameter families of solutions of (13.1) correspond simply to the ordinary one-parameter families of orbits (see Sect. 2.3).

We recall that we consider only the case where the starting point is in P (Sect. 12.1.4); as a consequence,

$$s_i = (-1)^i\,. \qquad (13.2)$$

This convention is identical with the one made in Sect. 8.4.1.

13.1 Properties

1. The quantities Y_i for $i = 1, \ldots, n-1$ can never vanish, because of (13.1b). Therefore each Y_i has a constant sign along a family. We recover the principle of the invariance of the side of passage (Broucke's principle), which was used in Chap. 8.
2. For any solution, by applying the fundamental symmetry Σ of the restricted problem (Sect. 2.7), we obtain another solution:

   $$\Sigma: \qquad (Y_i, X_i, s_i) \mapsto (Y_{n-i}, X_{n+1-i}, s_{n+1-i})\,. \qquad (13.3)$$

3. The equations(13.1) show that for any solution, there exists a symmetrical solution obtained by changing the signs of all variables Y_i, X_i, W. We call this symmetry Σ':

   $$\Sigma': \qquad (Y_i, X_i, W) \mapsto (-Y_i, -X_i, -W)\,. \qquad (13.4)$$

Therefore, for any family, there exists a symmetrical family, and it will be sufficient to study one of them. We do this by postulating that

$$Y_1 > 0 \,. \tag{13.5}$$

This convention is identical with the one made in Sect. 8.4.1. Σ' corresponds in fact to the second internal symmetry which was used in Sect. 8.4.1.

4. We can eliminate the X_i by extracting them from (13.1a) and substituting in (13.1b); we obtain a relation involving three successive values of Y_i:

$$(1 + Ks_i)Y_{i+1} + (1 - Ks_i)Y_{i-1} + 2Y_i + \frac{1 - K^2}{Y_i} - 2W = 0 \,, \quad i = 1, \dots, n-1 \,. \tag{13.6}$$

5. Let us define

$$\xi_k = Y_{2k+1} \,, \qquad \eta_k = Y_{2k+2} \,. \tag{13.7}$$

From (13.6), using (12.35), we have then

$$\begin{aligned} \xi_{k+1} &= \frac{2W - (1 - K)\xi_k - 2\eta_k}{1 + K} - \frac{1 - K}{\eta_k} \,, \\ \eta_{k+1} &= \frac{2W - (1 + K)\eta_k - 2\xi_{k+1}}{1 - K} - \frac{1 + K}{\xi_{k+1}} \,. \end{aligned} \tag{13.8}$$

Thus, the problem is formally equivalent to the study of a plane mapping. As is easily shown, that mapping is area-preserving. Numerical explorations (Hénon, unpublished) exhibit the mixture of regular and chaotic orbits characteristic of non-integrable systems. This supports the conjecture that the system of equations (13.1) is not explicitly solvable in general.

6. By comparing (12.111), (12.112), (12.31), and the definition (8.22) of s, we find

$$s = \text{sign}(W) \,. \tag{13.9}$$

13.1.1 Asymptotic Branches for $W \to \pm\infty$

In the limit $W \to \pm\infty$, we enter the region $|\Delta C| \gg \mu^{1/2}$, or $\nu < 1/2$, and the equations of Sect. 12.4 are applicable. The bifurcation arc or orbit can again be decomposed into a sequence of T and S arcs (or, as a special case, a first species orbit). By comparing the equations (12.63), (12.64), (12.111), (12.112), we obtain the correspondence relations

$$Y_i = W\bar{Y}_i \,, \qquad X_i = W\bar{X}_i \,. \tag{13.10}$$

From (12.75), (12.91), (12.92), (12.93), we obtain thus the following asymptotic equations:

– First species orbit:

$$X_1 = O(W^{-1}) , \qquad X_2 = O(W^{-1}) ,$$
$$Y_0 = \frac{W}{2} + O(W^{-1}) , \qquad Y_1 = \frac{W}{2} + O(W^{-1}) . \tag{13.11}$$

– Node:

$$Y_i = O(W^{-1}) . \tag{13.12}$$

– T-arc:

$$X_{i+1} = O(W^{-1}) , \qquad X_{i+2} = O(W^{-1}) , \qquad Y_{i+1} = W + O(W^{-1}) . \tag{13.13}$$

– S-arc:

$$X_{i+j} = \frac{(-1)^j}{m - Ks_i} W + O(W^{-1}) ,$$
$$Y_{i+j} = \begin{cases} \dfrac{m-j}{m-Ks_i} W + O(W^{-1}) & \text{if } j \text{ is odd,} \\ \dfrac{j}{m-Ks_i} W + O(W^{-1}) & \text{if } j \text{ is even.} \end{cases} \tag{13.14}$$

These results can be strenghtened. We have always

$$X_i = O(W) . \tag{13.15}$$

Therefore, from (13.1b) and (13.12) we obtain for a node i

$$Y_i = \Theta(W^{-1}) , \qquad X_i + X_{i+1} = \Theta(W) . \tag{13.16}$$

At an antinode, we have always

$$Y_i = \Theta(W) . \tag{13.17}$$

13.1.2 Variational Equations for $W \to \pm\infty$

We derive here some equations which will be needed in the next two chapters. We consider again the limit $W \to \pm\infty$. We consider the case where in this limit the bifurcating arc or orbit becomes a sequence of T and S arcs (i.e. we ignore the particular case of a first species orbit). In a total bifurcation, for each given free branch, we select one rooted branch for which the origin is a node. (There is always at least one such branch; otherwise all nodes would be in Q and the orbit would consist only of T arcs, which is impossible.)

We introduce some new notations. We call N the number of arcs which make up the bifurcating arc or orbit. Starting from the origin, we call the successive arcs $U_1, U_2, \ldots, U_\alpha, \ldots, U_N$. Each U_α arc is either a T-arc or a S-arc. We call $i_0, i_1, \ldots, i_\alpha, \ldots, i_N$ the positions of the nodes, with $i_0 = 0$ and $i_N = n$, so that the arc U_α extends from $i_{\alpha-1}$ to i_α. The number of basic arcs in arc U_α is $m_\alpha = i_\alpha - i_{\alpha-1}$.

From the two initial values Y_0 and X_1, using alternatively (13.1a) and (13.1b), we can compute successively Y_1, X_2, Y_2, ... We compute now the corresponding variational equations: we assume arbitrary infinitesimal variations $\mathrm{d}Y_0$ and $\mathrm{d}X_1$ and we compute the corresponding variations $\mathrm{d}Y_1$, $\mathrm{d}X_2$, ... (In general the perturbed values no longer correspond to a bifurcating arc.)

From (13.1) we have

$$\mathrm{d}Y_i = -\mathrm{d}Y_{i-1} + (1 + Ks_i)\,\mathrm{d}X_i\,, \qquad \mathrm{d}X_{i+1} = -\mathrm{d}X_i + \frac{1}{Y_i^2}\,\mathrm{d}Y_i\,. \quad (13.18)$$

As a preparatory step, we compute the variations inside a particular T arc, extending from i to $i+2$, as a function of its initial variations $\mathrm{d}Y_i$ and $\mathrm{d}X_{i+1}$. From (13.18) we obtain

$$\mathrm{d}Y_{i+1} = -\mathrm{d}Y_i + (1 - Ks_i)\,\mathrm{d}X_{i+1}\,, \quad (13.19)$$

$$\mathrm{d}X_{i+2} = -\mathrm{d}X_{i+1} + \frac{1}{Y_{i+1}^2}(-\mathrm{d}Y_i + (1 - Ks_i)\,\mathrm{d}X_{i+1})\,. \quad (13.20)$$

$i+1$ is an antinode; therefore, from (13.17), $Y_{i+1} = \Theta(W)$, and we can write the last relation as

$$\mathrm{d}X_{i+2} = O(W^{-2})\,\mathrm{d}Y_i - \mathrm{d}X_{i+1}\left[1 + O(W^{-2})\right]\,. \quad (13.21)$$

Finally, using (13.18a) again, we have

$$\mathrm{d}Y_{i+2} = \mathrm{d}Y_i\left[1 + O(W^{-2})\right] - 2\mathrm{d}X_{i+1}\left[1 + O(W^{-2})\right]\,. \quad (13.22)$$

We do a similar computation for a S-arc, extending from i to $i+m$. Proceeding in the same way as for the T-arc, we obtain generally

$$\begin{aligned}
&\mathrm{d}Y_{i+2j} = \mathrm{d}Y_i\,[\,] - (2j)\,\mathrm{d}X_{i+1}[\,] \qquad (0 \leqq 2j \leqq m-1)\,,\\
&\mathrm{d}Y_{i+2j+1} = -\mathrm{d}Y_i\,[\,] + (2j + 1 - Ks_i)\,\mathrm{d}X_{i+1}[\,] \qquad (1 \leqq 2j+1 \leqq m)\,,\\
&\mathrm{d}X_{i+2j} = O(W^{-2})\,\mathrm{d}Y_i - \mathrm{d}X_{i+1}\,[\,] \qquad (2 \leqq 2j \leqq m-1)\,,\\
&\mathrm{d}X_{i+2j+1} = O(W^{-2})\,\mathrm{d}Y_i + \mathrm{d}X_{i+1}\,[\,] \qquad (1 \leqq 2j+1 \leqq m)\,, \qquad (13.23)
\end{aligned}$$

where we have used the abbreviation $[\,] = \left[1 + O(W^{-2})\right]$.

In particular we have

$$\begin{aligned}
&\mathrm{d}X_{i+m} = O(W^{-2})\,\mathrm{d}Y_i + \mathrm{d}X_{i+1}\,[\,]\,,\\
&\mathrm{d}Y_{i+m} = -\mathrm{d}Y_i\,[\,] + (m - Ks_i)\,\mathrm{d}X_{i+1}\,[\,]\,. \qquad (13.24)
\end{aligned}$$

We can regroup (13.21), (13.22), (13.24) into general equations relating the initial and final variations in arc U_α:

$$\begin{aligned}
&\mathrm{d}X_{i_\alpha} = O(W^{-2})\,\mathrm{d}Y_{i_{\alpha-1}} - (-1)^{m_\alpha}\mathrm{d}X_{i_{\alpha-1}+1}\,[\,]\,,\\
&\mathrm{d}Y_{i_\alpha} = (-1)^{m_\alpha}\mathrm{d}Y_{i_{\alpha-1}}\,[\,] + u_\alpha\,\mathrm{d}X_{i_{\alpha-1}+1}\,[\,]\,, \qquad (13.25)
\end{aligned}$$

where we have introduced

$$u_\alpha = \begin{cases} -2 & \text{for a } T \text{ arc,} \\ m_\alpha + Ks_{i_\alpha} & \text{for a } S \text{ arc.} \end{cases} \tag{13.26}$$

m_α equals $-k_\alpha s_{i_\alpha}$ (see Sect. 12.4.3). Using Definition 8.2.1, we find that

$$\begin{cases} u_\alpha > 0 & \text{if } U_\alpha \text{ is a normal } S\text{-arc,} \\ u_\alpha < 0 & \text{if } U_\alpha \text{ is an abnormal } S\text{-arc or a } T\text{-arc.} \end{cases} \tag{13.27}$$

We compute now the initial variations in the next arc $U_{\alpha+1}$. We already know $\mathrm{d}Y_{i_\alpha}$. Since i_α is a node, we have from (13.16): $Y_{i_\alpha} = \Theta(W^{-1})$. From (13.18b), we obtain

$$\mathrm{d}X_{i_\alpha+1} = \frac{(-1)^{m_\alpha}}{Y_{i_\alpha}^2}\,\mathrm{d}Y_{i_{\alpha-1}}\,[\,] + \frac{u_\alpha}{Y_{i_\alpha}^2}\,\mathrm{d}X_{i_{\alpha-1}+1}\,[\,]\,. \tag{13.28}$$

The equations (13.25b) and (13.28) give the initial variations of $U_{\alpha+1}$ as functions of the initial variations of U_α. We can iterate these equations and we obtain

$$\begin{aligned} \mathrm{d}Y_{i_\alpha} &= \frac{(-1)^{m_1} u_2 u_3 \ldots u_\alpha}{Y_{i_1}^2 Y_{i_2}^2 \ldots Y_{i_{\alpha-1}}^2}\,\mathrm{d}Y_0\,[\,] + \frac{u_1 u_2 \ldots u_\alpha}{Y_{i_1}^2 Y_{i_2}^2 \ldots Y_{i_{\alpha-1}}^2}\,\mathrm{d}X_1\,[\,]\,, \\ \mathrm{d}X_{i_\alpha+1} &= \frac{(-1)^{m_1} u_2 u_3 \ldots u_\alpha}{Y_{i_1}^2 Y_{i_2}^2 \ldots Y_{i_\alpha}^2}\,\mathrm{d}Y_0\,[\,] + \frac{u_1 u_2 \ldots u_\alpha}{Y_{i_1}^2 Y_{i_2}^2 \ldots Y_{i_\alpha}^2}\,\mathrm{d}X_1\,[\,]\,. \end{aligned} \tag{13.29}$$

The orders of magnitude are

$$\begin{aligned} &\frac{\partial Y_{i_\alpha}}{\partial Y_0} = \Theta(W^{2(\alpha-1)})\,, && \frac{\partial Y_{i_\alpha}}{\partial X_1} = \Theta(W^{2(\alpha-1)})\,, \\ &\frac{\partial X_{i_\alpha+1}}{\partial Y_0} = \Theta(W^{2\alpha})\,, && \frac{\partial X_{i_\alpha+1}}{\partial X_1} = \Theta(W^{2\alpha})\,. \end{aligned} \tag{13.30}$$

The variations are strongly amplified after each node.

13.1.3 Jacobian

The fundamental equations (13.1) can be written

$$\begin{aligned} f_1 &= Y_1 - W - (1-K)X_1 = 0\,, \\ f_2 &= X_2 + X_1 + \frac{1}{Y_1} = 0\,, \\ &\ldots \\ f_{2i-1} &= Y_i + Y_{i-1} - W - (1 + Ks_i)X_i = 0\,, \\ f_{2i} &= X_{i+1} + X_i + \frac{1}{Y_i} = 0\,, \\ &\ldots \\ f_{2n-1} &= Y_n + Y_{n-1} - W - (1 + Ks_n)X_n = 0\,, \\ f_{2n} &= Y_n = 0\,. \end{aligned} \tag{13.31}$$

We have eliminated Y_0, substituting $Y_0 = 0$ in the first equation, but we have kept Y_n and the last equation $Y_n = 0$.

For a given value of W, this is a system of $2n$ equations for $2n$ variables. The Jacobian is

$$|J| = \frac{\partial(f_1,\ldots,f_{2n})}{\partial(X_1,Y_1,X_2,\ldots,X_n,Y_n)}$$

$$= \begin{vmatrix} -(1-K) & 1 & & & & & & \\ 1 & -\frac{1}{Y_1^2} & 1 & & & & & \\ & 1 & -(1+K) & 1 & & & & \\ & & & & \cdots & & & \\ & & & & & 1 & -(1+Ks_n) & 1 \\ & & & & & & & 1 \end{vmatrix} \tag{13.32}$$

which reduces to

$$|J| = \frac{\partial(f_1,\ldots,f_{2n-1})}{\partial(X_1,Y_1,X_2,\ldots,Y_{n-1},X_n)}\,. \tag{13.33}$$

13.1.4 Relation with Stability

We show now that there exists an intimate relation between the Jacobian $|J|$ and the *stability* of the orbit (Sect. 2.8).

In general, a second species orbit is "infinitely unstable" (Hénon and Guyot 1970, p. 364). For μ small but non-zero, the stability index z takes large positive or negative values. However, when a family of second species orbits is followed, the stability index occasionally jumps from $+\infty$ to $-\infty$, or vice versa. For $\mu > 0$, the stability index jumps then from large positive values to large negative values (or vice versa). There exists a short interval of stable orbits.

Such a jump happens when, for one of the arcs, the following condition is realized: "perturbations in the direction of the departure velocity have no effect on the impact parameter at the next encounter" (Hitzl and Hénon 1977b, p. 1029). These particular arcs, or *critical arcs*, have been computed in Hitzl and Hénon 1977b; they correspond to extremums in C along arc families (see Sect. 4.6).

The value of C is determined at a partial bifurcation. In general, the arcs which do not belong to the bifurcating arc are not critical arcs for this particular value of C (see Bruno 1973; 1994, Chap. IV). For the bifurcating arc itself, however, the above condition can be realized. We consider the system of equations (13.31) for a given value of W, with the last equation deleted. We have then a system of $2n-1$ equations for $2n$ variables. Starting from a given value X_1 and using the equations one by one, we obtain successively Y_1, X_2, $\ldots$, Y_n. In the same way, starting from a variation $\mathrm{d}X_1$, we can compute successively $\mathrm{d}Y_1, \mathrm{d}X_2, \ldots, \mathrm{d}Y_n$. The direction of the initial velocity is a function of X_1, and the final impact parameter is proportional to Y_n. Therefore the above condition is simply written

$$\frac{\mathrm{d}Y_n}{\mathrm{d}X_1} = 0 \,. \tag{13.34}$$

A bifurcating arc which verifies this condition will be called a *critical bifurcating arc.*

We have the variational equations

$$\begin{aligned}
&\frac{\partial f_1}{\partial X_1}\mathrm{d}X_1 + \ldots + \frac{\partial f_1}{\partial Y_n}\mathrm{d}Y_n = 0 \,, \\
&\ldots \\
&\frac{\partial f_{2n-1}}{\partial X_1}\mathrm{d}X_1 + \ldots + \frac{\partial f_{2n-1}}{\partial Y_n}\mathrm{d}Y_n = 0 \,.
\end{aligned} \tag{13.35}$$

We can solve this system to obtain

$$\frac{\mathrm{d}Y_n}{\mathrm{d}X_1} = -\frac{\dfrac{\partial(f_1,\ldots,f_{2n-1})}{\partial(X_1,Y_1,X_2,\ldots,X_n)}}{\dfrac{\partial(f_1,\ldots,f_{2n-1})}{\partial(Y_1,X_2,\ldots,X_n,Y_n)}} \,. \tag{13.36}$$

As is seen from (13.31), the denominator reduces to 1 (all elements above the diagonal are 0 and all elements in the diagonal are 1) and we have

$$\frac{\mathrm{d}Y_n}{\mathrm{d}X_1} = -|J| \,. \tag{13.37}$$

We have thus shown

Proposition 13.1.1. *In a partial bifurcation of type 1, the Jacobian vanishes if and only if the bifurcating arc is critical.*

We have from (13.31)

$$\frac{\mathrm{d}X_{i+1}}{\mathrm{d}X_1} = \frac{1}{Y_i^2}\frac{\mathrm{d}Y_i}{\mathrm{d}X_1} - \frac{\mathrm{d}X_i}{\mathrm{d}X_1} \,, \qquad \frac{\mathrm{d}Y_i}{\mathrm{d}X_1} = (1+Ks_i)\frac{\mathrm{d}X_i}{\mathrm{d}X_1} - \frac{\mathrm{d}Y_{i-1}}{\mathrm{d}X_1} \,, \tag{13.38}$$

from which we can compute the successive derivatives

$$\begin{aligned}
&\frac{\mathrm{d}Y_1}{\mathrm{d}X_1} = 1 - K \,, \\
&\frac{\mathrm{d}Y_2}{\mathrm{d}X_1} = \frac{1-K^2}{Y_1^2} - 2 \,, \\
&\frac{\mathrm{d}Y_3}{\mathrm{d}X_1} = \frac{(1-K^2)(1-K)}{Y_1^2Y_2^2} - \frac{2(1-K)}{Y_1^2} - \frac{2(1-K)}{Y_2^2} + 3 - K \,, \\
&\ldots
\end{aligned} \tag{13.39}$$

For given values of X_1 and W, one can compute successively Y_1, X_2, Y_2, ..., Y_n. Therefore Y_n is an analytical function of X_1 and W, which can be represented as a surface in a 3-dimensional space (W, X_1, Y_n). The characteristics of the bifurcating arcs of order n, in the (W, X_1) plane, are given by the intersection of that surface with the horizontal plane $Y_n = 0$.

We consider now the plane tangent to the surface at a point A representing a critical bifurcating arc. We have then $\partial Y_n/\partial X_1 = 0$, and two cases are possible:

1. The tangent plane is not horizontal. It intersects then the horizontal plane in a straight line parallel to the X_1 axis. This line is the tangent to the characteristic; therefore: *the characteristic has an extremum in W at the point A.*
2. The tangent plane is horizontal. The surface $Y_n(W, X_1)$ has then a maximum, a minimum, or a saddle at A. Since this point is on a characteristic, only the case of a saddle is possible. The intersection of the surface with the horizontal plane consists then in general of two curves which intersect at A: *the characteristic intersects another characteristic at the point A.*

Thus a critical bifurcating arc corresponds to either an extremum in W or an intersection with another family. Conversely, in an extremum in W or an intersection with another family, there is $\mathrm{d}Y_n/\mathrm{d}X_1 = 0$ and therefore a critical bifurcating arc.

All this has been shown to hold more generally for the restricted problem (Sect. 2.8).

Examples of both cases will be found below. The first case (extremum) does not correspond to a true singularity; we can obtain a non-zero Jacobian by changing the independent variable used as parameter of the characteristic, for instance by taking X_1 instead of W. The second case (intersection) corresponds to a true singularity: the implicit function theorem cannot be used, and the general method described in Sect. 11.4 fails. The reason is clear: for $\mu \neq 0$, the four branches arriving at the intersection can be joined in different ways. This is the same situation as a bifurcation involving two families of generating orbits (Fig. 1.1).

It will frequently be possible, however, to save the situation and establish the junctions by having recourse to symmetry considerations and to Restriction 7.3.1 (see Sect. 13.3.3.2).

The intersections which are mentioned here should not be confused with bifurcations. A bifurcation (see Chap. 1) corresponds to the intersection of two or more families of generating orbits, i.e. to an intersection which exists *after* the limit $\mu \to 0$ has been taken. On the contrary, here the characteristics in the (W, Y_1) plane represent families of periodic orbits, for μ small but not zero, i.e. *before* the limit has been taken. We are concerned with the intersections between two such families. Examples can be seen in the two upper cases of Fig. 13.2.

If we go back from the W, Y_1 variables to the physical variables ΔC, y_1 and let $\mu \to 0$, the whole figure in the $(\Delta C, y_1)$ plane shrinks to a point, which corresponds to the bifurcation orbit. We see thus that the present intersections take place in a finer level of description than the bifurcation orbits.

13.1.5 Asymptotic Behaviour for $|W| \to \infty$

In that case we can use the computations of Sect. 13.1.2. From (13.29), with $\mathrm{d}Y_0 = 0$, $\alpha = N$, $i_\alpha = n$ we obtain

$$\frac{\mathrm{d}Y_n}{\mathrm{d}X_1} = \frac{u_1 u_2 \ldots u_N}{Y_{i_1}^2 Y_{i_2}^2 \ldots Y_{i_{N-1}}^2} . \tag{13.40}$$

Thus, $\mathrm{d}Y_n/\mathrm{d}X_1 = \Theta(W^{2N-2})$: the Jacobian never vanishes in this limit.

13.2 Small Values of n

It will be simpler to use only the Y_i variables, and the equations (13.6) (with the boundary conditions (13.1c), (13.1d)). If desired, the X_i can then be deduced using (13.1a). The characteristics will be represented in the (W, Y_1) plane.

13.2.1 $n = 2$

We have a single equation

$$2Y_1 + \frac{1 - K^2}{Y_1} - 2W = 0 . \tag{13.41}$$

Depending on the value of K, the characteristic has one of the shapes represented on Fig. 13.1. It is a hyperbola; its asymptotes are the straight lines $Y_1 = 0$ and $Y_1 = W$ (dashed line). In accordance with (13.5), we represent only the upper half-plane $Y_1 > 0$; the lower half-plane is symmetrical with respect to the origin.

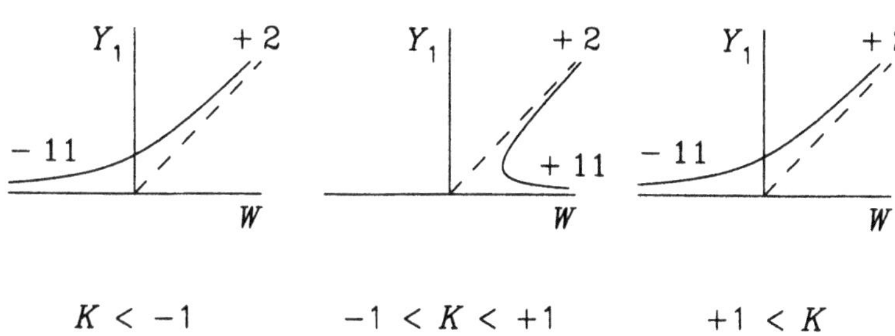

Fig. 13.1. Characteristics for 1P2 bifurcations.

The branches are easily identified. (13.9) shows that their sign s is the sign of W. Table 6.8 (under 1P2) shows that the bifurcating arc can consist either of one T-arc (branches ± 2) or two basic S-arcs (branches ± 11). From the equations of Sect. 13.1.1, we easily derive that the first case corresponds to the branches of the characteristic asymptotic to $Y_1 = W$, while the second case corresponds to the branches asymptotic to $Y_1 = 0$. We can then label the branches (Fig. 13.1). We recover the junctions established in Sect. 8.4.1 from Broucke's principle.

13.2.1.1 Stability and Jacobian. A critical bifurcating arc corresponds to $\mathrm{d}Y_2/\mathrm{d}X_1 = 0$. From (13.39) and (13.41) we find that this happens for

$$Y_1 = \sqrt{\frac{1-K^2}{2}}\,, \qquad W = \sqrt{2(1-K^2)}\,. \tag{13.42}$$

This is possibe only if $-1 < K < 1$. It corresponds then to the extremum in W (Fig.13.1b).

Using Proposition 13.1.1, we find thus that the Jacobian is generally non-zero.

13.2.2 $n = 3$

We have two equations

$$\begin{aligned}(1-K)Y_2 + 2Y_1 + \frac{1-K^2}{Y_1} - 2W &= 0\,,\\ (1-K)Y_1 + 2Y_2 + \frac{1-K^2}{Y_2} - 2W &= 0\,.\end{aligned} \tag{13.43}$$

Subtracting, multiplying by Y_1Y_2, and dividing by $1+K$, we obtain

$$(Y_1 - Y_2)(Y_1Y_2 + K - 1) = 0\,. \tag{13.44}$$

Thus we have two distinct families. The first corresponds to $Y_1 = Y_2$, which gives the characteristic

$$2W = (3-K)Y_1 + \frac{1-K^2}{Y_1}\,. \tag{13.45}$$

The second corresponds to $Y_1Y_2 = 1-K$ and has the characteristic

$$W = Y_1 + \frac{1-K}{Y_1}\,. \tag{13.46}$$

These curves are again hyperbolas, having as asymptotes the lines $Y_1 = 0$, $Y_1 = W$, $Y_1 = 2W/(3-K)$. A simple analysis shows that, depending on the value of K, the characteristics have one of the shapes represented on Fig. 13.2.

For $K < 1$, the two curves intersect at the point

$$Y_1 = \sqrt{1-K}\,, \qquad W = 2\sqrt{1-K}\,, \tag{13.47}$$

and this point is an extremum in W for the second family.

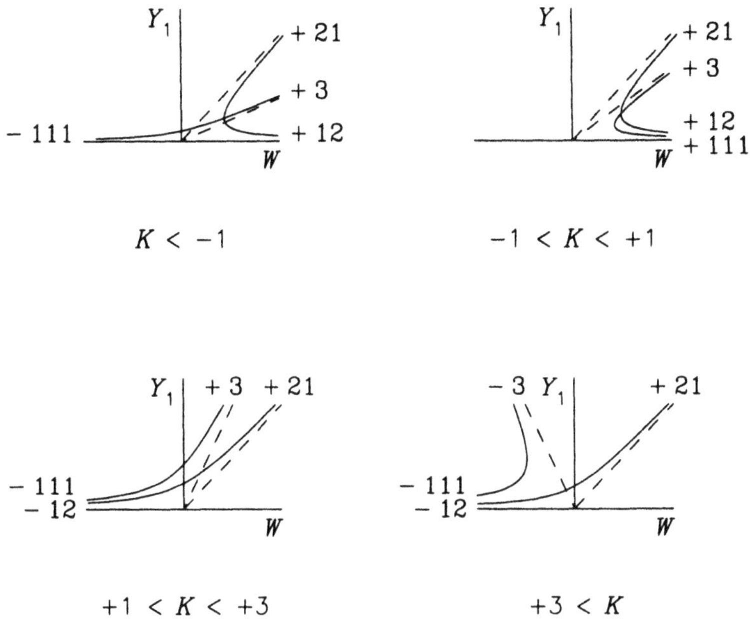

Fig. 13.2. Characteristics for 1P3 bifurcations.

The branches are listed in Table 6.8 (under 1P3). They are identified on Fig. 13.2 by looking at the asymptotic values of Y_1 and Y_2 and comparing them with the expressions of Sect. 13.1.1.

For $K > 1$, the two hyperbolas do not intersect, and the junctions are completely determined; we recover the junctions established in Sect. 8.4.1 on the basis of Broucke's principle (Table 8.4 to 8.11).

For $K < 1$, the two hyperbolas intersect, and the situation is more complicated. The characteristics of Fig. 13.2 are based on the asymptotic equations (13.1), which are correct only within $O(\mu^{1/2})$. So, for $\mu \neq 0$, the true characteristics are slightly different from those represented here. This may be sufficient to change the junctions between the four branches at the intersection point.

We remark however that in the first family, corresponding to $Y_1 = Y_2$, the bifurcating arc is symmetric: $Y_1 = Y_2$, $X_1 = X_3$. By virtue of Restriction 7.3.1, the whole orbit is then symmetric, and we can, as in Chap. 7, use the fact that the symmetry property is invariant along a family. In the second family, the bifurcating arc is asymmetric: $Y_1 \neq Y_2$, except at the intersection. Therefore the branch ± 3 is necessarily joined to the branch ± 111. We recover the results of Sect. 7.3.1.1.

If Restriction 7.3.1 is not satisfied, i.e. if the complement is not symmetric, then the quantitative approach is unable to establish the junctions for $K < 1$. It seems likely that this difficulty corresponds to a real fact, namely that the junctions do depend on what happens outside of the bifurcating arc, i.e. in the complement, through higher-order perturbations exerted at the two ends of the bifurcating arc.

13.2.2.1 Stability and Jacobian. A critical bifurcating arc corresponds to $dY_3/dX_1 = 0$. From (13.39) we find for the first family ($Y_1 = Y_2$)

$$(3-K)Y_1^4 - 4(1-K)Y_1^2 + (1-K^2)(1-K) = 0\,. \tag{13.48}$$

Using also (13.45), we find the two solutions

$$Y_1 = \sqrt{1-K}\,, \qquad W = 2\sqrt{1-K}\,, \tag{13.49}$$

and

$$Y_1 = \sqrt{\frac{(1-K^2)}{(3-K)}}\,, \qquad W = \operatorname{sign}(3-K)\sqrt{(1-K^2)(3-K)}\,. \tag{13.50}$$

The first solution exists only if $K < 1$; it corresponds to the point of intersection with the other family. The second solution exists only if $-1 < K < 1$ or $+3 < K$; it corresponds to the extremum in K along the family.

For the second family, we obtain

$$Y_1^4 - 2(1-K)Y_1^2 + (1-K)^2 = 0\,. \tag{13.51}$$

Using also (13.46), we find twice the solution (13.49), which exists for $K < 1$. This solution corresponds to the point of intersection with the other family, and is also an extremum in K along the family.

Using Proposition 13.1.1, we find that the Jacobian is generally non-zero.

13.2.3 $n = 4$

We have three equations

$$\begin{aligned}
&2Y_1 + (1-K)Y_2 + \frac{1-K^2}{Y_1} - 2W = 0\,,\\
&(1-K)Y_1 + 2Y_2 + (1+K)Y_3 + \frac{1-K^2}{Y_2} - 2W = 0\,,\\
&(1+K)Y_2 + 2Y_3 + \frac{1-K^2}{Y_3} - 2W = 0\,.
\end{aligned} \tag{13.52}$$

This system does not seem to be amenable to an analytical solution. Figs. 13.3 to 13.7 show the numerically computed characteristics for $K = -4, -2, 0, 2, 4$. Solid lines represent the characteristics for $n = 4$, i.e. the solutions of (13.52). Characteristics are also represented for $n = 3$ (dashed lines) and $n = 2$ (dotted lines).

The branches are listed in Table 6.8 (under 1P4). They are identified by looking at the asymptotic values of Y_1, Y_2, Y_3 (obtained from the numerical computation): an asymptotic behaviour $O(W^{-1})$ corresponds to a node, while an asymptotic behaviour $O(W)$ corresponds to an antinode (Sect. 13.1.1).

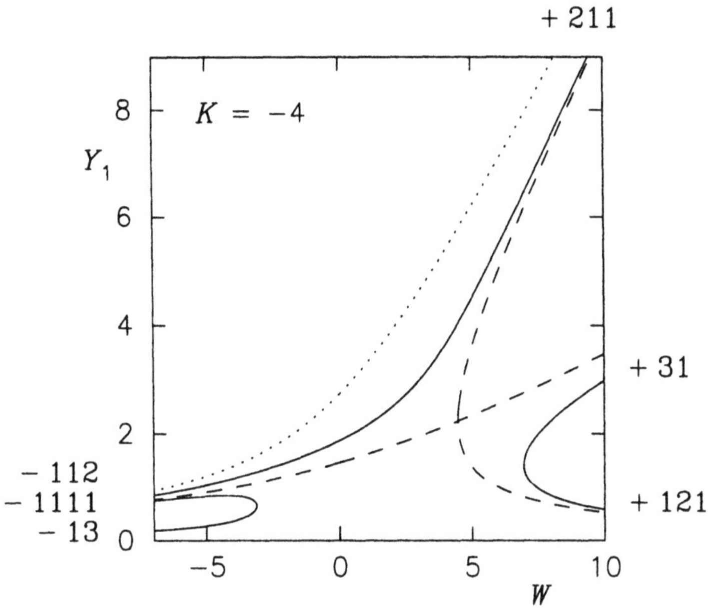

Fig. 13.3. Characteristics for 1P4 bifurcations, for $K = -4$ (solid lines). Also shown: 1P3 bifurcations (dashed lines), 1P2 bifurcation (dotted line).

Computations for other values of K indicate that the figure remains the same qualitatively inside each of the five intervals $K < -3$, $-3 < K < -1$, $-1 < K < +1$, $+1 < K < +3$, $+3 < K$. A proof of this fact will be given in Section 13.3.

The junctions are now fully determined, for all values of K. By contrast, the use of symmetry and of Broucke's principle left some junctions undetermined for $n = 4$ (Sect. 8.4.1). So we have here a first instance where the quantitative approach allows us to make progress.

13.3 Positional Method

13.3.1 Principle

In principle the numerical computation can be used to determine the junctions for any value of n (see for instance Sect. 13.3.3.5 where some characteristics are computed for $n = 6$). But in practice this approach cannot be pursued very far. The computations become time-consuming when n increases; they must be done for more and more values of K; and numerical problems arise because some branches lie very close to each other (see below).

However, the results obtained up to $n = 4$ suggest a new approach. We observe on Figs. 13.3 to 13.7 that *two characteristics which correspond to*

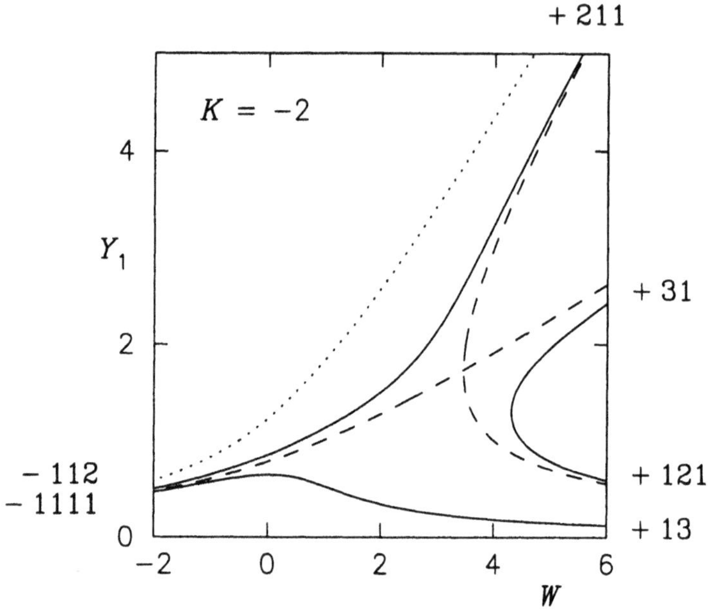

Fig. 13.4. Characteristics for 1P4 bifurcations, for $K = -2$.

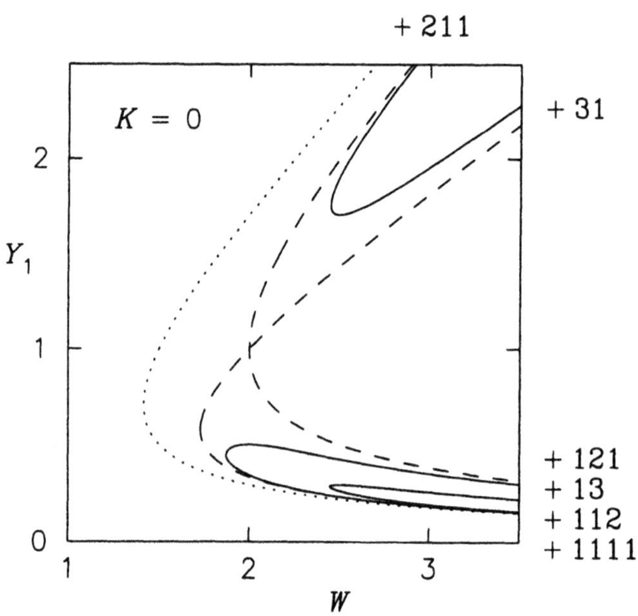

Fig. 13.5. Characteristics for 1P4 bifurcations, for $K = 0$.

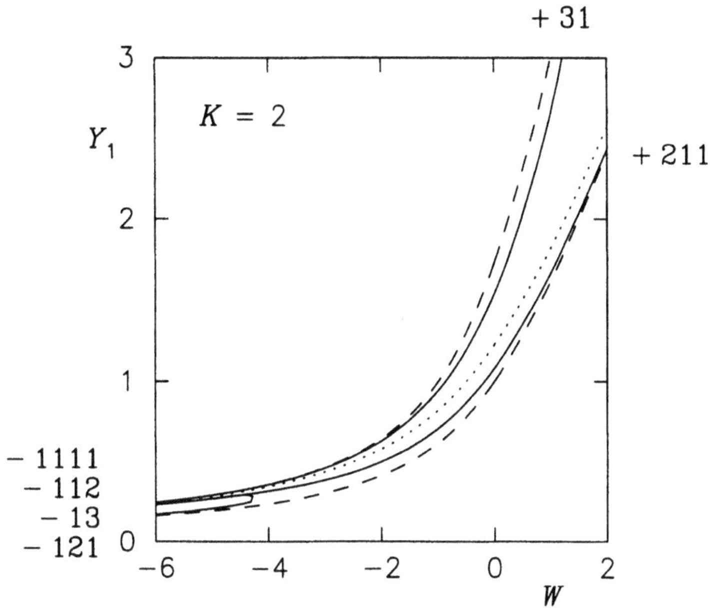

Fig. 13.6. Characteristics for 1P4 bifurcations, for $K = 2$.

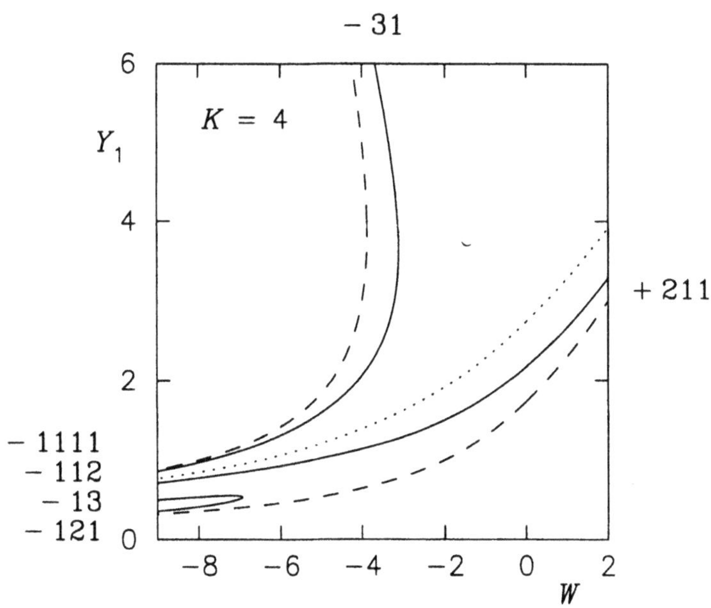

Fig. 13.7. Characteristics for 1P4 bifurcations, for $K = 4$.

different values of n never intersect. It is easy to prove that this is true generally. Let n and n' be the orders for the two characteristics, with $n < n'$. We have then, for the first characteristic: $Y_n = 0$ from (13.1c), and for the second characteristic: $Y_n \neq 0$ from (13.1b). But Y_n is a definite, single-valued function of W and Y_1. It follows that no point of the (W, Y_1) plane can belong to both characteristics.

Suppose now that we already know the shape of the characteristics for orders up to $n-1$, and we are trying to determine the junctions for order n. The previous characteristics divide the (W, Y_1) plane into several *regions*; and each characteristic for order n must lie entirely within one of these regions. In other words,

Proposition 13.3.1. *Two branches can be joined only if they lie in the same region.*

We can do better than that. Instead of defining an orbit by W and Y_1, we can just as well define it symmetrically by W and Y_{n-1}, and compute the other values backwards. The characteristics up to $n-1$ again divide the plane (W, Y_{n-1}) into several regions, and two branches can be joined only if their characteristics in that plane are in the same region. Using the symmetry Σ (see Sect. 13.1), we can reduce the consideration of the (W, Y_{n-1}) plane to the consideration of the (W, Y_1) plane, and we obtain

Proposition 13.3.2. *Two branches can be joined only if the two symmetrical branches lie in the same region of the (W, Y_1) plane, for the same value of K if n is odd and for the opposite value of K if n is even.*

These two propositions, taken together, constitute a powerful criterion for the determination of the junctions.

13.3.2 Branch Order

To apply the method, we need to determine the region in which a branch lies. As we will see now, this can be done by studying the relative position of the asymptotic branches for $W \to \pm\infty$.

13.3.2.1 Variations. We return here to the full set of equations (13.1) for the X_i and the Y_i. For a given large value of $|W|$, using the condition $Y_0 = 0$ and starting from any given value of X_1, we can compute the values of Y_1, X_2, Y_2, ..., successively by applying (13.1a) and (13.1b) alternatively. We consider the value of X_1 which corresponds to some particular branch of order n (so that the computation ends in $Y_n = 0$). We apply now a small variation $\mathrm{d}X_1$, and we compute the orders of magnitude of the corresponding variations of Y_1, X_2, ... (There are two small parameters in this problem: $\mathrm{d}X_1$ and W^{-1}.) We can use the computations made in Sect. 13.1.2, setting $\mathrm{d}Y_0 = 0$. We obtain in particular

$$\frac{\mathrm{d}Y_{i_\alpha}}{\mathrm{d}X_1} = \frac{u_1 u_2 \dots u_\alpha}{Y_{i_1}^2 Y_{i_2}^2 \dots Y_{i_{\alpha-1}}^2}[1 + O(W^{-2})] \, . \tag{13.53}$$

We will need the order of magnitude of this quantity

$$\frac{\mathrm{d}Y_{i_\alpha}}{\mathrm{d}X_1} = \Theta(W^{2(\alpha-1)}) \tag{13.54}$$

and its sign

$$\mathrm{sign}\left(\frac{\mathrm{d}Y_{i_\alpha}}{\mathrm{d}X_1}\right) = \mathrm{sign}(u_1 u_2 \dots u_\alpha) = (-1)^\varsigma \tag{13.55}$$

where ς is defined as the total number of T-arcs and abnormal S-arcs in the set $(U_1, \dots, U_\alpha)$, as shown by (13.26).

We have referred the variations to $\mathrm{d}X_1$. On the figures showing the characteristics, however, Y_1 is used. From (13.1a) with $i = 1$ we find that these two variables are related by

$$\mathrm{d}Y_1 = (1 - K)\mathrm{d}X_1 \, . \tag{13.56}$$

Therefore (13.54) is still true if we replace $\mathrm{d}X_1$ by $\mathrm{d}Y_1$. On the other hand, (13.55) becomes

$$\mathrm{sign}\left(\frac{\mathrm{d}Y_{i_\alpha}}{\mathrm{d}Y_1}\right) = (-1)^\varsigma \mathrm{sign}(1 - K) \, . \tag{13.57}$$

13.3.2.2 Relative Positions of Two Branches with Initial Common Arcs. We consider now two branches for which the arcs U_1 to U_α are the same, but the continuation is different: either the arcs $U_{\alpha+1}$ are different in the two branches, or the arc $U_{\alpha+1}$ does not exist in one of the two branches (the bifurcating arc ends in i_α).

The asymptotic expressions of $X_1, \dots, X_{i_\alpha}$ and $Y_1, \dots, Y_{i_\alpha}$ given in Sect. 13.1.1 are the same for the two branches; therefore these quantities differ by $O(W^{-1})$ between the two branches. For $|W|$ large enough, this can be made as small as desired. Therefore these differences will be called $\mathrm{d}X_1$, etc., and the above results on small variations can be applied.

From (13.13) and (13.14) we have

$$X_{i_\alpha} = \begin{cases} O(W^{-1}) & \text{if } U_\alpha \text{ is a } T\text{-arc,} \\ -\dfrac{W}{m_\alpha + K s_{i_\alpha}} + O(W^{-1}) & \text{if } U_\alpha \text{ is a } S\text{-arc,} \end{cases} \tag{13.58}$$

$$X_{i_\alpha+1} = \begin{cases} O(W^{-1}) & \text{if } U_{\alpha+1} \text{ is a } T\text{-arc,} \\ -\dfrac{W}{m_{\alpha+1} - K s_{i_\alpha}} + O(W^{-1}) & \text{if } U_{\alpha+1} \text{ is a } S\text{-arc.} \end{cases} \tag{13.59}$$

We have generally (see Sect. 12.4.3)

$$m_\alpha = -s_{i_\alpha} k_\alpha \, . \tag{13.60}$$

We find then from Definition 8.2.1 that the quantity $m_\alpha + Ks_{i_\alpha}$ is positive if U_α is a normal S-arc, negative if it is an abnormal S-arc. Similarly, we have

$$m_{\alpha+1} = s_{i_\alpha} k_{\alpha+1} \tag{13.61}$$

and the quantity $m_{\alpha+1} - Ks_{i_\alpha}$ is positive if $U_{\alpha+1}$ is a normal S-arc, negative if it is an abnormal S-arc.

We consider first the case $W > 0$.

Suppose first that U_α is a T-arc. If $U_{\alpha+1}$ exists, it must be a S-arc. We obtain from (13.1b)

$$Y_{i_\alpha} = (m_{\alpha+1} - Ks_{i_\alpha})W^{-1}\left[1 + O(W^{-2})\right] . \tag{13.62}$$

This is an increasing function of $m_{\alpha+1}$. Therefore, the order of the values of Y_{i_α} for the two branches is given by the order of the two values of $m_{\alpha+1}$ in the sequence

$$1, 3, 5, \ldots \tag{13.63}$$

We should also consider the case where the bifurcating arc ends in i_α. We represent this case symbolically by $m_{\alpha+1} = 0$. In that case $Y_{i_\alpha} = 0$. The expression (13.62) is positive if $U_{\alpha+1}$ is normal, negative if it is abnormal. Thus, this case should be positioned in the sequence (13.63) so that abnormal $U_{\alpha+1}$ arcs (if any) are on its left and normal arcs are on its right. For instance if $1 < Ks_{i_\alpha} < 3$, the sequence is

$$1;\ 0;\ 3, 5, 7, \ldots \tag{13.64}$$

Suppose next that U_α is a S-arc. If $U_{\alpha+1}$ is a S-arc, we obtain

$$Y_{i_\alpha} = \frac{(m_\alpha + Ks_{i_\alpha})(m_{\alpha+1} - Ks_{i_\alpha})}{m_\alpha + m_{\alpha+1}} W^{-1}\left[1 + O(W^{-2})\right] . \tag{13.65}$$

This can also be written

$$Y_{i_\alpha} = \left[m_\alpha + Ks_{i_\alpha} - \frac{(m_\alpha + Ks_{i_\alpha})^2}{m_\alpha + m_{\alpha+1}}\right] W^{-1}\left[1 + O(W^{-2})\right] . \tag{13.66}$$

This is again an increasing function of $m_{\alpha+1}$.

If $U_{\alpha+1}$ is a T-arc, we have

$$Y_{i_\alpha} = (m_\alpha + Ks_{i_\alpha})W^{-1}\left[1 + O(W^{-2})\right] . \tag{13.67}$$

This is equal to the limit of (13.66) for $m_{\alpha+1} \to +\infty$. Therefore, the order of the values of Y_{i_α} for the two branches is given by the order of the two values of $m_{\alpha+1}$ in the following sequence

$$1, 3, 5, \ldots;\ 2 . \tag{13.68}$$

The symbol 2 represents the case of a T-arc.

Finally, if the bifurcating arc ends in i_α, we have $Y_{i_\alpha} = 0$. The expression (13.65) is positive if U_α and $U_{\alpha+1}$ are both normal, negative if one of them is abnormal. So:

- If the U_α arc is normal, this case should be positioned in the sequence (13.68) so that abnormal $U_{\alpha+1}$ arcs (if any) are on its left and normal arcs are on its right. For instance if $3 < Ks_{i_\alpha} < 5$:

$$1,3;\ 0;\ 5,7,\ldots;\ 2\,. \tag{13.69}$$

- If the U_α arc is abnormal, this case should be positioned at the right of the sequence (13.68):

$$1,3,5,\ldots;\ 2;\ 0\,. \tag{13.70}$$

Having found the order of the values of Y_{i_α}, it is a simple matter to determine the order of the values of Y_1 for the two branches with the help of (13.57).

13.3.2.3 Relative Positions of Two Branches with Different First Arcs. The method of the previous section does not work when the two branches differ already in their first arc (this can be taken to correspond to $\alpha = 0$). In that case we compare directly the values of Y_1. If the first arc is a S-arc, we have from (13.14)

$$Y_1 = \frac{m_1 - 1}{m_1 - K}W + O(W^{-1}) = \left(1 + \frac{K-1}{m_1 - K}\right)W + O(W^{-1})\,. \tag{13.71}$$

If the first arc is a T-arc, we have from (13.13)

$$Y_1 = W + O(W^{-1})\,. \tag{13.72}$$

This is equal to the limit of (13.71) for $m_1 \to +\infty$. If $K < 1$, the denominator in (13.71) is always positive, and Y_1 is an increasing function of m_1; the order is

$$1,3,5,\ldots;\ 2 \tag{13.73}$$

If $1 < K$, then for $m_1 < K$ (abnormal arcs), Y_1 is smaller than 1, and is a decreasing function of m_1; for $m_1 > K$ (normal arcs), Y_1 is larger than 1, and is again a decreasing function of m_1. Thus, the order is: for $1 < K < 3$

$$1;\ 2;\ \ldots,7,5,3\,, \tag{13.74}$$

for $3 < K < 5$

$$3,1;\ 2;\ \ldots,9,7,5\,, \tag{13.75}$$

and so on.

13.3.2.4 The Case $W < 0$. In the previous two sections, we have supposed $W > 0$; we must now consider the opposite case $W < 0$. We notice that all equations used to determine the order of the branches, such as (13.62) for instance, are odd functions of W. Therefore, in the case $W < 0$, all sequences such as (13.63) are simply inverted. This is also a consequence of the symmetry Σ' of the figure in the (W, Y_1) plane (Sect. 13.1).

13.3.2.5 Packets. The results of Sect. 13.3.2.2 show that, for $\alpha > 0$,

$$\mathrm{d}Y_{i_\alpha} = \Theta(W^{-1}) . \tag{13.76}$$

From (13.54) and (13.56) we have then

$$\mathrm{d}Y_1 = \Theta(W^{1-2\alpha}) . \tag{13.77}$$

For $\alpha = 0$ (Sect. 13.3.2.3), the difference between the values of Y_1 for the two branches is $\Theta(W)$. Therefore (13.77) covers this case also, with the proviso that $\mathrm{d}Y_1$ is not a small quantity anymore.

This equation shows that for $|W|$ large, the branches are organized hierarchically. The characteristics of two branches which differ in their first arc lie at a distance $\Theta(W)$ from each other in the (W, Y_1) plane; they diverge for $|W| \to +\infty$. The characteristics of all branches having a given first arc in common are at a distance $O(W^{-1})$ of each other; they tend towards each other for $W \to +\infty$. We will say that they form a *first-order packet*. Inside such a packet, the characteristics of all branches having in common their first two arcs are at a distance $O(W^{-3})$ of each other; they tend towards each other even more rapidly, and form a *second-order packet*; and so on. This phenomenon is clearly seen for instance on Figs. 13.3 to 13.7.

Incidentally, the fact that the characteristics of two branches converge quickly towards one another for $|W| \to \infty$ as soon as α reaches a few units is one reason for the numerical difficulties alluded to above.

13.3.2.6 An Example. The rules which have been established in the preceding sections allow us to determine completely how the characteristics of all branches are ordered in Y_1. As an example, consider the case $1 < K < 3$, $W \to +\infty$, and the branches up to $n = 5$. The first arc can be 1, 3, 5, or 2. According to (13.74), the corresponding packets are arranged for increasing Y_1 in the order 1, 2, 5, 3. This is shown by Fig. 13.8, left column. (In this figure, Y_1 increases from bottom to top, as in Figs. 13.1 to 13.7, and packets of first order, second order, etc. have been separated into different columns, with their filiation indicated by line segments.) Next we consider the packet associated with a first arc 1. This arc is abnormal, therefore the order of the second arcs is given by (13.70) and the corresponding second-order packets are positioned as shown in Fig. 13.8. The first arc 3 is normal, and the next arc is also always normal, so the values of Y_3 increase along the sequence 0, 1, 2. In (13.57) we have $\zeta = 0$, $\mathrm{sign}(\mathrm{d}Y_{i_\alpha}/\mathrm{d}Y_1) = -1$, so that the order is reversed for the values of Y_1, as shown in Fig. 13.8. Finally, if the first arc is 2, the values of Y_2 increase along the sequence 1, 0, 3 as shown by (13.64). In (13.57) we have $\zeta = 1$, $\mathrm{sign}(\mathrm{d}Y_{i_\alpha}/\mathrm{d}Y_1) = +1$, so that the order is the same for the values of Y_1.

In the same way we find the order of the third-order packets, etc.

For $W \to -\infty$, because of the symmetry Σ', we have the same branches with the + sign replaced by -, and the ordering in Y_1 is inverted.

We apply now the convention (13.5): we consider only the half-plane $Y_1 > 0$. We remark also that the branch labelled + 1 on Fig. 13.8 corresponds to

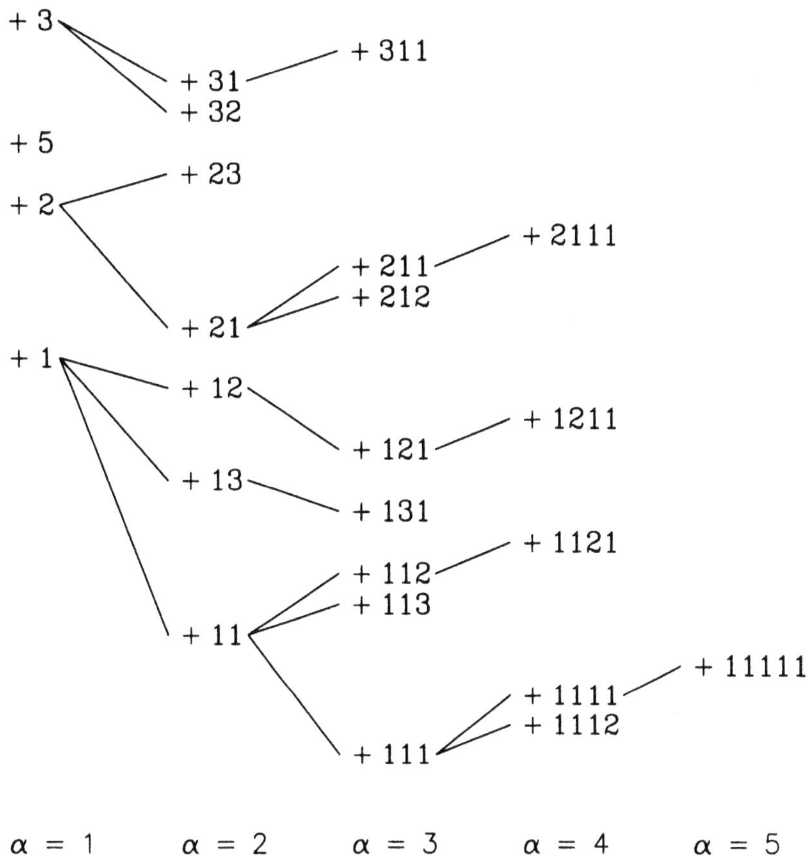

Fig. 13.8. Order of the branches for $1 < K < 3$, $W \to +\infty$, $n \leqq 5$. (Y_1 increases upwards.)

a single basic arc, and therefore to $Y_1 = 0$. (Incidentally, that branch does not participate in a true bifurcation: for $n = 1$ there are only two branches + 1 and - 1 smoothly joined, and therefore only one family (see Sect. 6.8).) Therefore we obtain a full picture of the position of the branches in the half-plane $Y_1 > 0$ as follows: (i) we take that part of the sequence of branches in Fig. 13.8 which lies above + 1, and we copy it on the right (W large and positive); (ii) we take the part of the sequence of branches in Fig. 13.8 which lies below + 1, we invert it, and we copy it on the left (W large and negative). Fig. 13.9 shows the result.

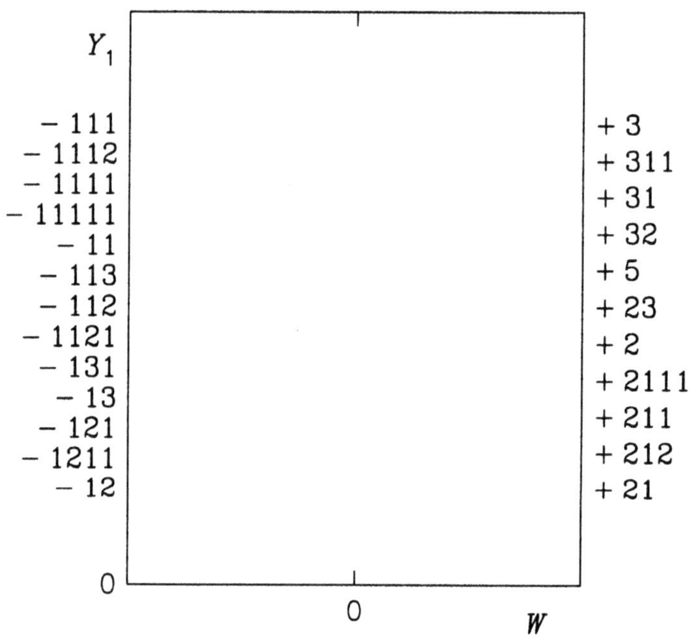

Fig. 13.9. Position of the branches in the half-plane $Y_1 > 0$ for $1 < K < 3$, $n \leqq 5$.

13.3.3 Results

We use now the method systematically to determine the junctions. The computations were done by hand up to $n = 6$. A program was also written to compute the branch order automatically; this allowed a check of the results and also a computation for higher values of n.

13.3.3.1 $n = 2$. In the half-plane $Y_1 > 0$, there are only two branches + 2 and ± 11, which are therefore necessarily joined (see Fig. 13.1).

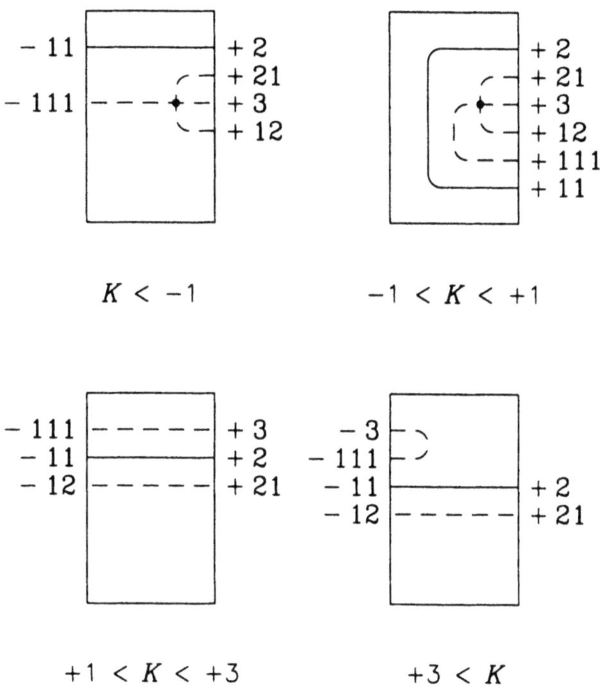

Fig. 13.10. Junctions for $n \leqq 3$.

13.3.3.2 $n = 3$. The junctions are represented on Fig. 13.10 for the various cases with respect to K. They are determined in three steps.

1. We determine first the position of the branches up to $n = 3$ in the half-plane $Y_1 > 0$, as explained above (see example of Fig. 13.9).
2. We represent the already known junction for $n = 2$ schematically, as a solid line. This junction divides the half-plane into two regions.
3. We consider now the branches for $n = 3$ and we use Proposition 13.3.1. In the two cases $+1 < K < +3$ and $+3 < K$, there are two branches in each region; therefore the junctions for $n = 3$ are immediately established. They are represented as dashed lines.
   For $K < -1$, all four branches are in the same region. We have here the first example of a situation which we will frequently encounter, and which we call a *trident*. Four branches lie in the same region. Two of them, H_1 and H_2, are made of symmetric orbits (in the present case: - 111 and + 3) (we use Restriction 7.3.1). The two other branches, H_3 and H_4, are made of asymmetric orbits, but are changed into each other by the symmetry (in the present case: + 21 and + 12). Then we can reason as in Chap. 7 and conclude that H_1 can be only joined to H_2, and H_3 to H_4.
   Consider two orbits Ω_3 and Ω_4, symmetrical of each other, belonging to H_3 and H_4, respectively. We let these two orbits move towards each other until they meet. The orbit Ω thus reached is symmetric. This symmetric orbit belongs to a family of symmetric orbits, which can only be the family H_1H_2. Thus *the two families H_1H_2 and H_3H_4 intersect in a common orbit Ω*. Moreover, the two symmetrical orbits H_3 and H_4 have the same value of W; it follows that *Ω is an extremum in W for the family H_3H_4*.
   In the (W, Y_1) plane, the four branches of the trident are disposed as shown in Fig. 13.10. The common orbit Ω is represented by a dot.
   Finally, for $-1 < K < +1$, we have again a trident. Here the stick of the trident is folded towards the right.
   In the present case of partial bifurcations of type 1, tridents can exist only when n is odd (see Sect. 7.3.1.1).

Thus, all junctions have been established by the positional method for $n = 3$. We recover the results obtained in Sect. 13.2.2.

13.3.3.3 $n = 4$. The junctions for $n = 4$ are determined in the same way. Fig. 13.11 shows the results. Cases corresponding to opposite values of K are shown side by side, because these cases are associated in Proposition 13.3.2.

First the position of the branches along the two sides is determined. The already known junctions up to $n = 3$ are drawn schematically as solid lines. Proposition 13.3.1 then immediately determines the new junctions for $n = 4$ (dashed lines) in several regions which contain only two branches. In the other regions, we use Proposition 13.3.2. Thus, for $-1 < K < +1$, + 211 is joined to + 31; therefore + 112 must be joined to + 13 (the change of sign

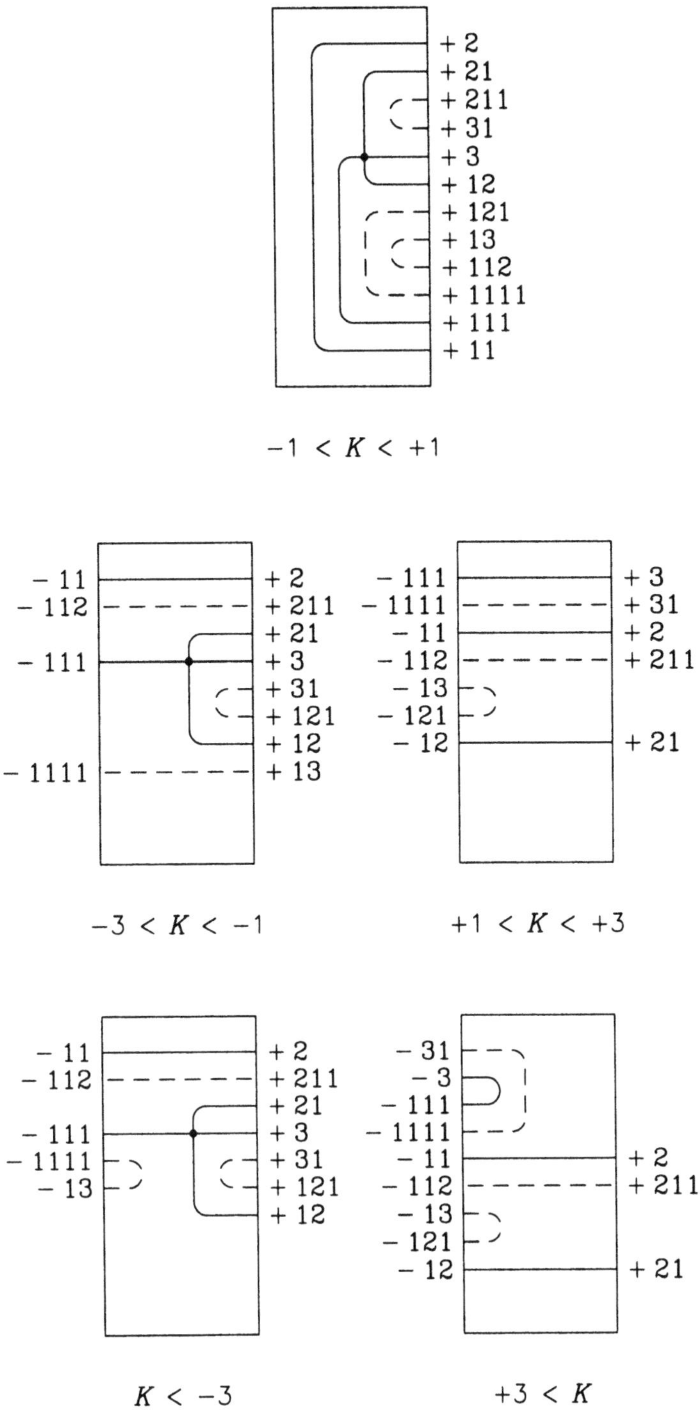

Fig. 13.11. Junctions for $n \leqq 4$.

of K leaves us in the same case). The two remaining branches + 121 and + 1111 must then also be joined. For $-3 < K < -1$, the branches + 31 and + 121 are joined; therefore, for $+1 < K < +3$, the branches + 13 and + 121 are joined. Using the symmetry Σ', we find that - 13 and - 121 are joined. Finally, the cases $K < -3$ and $+3 < K$ are treated in the same way.

All junctions have been established. We recover the results obtained numerically in Figs. 13.3 to 13.7.

13.3.3.4 $n = 5$. The junctions are shown in Figs. 13.12 to 13.14, in order of increasing K (here there is no particular reason to associate opposite values of K).

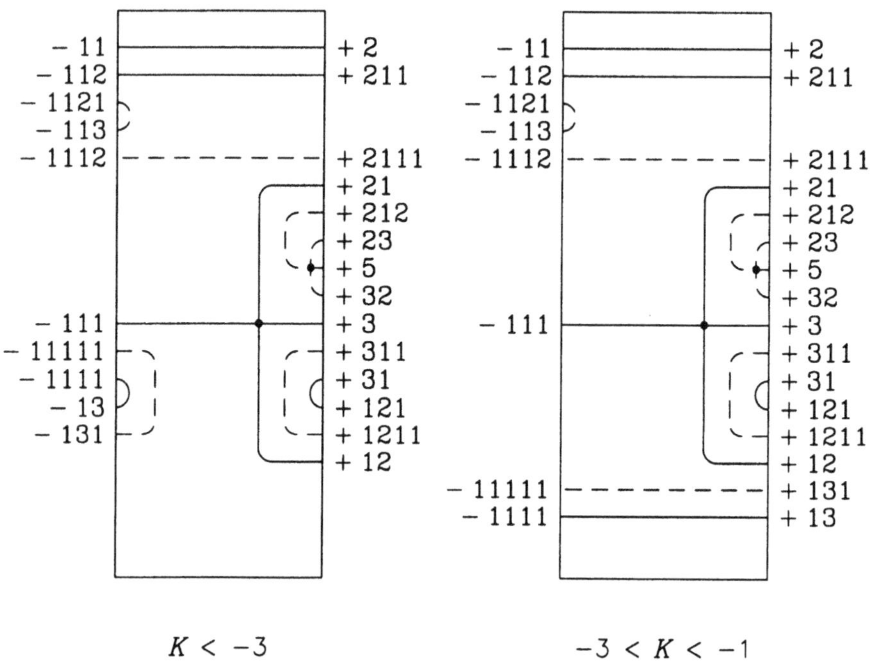

Fig. 13.12. Junctions for $n \leqq 5$.

Most junctions for $n = 5$ can be determined by the straightforward application of Propositions 13.3.1 and 13.3.2. New tridents appear; they are resolved in the same way as in Sect. 13.3.3.2. For $-1 < K < +1$, there are two regions containing 6 branches each, and a different reasoning is required. We consider first the upper region. It contains only two branches of symmetric orbits: + 212 and + 5; these branches must be joined. The branch + 23 cannot be joined to + 2111, because the symmetrical branches + 32 and + 1112 do not belong to the same region. In the same way, it cannot be joined to + 311. Therefore the branch + 23 must be joined to + 32 (incidentally, it

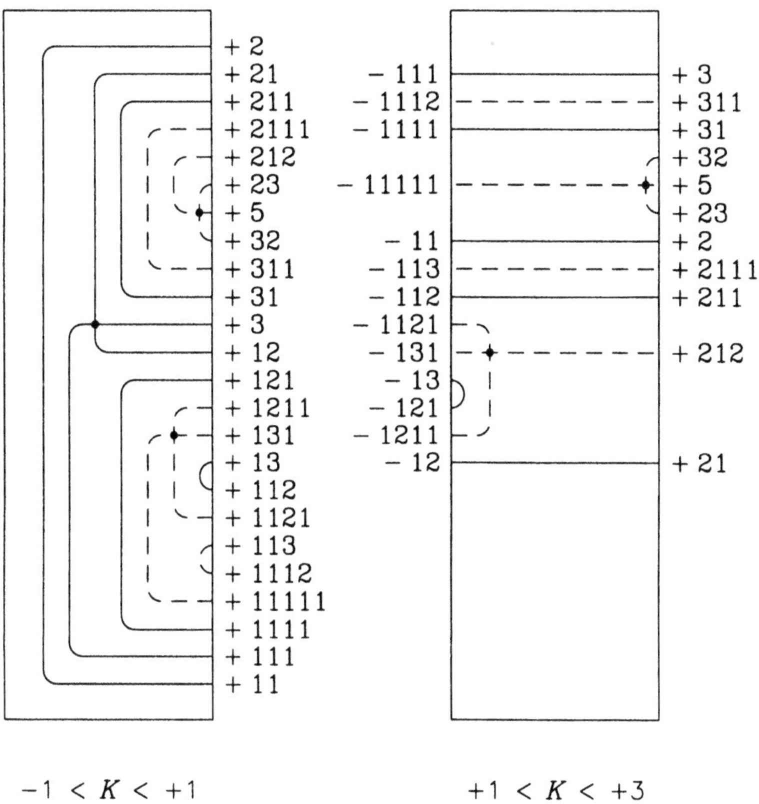

Fig. 13.13. Junctions for $n \leqq 5$ (continuation).

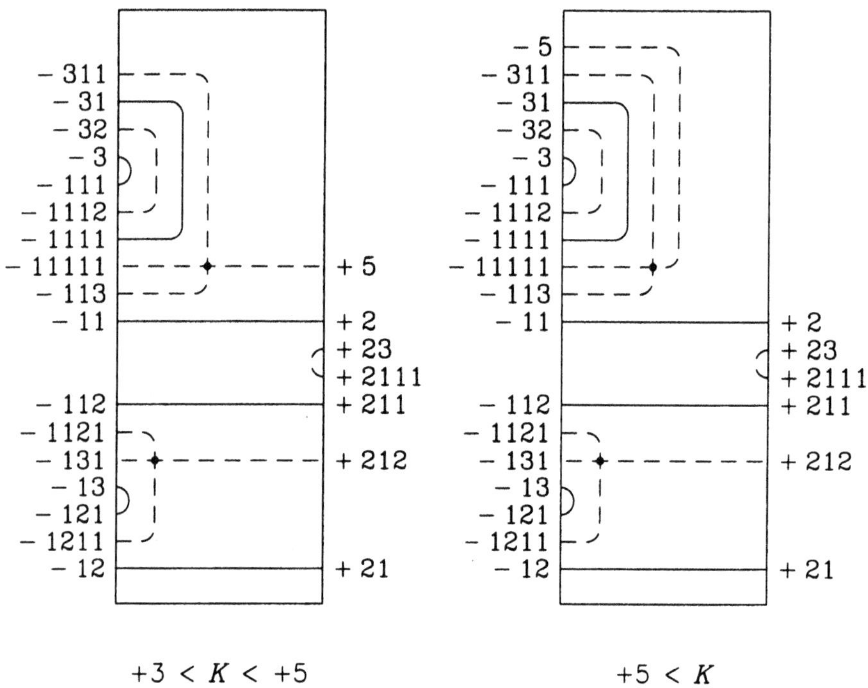

Fig. 13.14. Junctions for $n \leqq 5$ (continuation).

is part of a trident). The two remaining branches + 2111 and + 311 are then joined. A similar reasoning determines the junctions in the lower region.

Thus all junctions have been determined.

13.3.3.5 $n = 6$**.** The junctions are shown in Figs. 13.15 to 13.18. As in Sect. 13.3.3.3, the related cases corresponding to opposite values of K are shown side by side.

Many junctions for $n = 6$ are found by the use of Propositions 13.3.1 and 13.3.2. For $-3 < K < -1$ and $+1 < K < +3$, in fact, all junctions are found. In the other cases, however, we find that there exists a group of 4 branches for which the positional method is unable to establish the junctions. For $-1 < K < +1$, this group is: + 2112, + 213, + 33, + 312; for $+3 < K < +5$ and for $+5 < K$: - 33, - 321, - 11121, - 1113; for $-5 < K < -3$ and for $K < -5$: + 33, + 3111, + 12111, + 123.

There is a good reason for this breakdown of the positional method: the junctions between the 4 branches change inside the given interval of K. Fig. 13.19 shows the numerically computed characteristics for $K = -5.66$ and $K = -5.68$. The junctions between the branches change at about $K = -5.669369\ldots$. The positional method does not distinguish between these values of K, for which the normal and abnormal arcs are the same; therefore it cannot predict the junctions.

The characteristics were computed for a number of other values of K inside the intervals $K < -5$ and $-5 < K < -3$; the results indicate that there is no other change, i.e. the junctions remain as shown by Fig. 13.19, left, for $-5.669369\ldots < K < -3$, and by Fig. 13.19, right, for $K < -5.669369\ldots$.

Using Proposition 13.3.2, we find that there is a similar change in the junctions in the case $+5 < K$, at $K = +5.669369\ldots$.

Finally, in the interval $-1 < K < +1$, there is a change in the junctions at the value $K = 0$, as shown by Fig. 13.20. Computations for other values of K indicate that there is no other change in the interval $-1 < K < +1$, i.e. the junctions remain as shown by Fig. 13.20, left, for $-1 < K < 0$, and by Fig. 13.20, right, for $0 < K < 1$.

The fact that the change happens precisely at $K = 0$ can be explained. Consider the symmetry defined at the end of Sect. 13.1 for n even. In the particular case $K = 0$, this symmetry consists simply in inverting the sequences of the Y_i and the X_i. A solution of (13.1) can be invariant under that symmetry. This is indeed the case for the solutions of the branches + 2112 and + 33. The two other branches, + 213 and + 312, are changed into each other by the symmetry. Thus, for $K = 0$, we have a *trident* situation, and all four branches come to a common solution Ω (Fig. 13.20, center). As soon as $K \neq 0$, the four branches become joined two by two.

All junctions of branches have been thus determined up to $n = 6$. Results are in Tables 13.1 to 13.10. The format is similar to that of Tables 8.4 to 8.11. All cases left undecided in these tables are now solved.

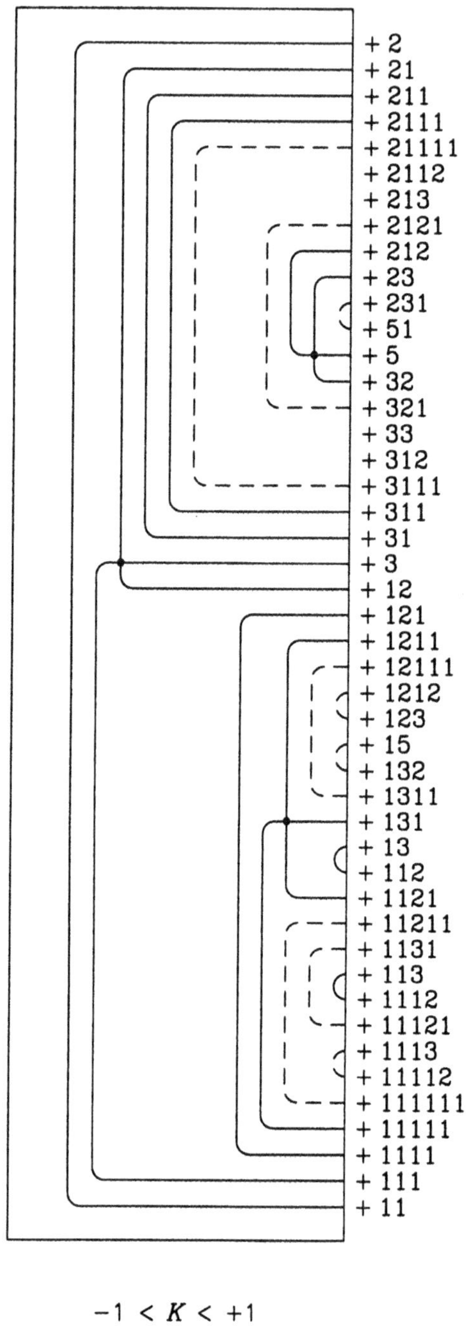

Fig. 13.15. Junctions for $n \leqq 6$.

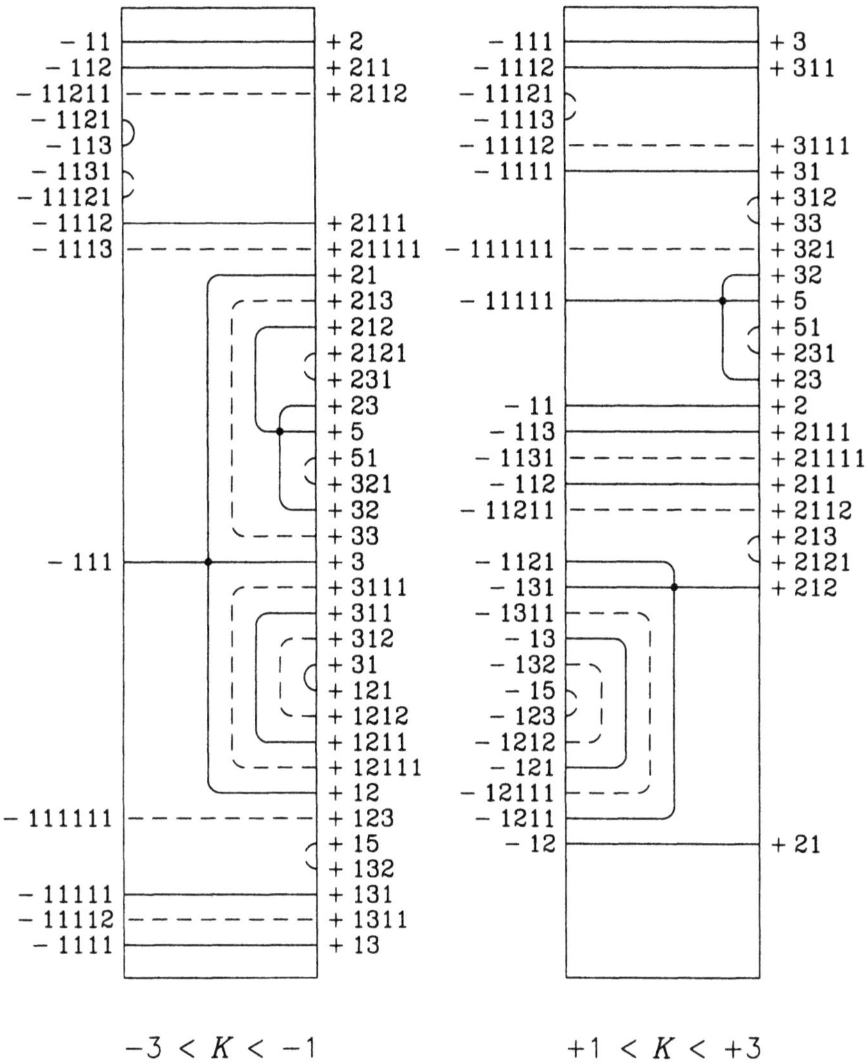

Fig. 13.16. Junctions for $n \leqq 6$ (continuation).

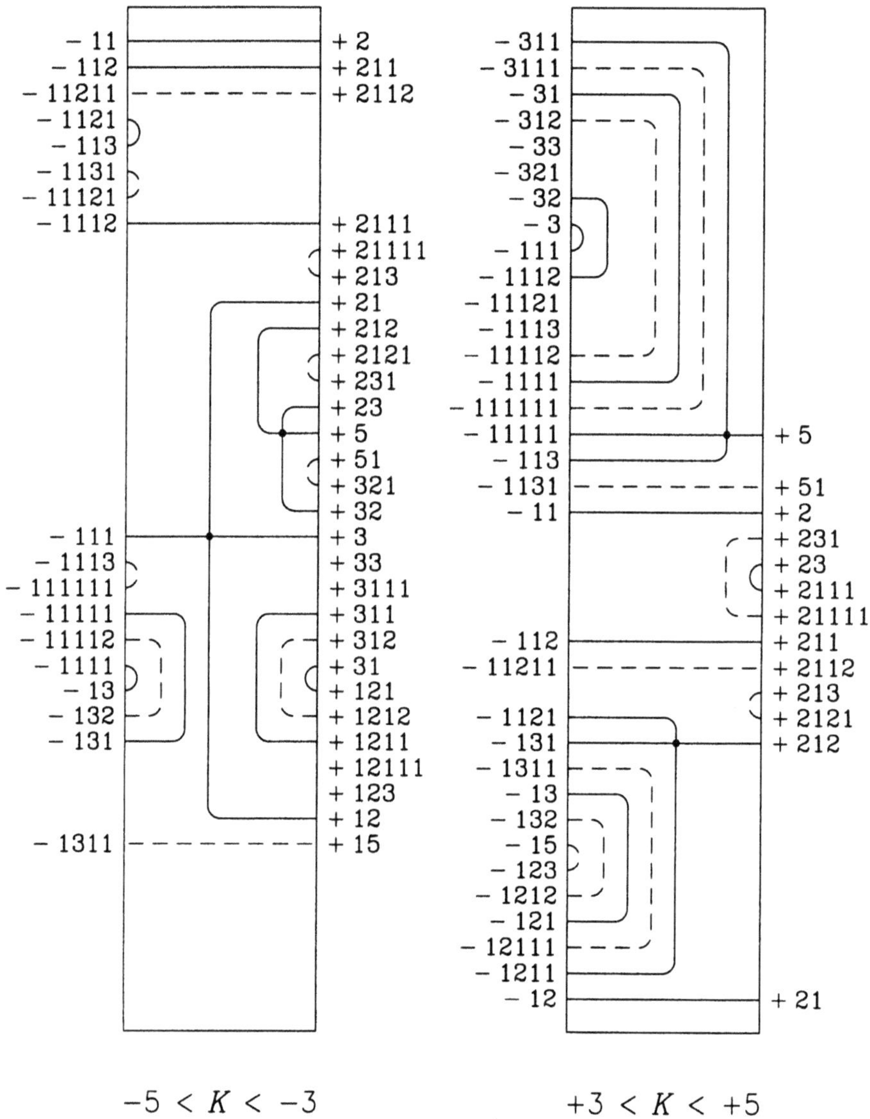

Fig. 13.17. Junctions for $n \leqq 6$ (continuation).

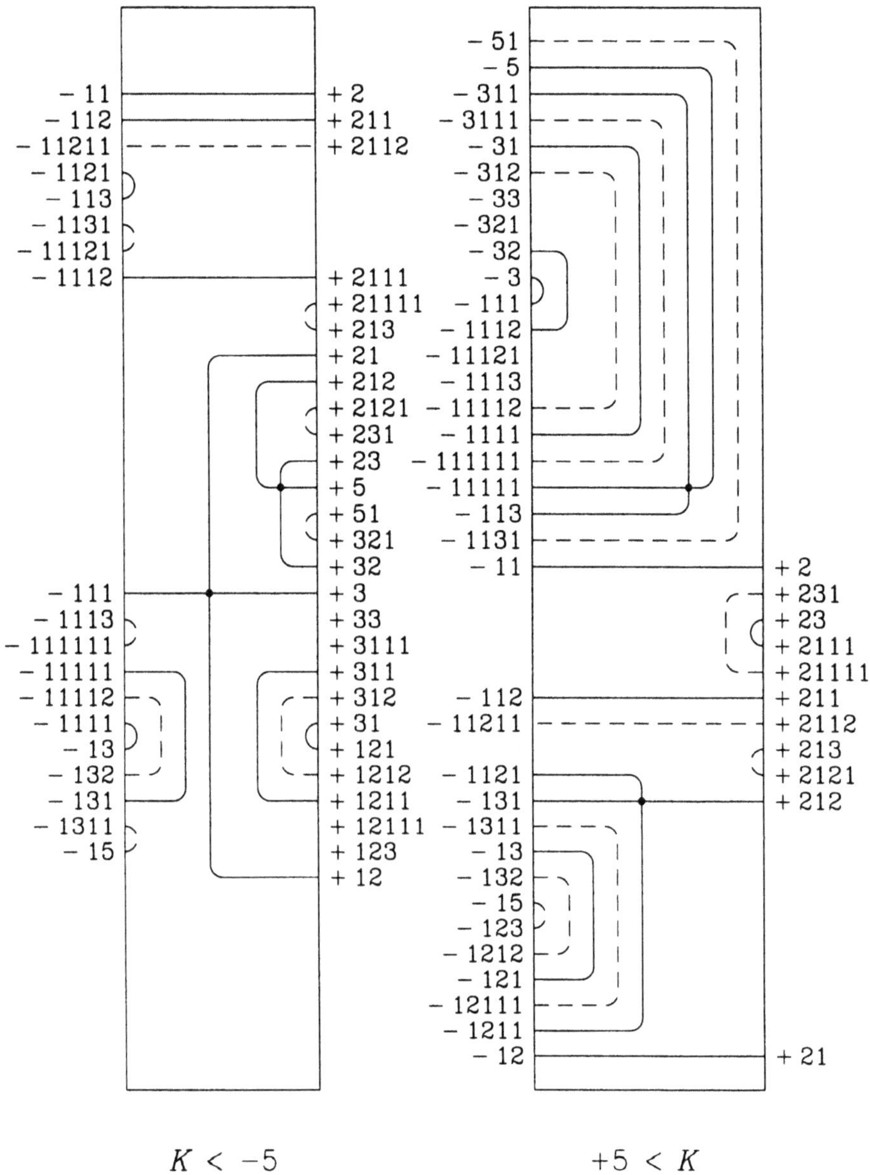

Fig. 13.18. Junctions for $n \leqq 6$ (continuation).

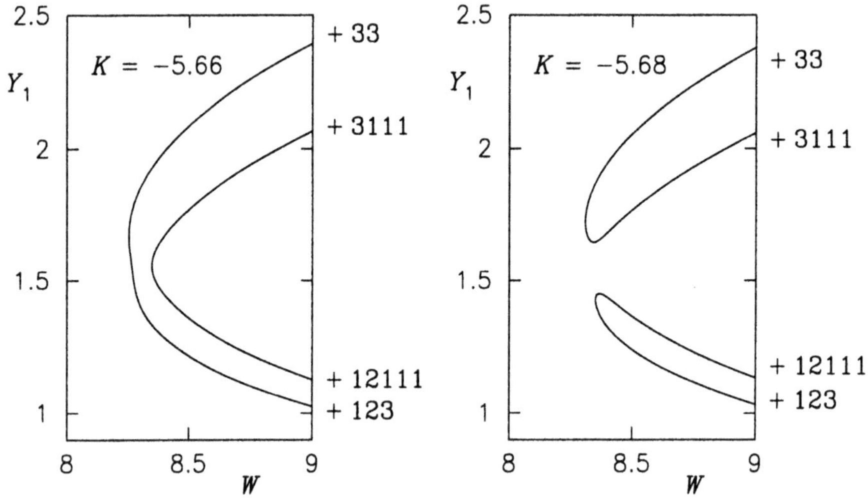

Fig. 13.19. Some characteristics of 1P6 bifurcations for $K = -5.66$ and $K = -5.68$.

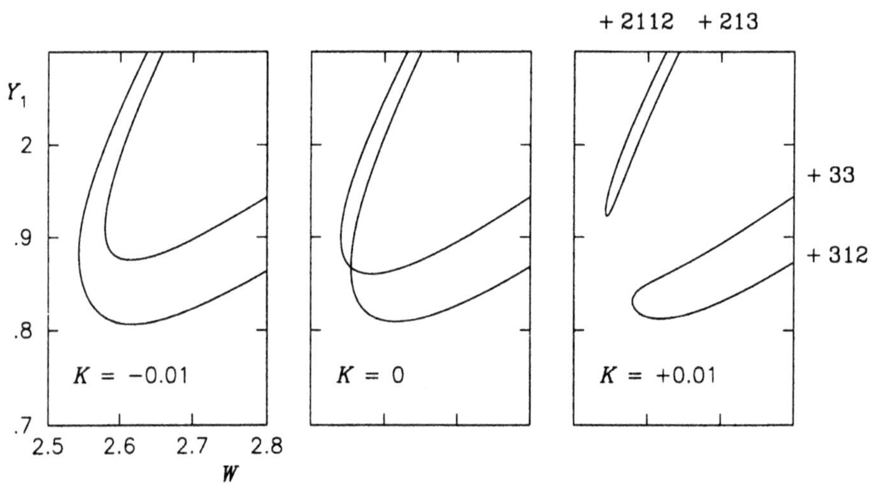

Fig. 13.20. Some characteristics of 1P6 bifurcations for $K = -0.01$, 0, +0.01.

Nothing prevents in principle the solution of higher values of n, using numerical computation. However, the amount of work grows exponentially.

13.3.3.6 $n > 6$. The positional method has also been applied for $n > 6$ in the two cases $-3 < K < -1$ and $+1 < K < +3$, in order to explore how far one can go. For $n = 7$, all junctions can still be determined. We do not show the schematic pictures of the junctions in the (W, Y_1) plane, which are too voluminous. However, the results are included in Tables 13.4 and 13.7.

For $n = 8$, in each of the two cases $-3 < K < -1$ and $+1 < K < +3$, there appear two groups of 4 branches which cannot be resolved by the positional method.

Table 13.1. Partial bifurcation, type 1: branch junctions determined by the quantitative study, for starting point in P and $K < -5.669369\ldots$. Other branches are joined as in Table 13.3.

1P6+++++A	1P6++---A
- $1\underline{5}$	+ $3\underline{3}$
- $1\underline{3}1\underline{1}$	+ $3\underline{1}1\underline{1}$
	+ $12\underline{3}$
	+ $12\underline{1}1\underline{1}$

Table 13.2. Partial bifurcation, type 1: branch junction determined by the quantitative study, for starting point in P and $-5.669369\ldots < K < -5$. Other branches are joined as in Table 13.3.

1P6+++++A
- $1\underline{5}$
- $1\underline{3}1\underline{1}$

Table 13.3. Partial bifurcation, type 1: branch junctions determined by the quantitative study, for starting point in P and $-5 < K < -3$.

1P2+A	
+	2
-	$1\underline{1}$

1P3++S	
+	3
-	$1\underline{1}1$

1P3++A	
+	21
+	12

1P4+++A	
-	$1\underline{3}$
-	$1\underline{1}1\underline{1}$

1P4++-A	
+	$3\underline{1}$
+	$12\underline{1}$
+	$21\underline{1}$
-	$1\underline{1}2$

1P5++++S	
+	5
+	212
-	$1\underline{3}1$
-	$1\underline{1}1\underline{1}1$

1P5++++A	
+	32
+	23

1P5++--A	
+	$3\underline{1}1$
+	$12\underline{1}1$
+	$21\underline{1}1$
-	$1\underline{1}12$
-	$1\underline{1}3$
-	$1\underline{1}21$

1P6+++++A	
+	15
-	$1\underline{3}1\underline{1}$
-	$1\underline{1}1\underline{3}$
-	$1\underline{1}1\underline{1}1\underline{1}$

1P6++++-A	
+	$5\underline{1}$
+	$32\underline{1}$
+	$23\underline{1}$
+	$212\underline{1}$
-	$1\underline{3}2$
-	$1\underline{1}1\underline{1}2$

1P6++--+A	
+	$3\underline{1}2$
+	$12\underline{1}2$
+	$21\underline{1}2$
-	$1\underline{1}21\underline{1}$
-	$1\underline{1}3\underline{1}$
-	$1\underline{1}12\underline{1}$

1P6++---A	
+	$3\underline{3}$
+	$12\underline{3}$
+	$3\underline{1}1\underline{1}$
+	$12\underline{1}1\underline{1}$
+	$21\underline{3}$
+	$21\underline{1}1\underline{1}$

Table 13.4. Partial bifurcation, type 1: branch junctions determined by the quantitative study, for starting point in P and $-3 < K < -1$.

1P2+A	
+ 2	- $1\underline{1}$

1P3++S	
+ 3	- $1\underline{1}1$

1P3++A	
+ 21	+ 12

1P4+++A	
+ 13	- $1\underline{1}1\underline{1}$

1P4++-A	
+ $3\underline{1}$	+ $12\underline{1}$
+ $21\underline{1}$	- $1\underline{1}2$

1P5++++S	
+ 5	+ 212
+ 131	- $1\underline{1}1\underline{1}1$

1P5++++A	
+ 32	+ 23

1P5++--A	
+ $3\underline{1}1$	+ $12\underline{1}1$
+ $21\underline{1}1$	- $1\underline{1}12$
- $1\underline{1}3$	- $1\underline{1}21$

1P6+++++A	
+ 33	+ 213
+ 15	+ 132
+ 123	- $1\underline{1}1\underline{1}1\underline{1}$

1P6++++-A	
+ $5\underline{1}$	+ $32\underline{1}$
+ $23\underline{1}$	+ $212\underline{1}$
+ $131\underline{1}$	- $1\underline{1}1\underline{1}2$

1P6++--+A	
+ $3\underline{1}2$	+ $12\underline{1}2$
+ $21\underline{1}2$	- $1\underline{1}21\underline{1}$
- $1\underline{1}3\underline{1}$	- $1\underline{1}12\underline{1}$

1P6++---A	
+ $3\underline{1}1\underline{1}$	+ $12\underline{1}1\underline{1}$
+ $21\underline{1}1\underline{1}$	- $1\underline{1}13$

1P7++++++S	
+ 7	+ 232
+ 151	- $1\underline{1}1\underline{1}1\underline{1}1$

1P7++++++A	
+ 52	+ 25
+ 331	+ 2131
+ 133	+ 1312
+ 1321	+ 1231

1P7++++--A	
+ $5\underline{1}1$	+ $32\underline{1}1$
+ $23\underline{1}1$	+ $212\underline{1}1$
+ $131\underline{1}1$	- $1\underline{1}1\underline{1}12$
- $1\underline{1}1\underline{1}3$	- $1\underline{1}1\underline{1}21$

1P7++--++S	
+ $3\underline{1}3$	+ $12\underline{1}21$
+ $21\underline{1}12$	- $1\underline{1}3\underline{1}1$

1P7++--++A	
+ $3\underline{1}21$	+ $12\underline{1}3$
+ $3\underline{1}12$	+ $12\underline{1}12$
+ $21\underline{1}3$	+ $21\underline{1}21$
- $1\underline{1}21\underline{1}1$	- $1\underline{1}12\underline{1}1$

1P7++----A	
+ $3\underline{1}1\underline{1}1$	+ $12\underline{1}1\underline{1}1$
+ $21\underline{1}1\underline{1}1$	- $1\underline{1}131$
- $1\underline{1}5$	- $1\underline{1}23$
- $1\underline{1}32$	- $1\underline{1}212$

Table 13.5. Partial bifurcation, type 1: branch junctions determined by the quantitative study, for starting point in P and $-1 < K < 0$.

1P2+A
+ 2 + 11

1P3++S
+ 3 + 111

1P3++A
+ 21 + 12

1P4+++A
+ 31 + 211
+ 13 + 112
+ 121 + 1111

1P5++++S
+ 5 + 212
+ 131 + 11111

1P5++++A
+ 32 + 23
+ 311 + 2111
+ 1211 + 1121
+ 113 + 1112

1P6+++++A
+ 51 + 231
+ 33 + 213
+ 312 + 2112
+ 321 + 2121
+ 3111 + 21111
+ 15 + 132
+ 1311 + 12111
+ 123 + 1212
+ 1131 + 11121
+ 11211 + 111111
+ 1113 + 11112

Table 13.6. Partial bifurcation, type 1: branch junctions determined by the quantitative study, for starting point in P and $0 < K < 1$. Other branches are joined as in Table 13.5.

1P6+++++A
+ 33 + 312
+ 213 + 2112

Table 13.7. Partial bifurcation, type 1: branch junctions determined by the quantitative study, for starting point in P and $1 < K < 3$.

1P2+A	+ 2 - $\underline{1}1$
1P3++S	+ 3 - $\underline{1}1\underline{1}$
1P3+-A	+ $2\underline{1}$ - $\underline{1}2$
1P4+++A	+ 31 - $\underline{1}1\underline{1}1$
1P4+--A	+ $2\underline{1}1$ - $\underline{1}12$
	- $\underline{1}3$ - $\underline{1}21$
1P5++++S	+ 5 - $\underline{1}1\underline{1}1\underline{1}$
1P5++++A	+ 32 + 23
1P5+++-A	+ $31\underline{1}$ - $\underline{1}1\underline{1}2$
1P5+--+S	+ $2\underline{1}2$ - $\underline{1}3\underline{1}$
1P5+--+A	- $\underline{1}21\underline{1}$ - $\underline{1}12\underline{1}$
1P5+---A	+ $2\underline{1}1\underline{1}$ - $\underline{1}13$
1P6+++++A	+ 51 + 231
	+ 33 + 312
	+ 321 - $\underline{1}1\underline{1}1\underline{1}1$
1P6+++--A	+ $31\underline{1}1$ - $\underline{1}1\underline{1}12$
	- $\underline{1}1\underline{1}3$ - $\underline{1}1\underline{1}21$
1P6+--++A	+ $2\underline{1}3$ + $2\underline{1}21$
	+ $2\underline{1}12$ - $\underline{1}12\underline{1}1$
	- $\underline{1}3\underline{1}1$ - $\underline{1}21\underline{1}1$
1P6+----A	+ $2\underline{1}1\underline{1}1$ - $\underline{1}131$
	- $\underline{1}5$ - $\underline{1}23$
	- $\underline{1}32$ - $\underline{1}212$
1P7++++++S	+ 7 + 232
	+ 313 - $\underline{1}1\underline{1}1\underline{1}1\underline{1}$
1P7++++++A	+ 52 + 25
1P7+++++-A	+ $51\underline{1}$ + $231\underline{1}$
	+ $33\underline{1}$ + $312\underline{1}$
	+ $321\underline{1}$ - $\underline{1}1\underline{1}1\underline{1}2$
1P7+++--+A	+ $31\underline{1}2$ - $\underline{1}1\underline{1}21\underline{1}$
	- $\underline{1}1\underline{1}3\underline{1}$ - $\underline{1}1\underline{1}12\underline{1}$
1P7+++---A	+ $31\underline{1}1\underline{1}$ - $\underline{1}1\underline{1}13$
1P7+--+++A	+ $2\underline{1}13$ - $\underline{1}12\underline{1}1\underline{1}$
	- $\underline{1}3\underline{1}1\underline{1}$ - $\underline{1}21\underline{1}1\underline{1}$
1P7+--++-A	+ $2\underline{1}3\underline{1}$ + $2\underline{1}12\underline{1}$
	+ $2\underline{1}21\underline{1}$ - $\underline{1}12\underline{1}2$
	- $\underline{1}3\underline{1}2$ - $\underline{1}21\underline{1}2$
1P7+----+S	+ $2\underline{1}1\underline{1}2$ - $\underline{1}131\underline{1}$
	- $\underline{1}5\underline{1}$ - $\underline{1}212\underline{1}$
1P7+----+A	- $\underline{1}32\underline{1}$ - $\underline{1}23\underline{1}$
1P7+-----A	+ $2\underline{1}1\underline{1}1\underline{1}$ - $\underline{1}123$
	- $\underline{1}33$ - $\underline{1}213$
	- $\underline{1}15$ - $\underline{1}132$

Table 13.8. Partial bifurcation, type 1: branch junctions determined by the quantitative study, for starting point in P and $3 < K < 5$.

Branch	Junction
1P2+A	+ 2, - 11
1P3++S	- 3, - 111
1P3+-A	+ 21, - 12
1P4+++A	- 31, - 1111
1P4+--A	+ 211, - 112
	- 13, - 121
1P5++++S	+ 5, - 11111
1P5++++A	- 311, - 113
1P5+++-A	- 32, - 1112
1P5+--+S	+ 212, - 131
1P5+--+A	- 1211, - 1121
1P5+---A	+ 23, + 2111
1P6+++++A	+ 51, - 1131
	- 3111, - 111111
1P6+++--A	- 33, - 321
	- 1113, - 11121
	- 312, - 11112
1P6+--++A	+ 213, + 2121
	+ 2112, - 11211
	- 1311, - 12111
1P6+----A	+ 231, + 21111
	- 15, - 123
	- 132, - 1212

Table 13.9. Partial bifurcation, type 1: branch junctions determined by the quantitative study, for starting point in P and $5 < K < 5.669369\ldots$. Other branches are joined as in Table 13.8.

Branch	Junction
1P5++++S	- 5, - 11111
1P6+++++A	- 51, - 1131

Table 13.10. Partial bifurcation, type 1: branch junctions determined by the quantitative study, for starting point in P and $5.669369\ldots < K$. Other branches are joined as in Table 13.8.

Branch	Junction
1P5++++S	- 5, - 11111
1P6+++++A	- 51, - 1131
1P6+++--A	- 33, - 1113
	- 321, - 11121

14. Total Bifurcation of Type 1

The equations (12.114) are in the case of a total bifurcation

$$Y_i + Y_{i-1} - W - (1 + Ks_i)X_i = 0 \,,$$
$$Y_i(X_{i+1} + X_i) + 1 = 0 \,, \tag{14.1}$$

where i is to be taken modulo n and takes all values from 1 to n.

These equations form a system of $2n$ equations for the $2n + 1$ variables W, Y_0 to Y_{n-1}, X_1 to X_n. As in the case of partial bifurcations (Sect. 13.1), we expect one-parameter families of solutions, which correspond to ordinary one-parameter families of orbits.

As in Sect. 8.5.1, we can use the two isomorphisms to consider only the case $K > 0$, and the case $Y_0 > 0$. The origin is taken in P.

14.1 Properties

The properties are essentially the same as for a partial bifurcation (Sect. 13.1).

In regard to property 5, the bifurcation orbit corresponds to a cycle of period n of the mapping.

14.1.1 Jacobian

The fundamental equations (14.1) can be written

$$f_1 = Y_1 + Y_0 - W - (1 - K)X_1 = 0 \,,$$
$$f_2 = X_2 + X_1 + \frac{1}{Y_1} = 0 \,,$$
$$\dots$$
$$f_{2i-1} = Y_i + Y_{i-1} - W - (1 + Ks_i)X_i = 0 \,,$$
$$f_{2i} = X_{i+1} + X_i + \frac{1}{Y_i} = 0 \,,$$
$$\dots$$
$$f_{2n-1} = Y_n + Y_{n-1} - W - (1 + K)X_n = 0 \,,$$
$$f_{2n} = X_{n+1} + X_n + \frac{1}{Y_n} = 0 \,,$$

$$
\begin{aligned}
f_{2n+1} &= Y_0 - Y_n = 0\,, \\
f_{2n+2} &= X_1 - X_{n+1} = 0\,.
\end{aligned}
\tag{14.2}
$$

The last two equations ensure periodicity.

For a given value of W, this is a system of $2n+2$ equations for $2n+2$ variables. The Jacobian is

$$
|J| = \frac{\partial(f_1, \ldots, f_{2n+2})}{\partial(Y_0, X_1, Y_1, X_2, \ldots, X_n, Y_n, X_{n+1})}
$$

$$
= \begin{vmatrix}
1 & -(1-K) & 1 & & & & & & & \\
 & 1 & -\frac{1}{Y_1^2} & 1 & & & & & & \\
 & & 1 & -(1+K) & 1 & & & & & \\
 & & & & & \cdots & & & & \\
 & & & & & & 1 & -(1+K) & 1 & \\
 & & & & & & & 1 & -\frac{1}{Y_n^2} & 1 \\
1 & & & & & & & & -1 & \\
 & 1 & & & & & & & & -1
\end{vmatrix}
\tag{14.3}
$$

We decompose with respect to the last two lines, obtaining four $n \times n$ determinants. Two of them are equal to 1; the third is obtained by removing the second column and the second from the last column; the fourth is obtained by removing the first and last columns. The result is

$$
|J| = 2 + \frac{\partial(f_1, \ldots, f_{2n})}{\partial(Y_0, Y_1, X_2, \ldots, X_n, X_{n+1})} - \frac{\partial(f_1, \ldots, f_{2n})}{\partial(X_1, Y_1, X_2, \ldots, X_n, Y_n)}\,.
\tag{14.4}
$$

14.1.2 Relation with Stability

Here again we will show that there exists an intimate relation between the Jacobian $|J|$ and the stability of the orbit.

For a total bifurcation, the stability index (Sect. 2.8) can be computed, since the whole orbit is known. We consider the system of equations (14.2), for a given value of W, with the last two equations deleted. We have then a system of $2n$ equations for $2n+2$ variables. Starting from given values of Y_0 and X_1 and applying the equations one by one, we obtain successively Y_1, X_2, $\ldots$, Y_n, X_{n+1}. In the same way, starting from variations $\mathrm{d}Y_0$ and $\mathrm{d}X_1$, we can compute successively $\mathrm{d}Y_1$, $\mathrm{d}X_2$, $\ldots$, $\mathrm{d}Y_n$, $\mathrm{d}X_{n+1}$.

For the whole orbit, we have then

$$
\begin{pmatrix} \mathrm{d}Y_n \\ \mathrm{d}X_{n+1} \end{pmatrix} = \begin{pmatrix} \dfrac{\partial Y_n}{\partial Y_0} & \dfrac{\partial Y_n}{\partial X_1} \\[2ex] \dfrac{\partial X_{n+1}}{\partial Y_0} & \dfrac{\partial X_{n+1}}{\partial X_1} \end{pmatrix} \begin{pmatrix} \mathrm{d}Y_0 \\ \mathrm{d}X_1 \end{pmatrix}\,.
\tag{14.5}
$$

The stability index is the trace of the matrix divided by 2:

$$z = \frac{1}{2}\left(\frac{\partial Y_n}{\partial Y_0} + \frac{\partial X_{n+1}}{\partial X_1}\right) . \tag{14.6}$$

We have the variational equations

$$\begin{aligned} &\frac{\partial f_1}{\partial Y_0}\mathrm{d}Y_0 + \ldots + \frac{\partial f_1}{\partial X_{n+1}}\mathrm{d}X_{n+1} = 0 , \\ &\ldots \\ &\frac{\partial f_{2n}}{\partial Y_0}\mathrm{d}Y_0 + \ldots + \frac{\partial f_{2n}}{\partial X_{n+1}}\mathrm{d}X_{n+1} = 0 , \end{aligned} \tag{14.7}$$

In order to compute $\partial Y_n/\partial Y_0$, we set $\mathrm{d}X_1 = 0$. The remaining system of $2n$ equations for $2n+1$ variables can then be solved to give

$$\frac{\partial Y_n}{\partial Y_0} = -\frac{\dfrac{\partial(f_1,\ldots,f_{2n})}{\partial(Y_0,Y_1,X_2,\ldots,X_n,X_{n+1})}}{\dfrac{\partial(f_1,\ldots,f_{2n})}{\partial(Y_1,X_2,\ldots,X_n,Y_n,X_{n+1})}} . \tag{14.8}$$

Similarly, to compute $\partial X_{n+1}/\partial X_1$, we set $\mathrm{d}Y_0 = 0$, and we find

$$\frac{\partial X_{n+1}}{\partial X_1} = \frac{\dfrac{\partial(f_1,\ldots,f_{2n})}{\partial(X_1,Y_1,X_2,\ldots,X_n,Y_n)}}{\dfrac{\partial(f_1,\ldots,f_{2n})}{\partial(Y_1,X_2,\ldots,X_n,Y_n,X_{n+1})}} . \tag{14.9}$$

We obtain the stability index

$$z = \frac{\dfrac{\partial(f_1,\ldots,f_{2n})}{\partial(X_1,Y_1,X_2,\ldots,X_n,Y_n)} - \dfrac{\partial(f_1,\ldots,f_{2n})}{\partial(Y_0,Y_1,X_2,\ldots,X_n,X_{n+1})}}{2\,\dfrac{\partial(f_1,\ldots,f_{2n})}{\partial(Y_1,X_2,\ldots,X_n,Y_n,X_{n+1})}} . \tag{14.10}$$

The denominator reduces to 1, and comparing with (14.4), we obtain

$$z = 1 - \frac{|J|}{2} . \tag{14.11}$$

We have thus shown

Proposition 14.1.1. *In a total bifurcation of type 1, the Jacobian vanishes if and only if the orbit is a critical orbit of the first kind (the stability index is $z = 1$).*

14.1.3 Asymptotic Behaviour for $|W| \to \infty$

In the general case, the bifurcation orbit becomes in this limit a sequence of T and S arcs, and we can use the computations of Sect. 13.1.2. (The particular case of the first species orbit E will be considered in Sect. 14.2.1.) From (13.29), with $\alpha = N$, $i_\alpha = n$ we obtain for the stability index

$$z = \frac{u_1 u_2 \dots u_N}{2Y_{i_1}^2 Y_{i_2}^2 \dots Y_{i_N}^2} [1 + O(W^{-2})] \,. \tag{14.12}$$

There is $z = \Theta(W^{2N})$: the orbit is strongly unstable. The Jacobian never vanishes in this limit. The sign is

$$\text{sign}(z) = (-1)^{\zeta} \tag{14.13}$$

where ζ is defined, as in Sect. 13.3.2.1, as the total number of T-arcs and abnormal S-arcs.

14.2 Small Values of n

We will represent the characteristics in the (W, Y_0) plane. In the present case, a point of this plane is not sufficient to define an orbit: the three values W, Y_0, X_1 are needed. As a consequence, the intersection of two characteristics on the figures does not necessarily correspond to a true intersection of the two families; it can be merely a projection effect.

Only the upper half $Y_0 > 0$ is represented, since here again the figure is symmetric with respect to the origin because of the symmetry Σ' (13.4).

14.2.1 $n = 2$

From (13.6) we have

$$2Y_0 + 2Y_1 + \frac{1 - K^2}{Y_1} - 2W = 0 \,, \qquad 2Y_1 + 2Y_0 + \frac{1 - K^2}{Y_0} - 2W = 0 \,, \tag{14.14}$$

from which we obtain $Y_0 = Y_1$, and

$$W = 2Y_0 + \frac{1 - K^2}{2Y_0} \,, \qquad X_1 = -\frac{1 + K}{2Y_0} \,. \tag{14.15}$$

Depending on the value of K, the characteristic has one of the shapes represented on Fig. 14.1. It is a hyperbola; its asymptotes are the straight lines $Y_0 = 0$ and $Y_0 = W/2$ (dashed line).

Table 6.4 (under 1T2) shows that the bifurcating orbit can consist either of a first species orbit (branches ± E) or two basic S-arcs (branches ± 11). From the equations of Sect. 13.1.1, we find that the first case corresponds to the branches of the characteristic asymptotic to $Y_0 = W/2$, while the second case corresponds to the branches asymptotic to $Y_0 = 0$. We recover the junctions established in Sect. 8.5.1 from Broucke's principle.

For $0 \leqq K < 1$, the characteristic has an extremum in W (dot on Fig. 14.1), with the coordinates

$$Y_0 = \frac{\sqrt{1 - K^2}}{2} \,, \qquad W = 2\sqrt{1 - K^2} \,. \tag{14.16}$$

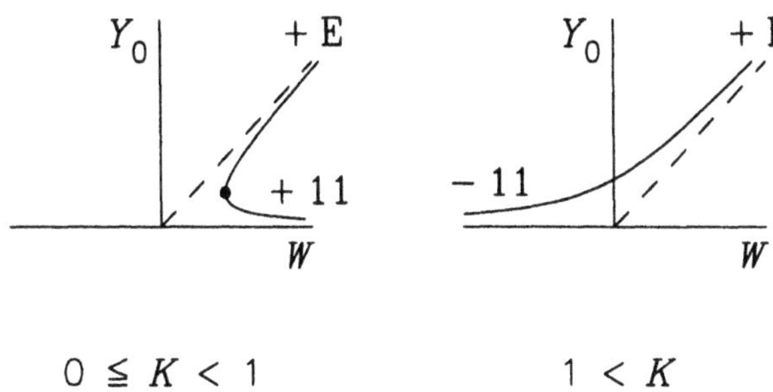

Fig. 14.1. Characteristics for 1T2 bifurcations.

We obtain the stability index from (14.6)

$$z = 1 - \frac{2}{Y_0^2} + \frac{1 - K^2}{2Y_0^4} . \tag{14.17}$$

For $0 \leqq K < 1$, there is $z = 1$ at the extremum.

On the branch + E one has $Y_0 \to +\infty$, and therefore $z \to 1_-$: the first species orbit is stable. This agrees with the numerical results of Bruno (1976; 1994, Chap. VII).

The total bifurcation 1T2 was already studied quantitatively by Guillaume (1971, pp. 112 to 119).

14.2.2 Numerical Computation: Method

For $n \geqq 4$, the system of 4 equations (13.6) does not seem to be explicitly solvable. (For $n = 4$, the *symmetric* orbits can be determined.) No equivalent of the positional method used for partial bifurcations (Sect. 13.3) was found in the present case. A program was written to solve the equations numerically. The computation of a given branch is begun at a large value of W. An iterative relaxation method, based on the asymptotic decomposition of the orbit into T- and S-arcs and on the asymptotic equations of Sect. 13.1.1, works well in that case. For a given value of W, the steps are as follows:

1. We take as initial approximation $Y_i = 0$ in nodes, $1/Y_i = 0$ in antinodes.
2. For each arc, we compute the values of the internal Y_i (antinodes) and the values of the X_i by solving the equations (14.1). In (14.1b), the value of Y_i from the previous iteration is first substituted; we have then a linear system of $2m - 1$ equations for $2m - 1$ unknowns, which is easily solved.
3. For each node, we recompute Y_i, using (14.1b).
4. We go back to step 2, until the solution has converged.

The branch is followed towards decreasing $|W|$. Usually at some point the iteration does not converge anymore, because the decomposition into T- and S-arcs ceases to be a good approximation. We shift then to a trial-and-error shooting method, which works well for moderate values of W where the branches are well separated: starting from approximate values of Y_0 and X_1, we compute successively $Y_1, X_2, \ldots, Y_n, X_{n+1}$, with (14.1). We compute also the variations, as functions of $\mathrm{d}Y_0$ and $\mathrm{d}X_1$, obtaining the 4 elements of the matrix in (14.5). This allows a correction of the initial values by the Newton-Raphson method, until a periodic orbit is obtained ($Y_n = Y_0$, $X_{n+1} = X_1$). It allows also a computation of the stability index.

14.2.3 $n = 4$

Figs. 14.2 to 14.4 show results for $K = 0.3$, 2.2, 4. Characteristics are also represented for $n = 2$ (dotted lines).

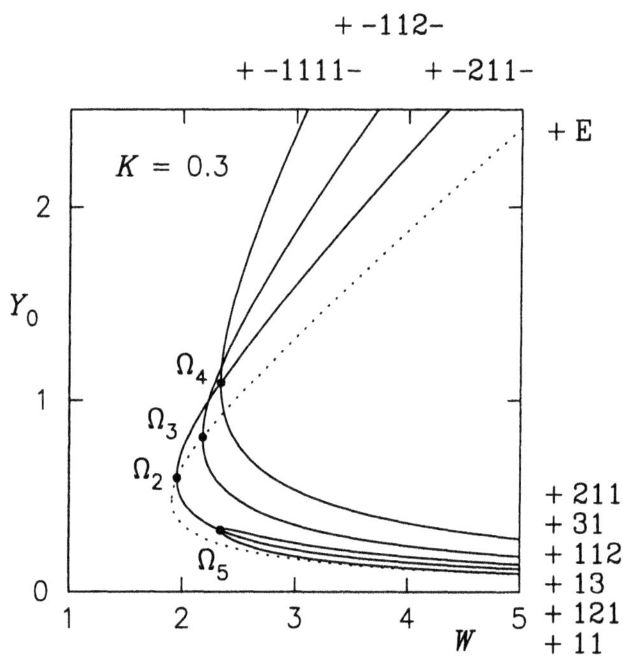

Fig. 14.2. Characteristics for 1T4 bifurcations, for $K = 0.3$ (solid lines). Also shown: 1T2 bifurcation (dotted line).

The branches are listed in Table 6.4 (under 1T4). They are automatically identified on the figures since the numerical procedure explicitly uses their asymptotic behaviour.

Computations for other values of K indicate that the figure remains the same qualitatively inside each of the three intervals $0 \leqq K < 1$, $1 < K < 3$, $3 < K$. The junctions are thus fully determined.

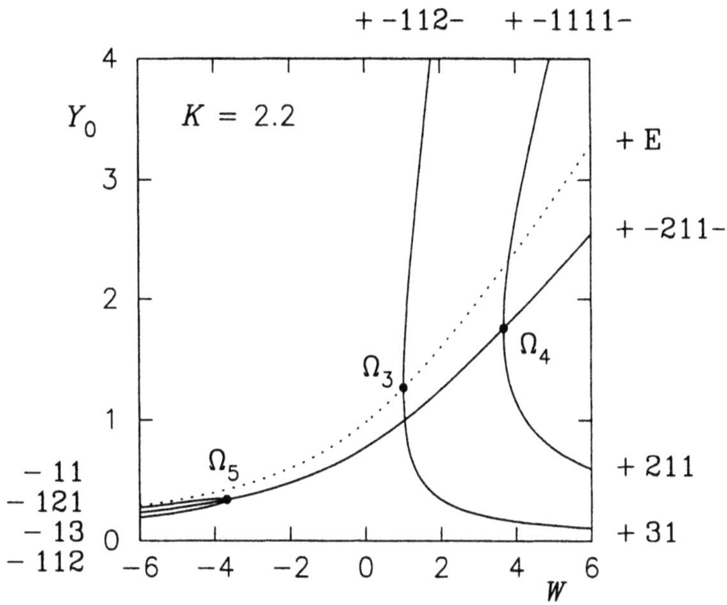

Fig. 14.3. Characteristics for 1T4 bifurcations, for $K = 2.2$.

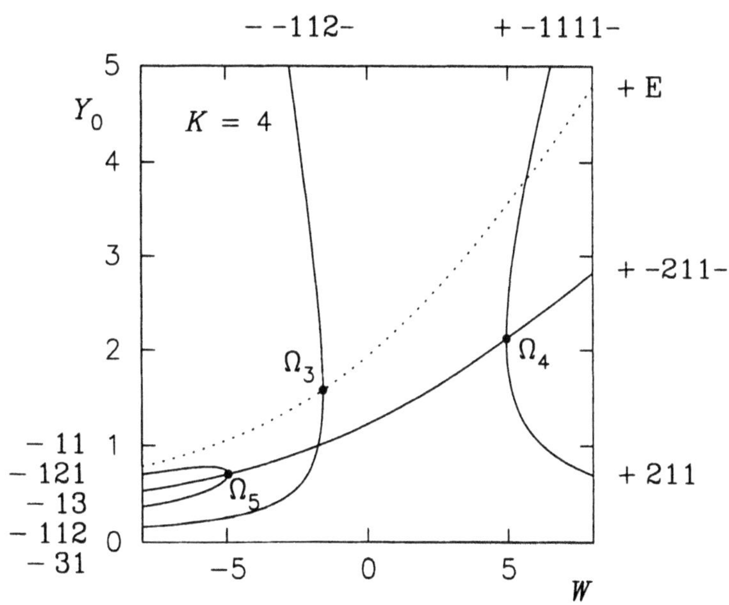

Fig. 14.4. Characteristics for 1T4 bifurcations, for $K = 4$.

Several intersections are significant on Figs. 14.2 to 14.4; they are indicated by dots and labels. The orbits Ω_2 and Ω_3 have as coordinates respectively

$$W = (3+K)\sqrt{\frac{1-K}{2}}\,, \qquad Y_0 = \sqrt{\frac{1-K}{2}}\,, \qquad X_1 = -\frac{1+K}{\sqrt{2(1-K)}} \tag{14.18}$$

and

$$W = (3-K)\sqrt{\frac{1+K}{2}}\,, \qquad Y_0 = \sqrt{\frac{1+K}{2}}\,, \qquad X_1 = -\sqrt{\frac{1+K}{2}}\,. \tag{14.19}$$

(Ω_2 exists only for $0 \leqq K < 1$.) Ω_2 and Ω_3 correspond to an intersection of a $n = 4$ family with a $n = 2$ family described twice. The intersection is also an extremum in W for the $n = 4$ family. The stability index at the intersection is $z = +1$ for the $n = 4$ family, $z = -1$ for the $n = 2$ family.

The orbits Ω_4 and Ω_5 exist for all K. They are respectively in

$$W = \frac{3+K}{\sqrt{2}}\,, \qquad Y_0 = \frac{1+\sqrt{K}}{\sqrt{2}}\,, \qquad X_1 = -\frac{1-\sqrt{K}}{\sqrt{2}(1+\sqrt{K})}\,, \tag{14.20}$$

and

$$W = \operatorname{sign}(1-K)\frac{3+K}{\sqrt{2}}\,, \qquad Y_0 = \frac{\left|1-\sqrt{K}\right|}{\sqrt{2}}\,, \qquad X_1 = -\frac{1+\sqrt{K}}{\sqrt{2}\left|1-\sqrt{K}\right|}\,. \tag{14.21}$$

They correspond to an intersection of two $n = 4$ families, which is also an extremum in W for one of the two families. Moreover, one of the families is made of asymmetric orbits and the other of symmetric orbits. This is another instance of the *trident* situation already found in the case of partial bifurcation (Sects. 13.2.2 and 13.3.3.2). The stability index at the intersection is $z = +1$ for both families.

14.2.4 $n = 6$

Numerical computation was also used to determine the junctions for $n = 6$. Showing characteristics in the (W, Y_0) plane would not be useful because the number of branches is too great and the figure is crowded. Instead, detailed print-outs of the branches were made, from which the junctions were found by inspection.

The computations were made for several values of K. They indicate that the figure remains the same qualitatively inside each of the three intervals $0 \leqq K < 1$, $1 < K < 3$, $3 < K < 5$. For $5 < K$, however, some junctions change at the value $K = 5.2612\ldots$; this is illustrated by Fig. 14.5.

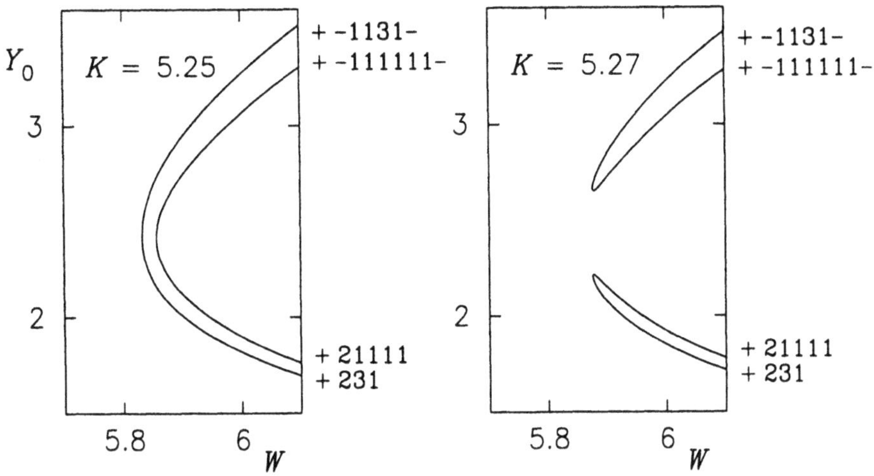

Fig. 14.5. Some characteristics of 1T6 bifurcations for $K = 5.25$ and $K = 5.27$.

14.3 Conclusions for Type 1

The study of bifurcations of type 1, begun in Chap. 12, is now complete. All junctions of branches have been determined at least up to $n = 6$. Results are in Tables 13.1 to 13.10 for partial bifurcations, and in Tables 14.1 to 14.5 for total bifurcations. These tables are in the same format as Tables 8.4 to 8.11 and 8.14 to 8.17 of Volume I. It can be verified that there is agreement for all branches which appear in both series of tables. In addition, all cases up to $n = 6$ which had been left undecided in the tables of Chap. 8 are now solved. This shows that the quantitative approach is much more powerful than the qualitative one.

Nothing prevents in principle the solution of higher values of n, using numerical computation. However, the amount of work grows exponentially.

Table 14.1. Total bifurcation, type 1, $n = 2$ to 6: branch junctions determined by the quantitative study, for $0 \leqq K < 1$.

```
1T2+++1
-------
+ E
+ 11
-------
1T4+++++1
-------
+ -211-
+ 13
-------
1T4+++++3
-------
+ 31
+ -112-
-------
1T4+++++A
-------
+ 211
+ -1111-
-------
+ 112
+ 121
-------

1T6+++++++1
-------
+ -312-
+ -231-
-------
+ 15
+ 1212
-------
+ 1131
+ -21111-
-------
1T6+++++++3
-------
+ -114-
+ 33
-------
+ -411-
+ -11211-
-------
+ 3111
+ 1113
-------

1T6+++++++5
-------
+ 51
+ -132-
-------
+ -213-
+ 2121
-------
+ -11112-
+ 1311
-------
1T6+++++++A
-------
+ 321
+ 231
-------
+ -1212-
+ 312
-------
+ -1131-
+ -1122-
-------

+ -2211-
+ -1311-
-------
+ 213
+ -2121-
-------
+ 132
+ 123
-------
+ 21111
+ -111111-
-------
+ 11112
+ 11121
-------
+ 11211
+ 12111
-------
```

Table 14.2. Total bifurcation, type 1, $n = 2$ to 6: branch junctions determined by the quantitative study, for $1 < K < 3$.

```
1T2+++1
-------
+ E
- 11
-------
1T4+++++3
-------
+ 31
+ -112-
-------
1T4++--+1
-------
+ -211-
- 13
-------
1T4++--+A
-------
+ 211
+ -1111-
-------
- 112
- 121
-------

1T6+++++++1
-------
+ -312-
+ -231-
-------
1T6+++++++3
-------
+ -114-
+ 33
-------
1T6+++++++5
-------
+ 51
+ -132-
-------
1T6+++++++A
-------
+ 321
+ 312
-------
+ -1212-
+ -1122-
-------
+ -1131-
+ 231
-------

1T6++++--+3
-------
+ -411-
+ -11211-
-------
+ 3111
- 1113
-------
1T6++++--+A
-------
+ -2211-
+ -1311-
-------
- 11112
- 11121
-------
1T6++--+++5
-------
+ -213-
+ 2121
-------
+ -11112-
- 1311
-------

1T6++--+++A
-------
+ 213
+ -2121-
-------
- 11211
- 12111
-------
1T6++----+1
-------
- 15
- 1212
-------
- 1131
+ -21111-
-------
1T6++----+A
-------
- 132
- 123
-------
+ 21111
+ -111111-
-------
```

Table 14.3. Total bifurcation, type 1, $n = 2$ to 6: branch junctions determined by the quantitative study, for $3 < K < 5$.

1T2+++1
+ E - 11

1T4+++++3
- 31 - -112-

1T4++--+1
+ -211- - 13

1T4++--+A
+ 211 + -1111-
- 112 - 121

1T6+++++++1
+ -312- - 1131

1T6+++++++3
+ -114- - 3111

1T6+++++++5
+ 51 - -11112-

1T6++++--+3
- 33 - 1113
+ -411- + -11211-

1T6++++--+A
- 321 - 312
+ -2211- + -1311-
- 11112 - 11121

1T6++--+++5
- -132- - 1311
+ -213- + 2121

1T6++--+++A
- -1212- - -1122-
+ 213 + -2121-
- 11211 - 12111

1T6++----+1
+ -231- + -21111-
- 15 - 1212

1T6++----+A
+ -1131- + 231
- 132 - 123
+ 21111 + -111111-

Table 14.4. Total bifurcation, type 1, $n = 2$ to 6: branch junctions determined by the quantitative study, for $5 < K < 5.26\ldots$.

1T2+++1
+ E
- 11

1T4+++++3
- 31
- -112-

1T4++--+1
+ -211-
- 13

1T4++--+A
+ 211
+ -1111-
- 112
- 121

1T6+++++++1
- -312-
- 1131

1T6+++++++3
- -114-
- 3111

1T6+++++++5
- 51
- -11112-

1T6++++--+3
- 33
- 1113
+ -411-
+ -11211-

1T6++++--+A
- 321
- 312
- 11112
- 11121
+ -2211-
+ -1311-

1T6++--+++5
- -132-
- 1311
+ -213-
+ 2121

1T6++--+++A
- -1212-
- -1122-
- 11211
- 12111
+ 213
+ -2121-

1T6++----+1
+ -231-
+ -21111-
- 15
- 1212

1T6++----+A
+ -1131-
+ 231
+ 21111
+ -111111-
- 132
- 123

Table 14.5. Total bifurcation, type 1, $n = 2$ to 6: branch junctions determined by the quantitative study, for $5.26\ldots < K$.

1T2+++1
+ E
- 11

1T4+++++3
- 31
- -112-

1T4++--+1
+ -211-
- 13

1T4++--+A
+ 211
+ -1111-

- 112
- 121

1T6+++++++1
- -312-
- 1131

1T6+++++++3
- -114-
- 3111

1T6+++++++5
- 51
- -11112-

1T6++++--+3
- 33
- 1113

+ -411-
+ -11211-

1T6++++--+A
- 321
- 11121

- 312
- 11112

+ -2211-
+ -1311-

1T6++--+++5
- -132-
- 1311

+ -213-
+ 2121

1T6++--+++A
- -1212-
- 12111

- -1122-
- 11211

+ 213
+ -2121-

1T6++----+1
+ -231-
+ -21111-

- 15
- 1212

1T6++----+A
+ -1131-
+ -111111-

+ 231
+ 21111

- 132
- 123

15. The Newton Approach

After the present monograph was completed (with the exception of the present chapter), there appeared a book by Bruno (1998, 2000). Chapter 1 of that book presents a general geometrical framework, based on the fundamental concept of *Newton polyhedra*, for the study of algebraic, ordinary differential, and partial differential equations in the vicinity of a singular point. Chapter 2 applies that framework to the case of a system of nonlinear equations.

This work is in principle directly applicable to the problem which we have considered in the preceding chapters. We have a system of nonlinear equations (12.32), relating the variables μ, ΔC, y_i, Δa_i, and we wish to investigate the solutions of that system for small values of these variables, i.e. in the vicinity of the origin. If the origin corresponds to a bifurcation, more than one family passes through it (see Definition 6.0.2) and the origin is a singular point.

We solved this problem in Chapters 12 to 14, using a somewhat intuitive, heuristic approach. Here we present a more rigorous approach, based on the use of Newton polyhedra as developed by Bruno. We will call it the *Newton approach*. Unfortunately, we will find that this approach is not applicable to the general case, and that in practice it can only be used for small values of n. However, the results of the preceding chapters will be confirmed and put on firmer ground.

In the present Chapter, a reference such as "Bruno 1.6" will mean: Bruno (2000), Chap. 1, Section 6.

15.1 Partial Bifurcation of Type 1: Variables and Equations

We will describe the Newton approach mostly in the case 1Pn, i.e. a partial bifurcation of type 1 in which the bifurcating arc is made of n basic arcs (see Sect. 6.2.2); the case of a total bifurcation of type 1 will be briefly considered in Sect. 15.8. There is (Sect. 6.2.2.1):

$$n \geqq 2 \, . \tag{15.1}$$

The equations which we have to consider are (12.32), and occasionally also the additional relations (12.33). We eliminate the two variables y_0 and y_n, using the last two equations (12.32). We are thus left with $2n+1$ variables: μ, ΔC, y_i ($i = 1, \ldots, n-1$), Δa_i ($i = 1, \ldots, n$). (We recall that G_1, G_2, G_3, K, $s_i = (-1)^i$ are constants.) It will be convenient to designate these variables by the uniform notations x_1, x_2, ..., as in Bruno: we write

$$\begin{aligned} &x_1 = \mu\,, \quad x_2 = \Delta C\,, \quad x_3 = \Delta a_1\,, \quad x_4 = y_1\,, \quad x_5 = \Delta a_2\,, \\ &x_6 = y_2\,, \quad \ldots\,, \quad x_{2i+1} = \Delta a_i\,, \quad x_{2i+2} = y_i\,, \quad \ldots\,, \\ &x_{2n-1} = \Delta a_{n-1}\,, \quad x_{2n} = y_{n-1}\,, \quad x_{2n+1} = \Delta a_n\,. \end{aligned} \tag{15.2}$$

Thus there are $2n+1$ variables. The number of variables is called n by Bruno. To avoid confusion, Bruno's variables will be written with a subscript B when needed; thus we have

$$n_B = 2n + 1. \tag{15.3}$$

The fundamental equations (12.32) become

$$\begin{aligned} &f_1 = x_4 - G_2 x_2 + G_3(1-K)x_3 + O(x_1) + O(x_2^2) + O(x_3^2) = 0\,, \\ &f_2 = x_4(x_5 - x_3) + G_1 x_1 + O(x_1 x_3) + O(x_1 x_4) + O(x_1 x_5) = 0\,, \\ &\ldots \\ &f_{2i-1} = x_{2i+2} - x_{2i} + G_2 s_i x_2 + G_3(1 + K s_i) x_{2i+1} + O(x_1) \\ &\qquad + O(x_2^2) + O(x_{2i+1}^2) = 0\,, \qquad i = 2, \ldots, n-1\,, \\ &f_{2i} = x_{2i+2}(x_{2i+3} - x_{2i+1}) + G_1 x_1 + O(x_1 x_{2i+1}) + O(x_1 x_{2i+2}) \\ &\qquad + O(x_1 x_{2i+3}) = 0\,, \qquad i = 2, \ldots, n-1\,, \\ &\ldots \\ &f_{2n-1} = -x_{2n} + G_2 s_n x_2 + G_3(1 + K s_n) x_{2n+1} \\ &\qquad + O(x_1) + O(x_2^2) + O(x_{2n+1}^2) = 0\,. \end{aligned} \tag{15.4}$$

The constants G_1, G_2, G_3 are defined in (12.31); we recall that

$$G_1 > 0\,, \quad G_2 \neq 0\,, \quad G_3 > 0\,. \tag{15.5}$$

We have thus reduced our problem to the form considered by Bruno (2, (5.1), (6.1), (8.1)). The number of equations is called m by Bruno; here we will designate it by m_B. This number is

$$m_B = 2n - 1. \tag{15.6}$$

There is $m_B = n_B - 2$; therefore we expect that the solutions of the system of equations (15.4) will lie on two-dimensional manifolds. One of the two dimensions corresponds to the variation of μ, while the other is the parameter along a family of periodic orbits, which can be taken to be ΔC.

The most studied case in the literature has been the particular case $m_B = n_B - 1$ (Bruno 2.8), for which the solutions lie on one-dimensional manifolds. Indeed, most examples in Bruno's book fall into this case, the only exception being the comparatively simple Example 4.1 in Chapter 2 which has $n_B = 3$,

$m_B = 1$. The present problem might therefore be of theoretical interest as a non-trivial example of the more general case.

15.1.1 Additional Relations

In some cases the system of equations becomes degenerate, as will be seen, and it is then necessary to substitute one or more additional relations (12.33) to some of the equations (15.4). (The total number of equations remains constant.) Therefore we write also these relations in Bruno's variables. Their number is $n - 1$. For definiteness we describe them with indices from $2n$ to $3n - 2$:

$$\begin{aligned}
f_{2n} &= x_6 + G_3(1-K)x_3 + G_3(1+K)x_5 \\
&\quad + O(x_1) + O(x_3^2) + O(x_5^2) = 0\,, \\
&\dots \\
f_{2n-1+i} &= -x_{2i} + x_{2i+4} + G_3(1+Ks_i)x_{2i+1} + G_3(1-Ks_i)x_{2i+3} \\
&\quad + O(x_1) + O(x_{2i+1}^2) + O(x_{2i+3}^2) = 0\,, \qquad i = 2,\dots,n-2\,, \\
&\dots \\
f_{3n-2} &= -x_{2n-2} + G_3(1+Ks_{n-1})x_{2n-1} + G_3(1-Ks_{n-1})x_{2n+1} \\
&\quad + O(x_1) + O(x_{2n-1}^2) + O(x_{2n+1}^2) = 0\,.
\end{aligned} \tag{15.7}$$

In the particular case $n = 2$, there is a single additional relation

$$f_4 = G_3(1-K)x_3 + G_3(1+K)x_5 + O(x_1) + O(x_3^2) + O(x_5^2) = 0\,. \tag{15.8}$$

15.2 Method of Solution

The method consists in looking for solutions where all variables can be expressed as powers of a parameter. More precisely, we consider solutions of the system (15.4) of the form (Bruno 2.2)

$$x_i = b_i \tau^{p_i}(1 + o(1))\,, \qquad i = 1,\dots,n_B\,, \tag{15.9}$$

with

$$b_i \neq 0\,, \quad p_i \neq 0\,, \qquad i = 1,\dots,n_B\,. \tag{15.10}$$

τ is a parameter, and we consider the limit

$$\tau \to +\infty\,. \tag{15.11}$$

The parameter ν which was defined in (11.80) and which played a fundamental role in Chap. 12 can be related to the quantities p_i introduced here. We use the first two relations (15.2) and we substitute the first two equations (15.9) in (11.80). We obtain

$$b_2\tau^{p_2}(1+o(1)) = \Theta[b_1^\nu \tau^{p_1\nu}(1+o(1))] \tag{15.12}$$

which reduces to

$$\tau^{p_2} = \Theta(\tau^{p_1\nu}) \tag{15.13}$$

from which we have

$$\nu = \frac{p_2}{p_1} . \tag{15.14}$$

(15.9) represents a one-parameter family of solutions. In the most studied case (Bruno 2.8) $m_B = n_B - 1$, the solutions form one-dimensional manifolds, or curves, passing through the origin. Then the expressions (15.9) provide a parametric representation of one of these curves. In our case, there is $m_B = n_B - 2$, and (15.9) represents a curve lying on one of the two-dimensional manifolds passing through the origin.

Bruno considers the general case where the p_i can be positive or negative, i.e. in the limit some variables tend to zero while others tend to infinity. In our problem, all variables tend to zero; there is

$$p_i < 0 , \qquad i = 1, \ldots, n_B . \tag{15.15}$$

The *cone of the problem* (Bruno 1.6) is simply (see also Bruno 2.6, between (6.5) and (6.6)):

$$\mathbf{K} = \{P\colon\ p_1 < 0, \ldots, p_{n_B} < 0\} . \tag{15.16}$$

We consider now one particular expression $f_i(X)$ in (15.4). For simplicity, we drop temporarily the subscript i and write simply $f(X)$. This expression is a power series with respect to the variables. Each term is of the form $x_1^{q_1} x_2^{q_2} \ldots$. It will be convenient to introduce the abbreviations: $X = (x_1, \ldots, x_{n_B})$, $Q = (q_1, \ldots, q_{n_B})$, $X^Q = x_1^{q_1} x_2^{q_2} \ldots$.

We consider a n_B-dimensional space $\mathbb{R}^{n_B}$ with coordinates $q_1, \ldots, q_{n_B}$. Each term in the equation corresponds to a given point Q in $\mathbb{R}^{n_B}$. We call $\mathbf{S}$ the set of these points for the equation $f(X) = 0$; $\mathbf{S}$ is called the *support* of $f(X)$. The equation can then be written in a compact way (Bruno 2.8)

$$f(X) = \sum_{Q \in \mathbf{S}} f_Q X^Q = 0 . \tag{15.17}$$

We assume that similar terms have been collected, and that all coefficients f_Q differ from zero.

We also introduce $P = (p_1, \ldots, p_{n_B})$, and we consider a given P. We consider a monomial X^Q. If we substitute (15.9), it becomes

$$X^Q = B^Q \tau^{\langle P,Q\rangle}(1+o(1)) , \tag{15.18}$$

where $B = (b_1, \ldots, b_{n_B})$ and $\langle P, Q\rangle$ is the scalar product

$$\langle P, Q\rangle = p_1 q_1 + \ldots + p_{n_B} q_{n_B} . \tag{15.19}$$

Each term of $f(X)$ is of the form (15.18); $f(X)$ is thus expressed as a power series of τ. Clearly, as $\tau \to +\infty$, terms with the largest exponents dominate. We define

$$c_P = \sup\langle P, Q\rangle \qquad \text{for} \qquad Q \in \mathbf{S} \, . \tag{15.20}$$

We call *dominant* the terms of $f(X)$ for which

$$\langle P, Q\rangle = c_P \, . \tag{15.21}$$

For $\tau \to +\infty$, non-dominant terms become negligible. If we keep only the dominant terms in $f(X)$, we obtain the *truncated function* or *truncation* to the order P, which we designate by $\widehat{f}(X)$. Correspondingly, we have the *truncated equation*

$$\widehat{f}(X) = 0 \, . \tag{15.22}$$

In that equation, we can factor out $\tau^{\langle P,Q\rangle} = \tau^{c_P}$. In the limit $\tau \to \infty$, we obtain an equation for the b_i:

$$\widehat{f}(B) = 0 \, . \tag{15.23}$$

These considerations are illuminated by a nice geometrical interpretation in the space $\mathbb{R}^{n_B}$. We consider the set of points $\mathbf{S}$. Its convex hull is called the *Newton polyhedron,* and is designated by $\mathbf{\Gamma}$. For a given vector P, the equation

$$\langle P, Q\rangle = c \tag{15.24}$$

defines a hyperplane orthogonal to P. If $c = c_P$, we obtain the *supporting hyperplane*, which touches the Newton polyhedron. Intuitively, this plane may be defined as follows: we start with a plane (15.24) at a large distance in the direction of the vector P, and we let it slide in the direction of $-P$ until it touches the Newton polyhedron.

The intersection of the supporting hyperplane with $\mathbf{S}$ is called a *boundary subset* and designated by $\mathbf{S}_P$. Each boundary subset corresponds to a *face* of the Newton polyhedron, which is the convex hull of $\mathbf{S}_P$. The term "face" is used here in a general, multi-dimensional sense: it can be a vertex of the polyhedron (its dimension is $d = 0$), an edge ($d = 1$), a face in the usual sense ($d = 2$), and so on.

Consider now a given boundary subset, which we call $\mathbf{S}'$. The set of the vectors P such that their boundary subset $\mathbf{S}_P$ coincides with $\mathbf{S}'$ is called the *normal cone* of the boundary subset $\mathbf{S}'$. The normal cones form a partition of the set of all vectors P.

In order to find the solutions of the form (15.9) which satisfy one of the equations (15.4), we can therefore use the following geometrical approach. First from $f(X)$ we construct the set of points $\mathbf{S}$. It defines the Newton polyhedron $\mathbf{\Gamma}$. We find the faces of $\mathbf{\Gamma}$. Next we consider each face in turn. We find the associated boundary subset, and the corresponding truncated function $\widehat{f}(X)$. We write down also the set of equalities and inequalities which define the normal cone of the boundary subset. (They derive immediately from the above definitions: $\langle P, Q\rangle$ must have the same value for all points of the boundary subset, and a lower value for all other points.) We have thus found a set of solutions of the form (15.9): it consists of all solutions such that (i) $\widehat{f}(B) = 0$ and (ii) P belongs to the normal cone.

A face of dimension $d = 0$ corresponds to a vertex of the Newton polyhedron; the corresponding boundary subset contains only one point. But in that case, the truncated equation (15.22) contains only one term and cannot be satisfied (see Bruno 2.8). (In other words, there must be at least two dominant terms.) Therefore, *we only need to consider the faces with a dimension* $d \geqq 1$.

15.3 Newton Polyhedra

The first step consists thus in determining, for each equation $f_i(X) = 0$ of (15.4) in turn, the associated Newton polyhedron $\mathbf{\Gamma}_i$, its faces $\mathbf{\Gamma}_{ik}^{(d)}$, the truncated equations $\widehat{f}_{ik}^{(d)}(X)$, and their normal cones $\mathbf{U}_{ik}^{(d)}$.

15.3.1 Encounter Equations

The $n-1$ encounter equations are the equations $f_{2i} = 0$ in (15.4), with $i = 1, \ldots, n-1$. All these equations are identical except for a shift of the subscripts; therefore we consider one of them for a fixed i.

The support $\mathbf{S}_{2i}$ is infinite, because the terms $O()$ hide a series expansion in the variables. However, $\mathbf{S}_{2i}$ can be replaced by its *minimal dominant subset* $\mathbf{S}'_{2i}$, which is finite (Bruno 1.8):
it contains the 3 points

$$
\begin{aligned}
Q_1^{2i} &= (1, \underbrace{0, \ldots, 0}_{2n}) = E_1 \;, \\
Q_2^{2i} &= (\underbrace{0, \ldots, 0}_{2i}, 1, 1, \underbrace{0, \ldots, 0}_{2n-2i-1}) = E_{2i+1} + E_{2i+2} \;, \\
Q_3^{2i} &= (\underbrace{0, \ldots, 0}_{2i+1}, 1, 1, \underbrace{0, \ldots, 0}_{2n-2i-2}) = E_{2i+2} + E_{2i+3} \;.
\end{aligned}
\tag{15.25}
$$

Here and in what follows, E_i represents the unit vector in the direction q_i. (For the Q points, we write the index $2i$ of the equation as an exponent, in the manner of Bruno 1.9.)

We note that going from the full support $\mathbf{S}_{2i}$ to the minimal dominant subset $\mathbf{S}'_{2i}$ is equivalent to omitting the $O()$ terms in the equation, which reduces to

$$
f'_{2i} = x_{2i+2}(x_{2i+3} - x_{2i+1}) + G_1 x_1 = 0 \;. \tag{15.26}
$$

In general, finding the Newton polyhedron and its faces for a given set of points $\mathbf{S}$ can be a complex operation (see Bruno 1.4, 1.5, 1.7). In our case, however, $\mathbf{S}'_{2i}$ contains only 3 points, which are not aligned. Thus the Newton polyhedron is simply a triangle. Its dimension is $d = 2$. It is therefore a *simplex* (Bruno 1.1). In such a case, finding the faces is very simple: every subset of $\mathbf{S}'_{2i}$ corresponds to one face.

We designate a face by $\mathbf{\Gamma}^{(d)}_{2i,k}$, where d is the dimension of the face, $2i$ identifies the equation, and k represents a numbering of the faces for each dimension (Bruno 2.8). We define a face by its associated boundary subset, as in Bruno 1.3. We obtain one face with dimension 2

$$\mathbf{\Gamma}^{(2)}_{2i,1} \supset \{Q^{2i}_1, Q^{2i}_2, Q^{2i}_3\}\,, \tag{15.27}$$

and three faces with dimension 1

$$\mathbf{\Gamma}^{(1)}_{2i,1} \supset \{Q^{2i}_1, Q^{2i}_2\}\,, \quad \mathbf{\Gamma}^{(1)}_{2i,2} \supset \{Q^{2i}_1, Q^{2i}_3\}\,, \quad \mathbf{\Gamma}^{(1)}_{2i,3} \supset \{Q^{2i}_2, Q^{2i}_3\}\,. \tag{15.28}$$

The corresponding truncations are

$$\begin{aligned} \hat{f}^{(2)}_{2i,1} &= x_{2i+2}(x_{2i+3} - x_{2i+1}) + G_1 x_1\,, \qquad & \hat{f}^{(1)}_{2i,1} &= -x_{2i+1}x_{2i+2} + G_1 x_1\,, \\ \hat{f}^{(1)}_{2i,2} &= x_{2i+2}x_{2i+3} + G_1 x_1\,, & \hat{f}^{(1)}_{2i,3} &= x_{2i+2}(x_{2i+3} - x_{2i+1})\,. \end{aligned} \tag{15.29}$$

We remark that $f^{(2)}_{2i,1} = f'_{2i}$: all terms participate in that case.

Finally, the normal cones are

$$\begin{aligned} &\mathrm{a}: \quad \mathbf{U}^{(2)}_{2i,1} = \{P\colon\ p_1 = p_{2i+1} + p_{2i+2} = p_{2i+2} + p_{2i+3}\}\,, \\ &\mathrm{c}: \quad \mathbf{U}^{(1)}_{2i,1} = \{P\colon\ p_1 = p_{2i+1} + p_{2i+2}\,, \quad p_{2i+1} > p_{2i+3}\}\,, \\ &\mathrm{d}: \quad \mathbf{U}^{(1)}_{2i,2} = \{P\colon\ p_1 = p_{2i+2} + p_{2i+3}\,, \quad p_{2i+3} > p_{2i+1}\}\,, \\ &\mathrm{b}: \quad \mathbf{U}^{(1)}_{2i,3} = \{P\colon\ p_{2i+1} = p_{2i+3}\,, \quad p_{2i+1} + p_{2i+2} > p_1\}\,. \end{aligned} \tag{15.30}$$

The letters a to d are labels which will be needed below. They are inherited from some older work (unpublished); this is why they are not in alphabetical order.

15.3.2 Arc Equations: General Case

The general case corresponds to an arc which is not one of the two end arcs and to one of the equations $f_{2i-1} = 0$ in (15.4), with $i = 2, \ldots, n-1$. Their number is $n-2$: they exist only if $n \geqq 3$. All these equations are identical except for a shift of the subscripts; therefore we consider one of them for a fixed i.

The term $O(x_1) = O(\mu)$ can be neglected, because it is negligible with respect to the terms $x_{2i} = y_{i-1}$ and $x_{2i+2} = y_i$ as shown by (11.29).

Here again, the support $\mathbf{S}_{2i-1}$ can be replaced by its minimal dominant subset $\mathbf{S}'_{2i-1}$, which is finite: it contains the 4 points

$$Q_1^{2i-1} = (0,1,0,\ldots,0) = E_2 \,,$$
$$Q_2^{2i-1} = (\underbrace{0,\ldots,0}_{2i-1},1,\underbrace{0,\ldots,0}_{2n-2i+1}) = E_{2i} \,,$$
$$Q_3^{2i-1} = (\underbrace{0,\ldots,0}_{2i},1,\underbrace{0,\ldots,0}_{2n-2i}) = E_{2i+1} \,,$$
$$Q_4^{2i-1} = (\underbrace{0,\ldots,0}_{2i+1},1,\underbrace{0,\ldots,0}_{2n-2i-1}) = E_{2i+2} \,. \tag{15.31}$$

This is equivalent to omitting the $O()$ terms in the equation, which reduces to

$$f'_{2i-1} = x_{2i+2} - x_{2i} + G_2 s_i x_2 + G_3(1+Ks_i)x_{2i+1} = 0 \tag{15.32}$$

The 4 points are not coplanar. The Newton polyhedron is a tetrahedron. Its dimension is $d = 3$. It is again a simplex: every subset of $\mathbf{S}'_{2i-1}$ corresponds to a face. We obtain a face with dimension 3:

$$\mathbf{\Gamma}^{(3)}_{2i-1,1} \supset \{Q_1^{2i-1}, Q_2^{2i-1}, Q_3^{2i-1}, Q_4^{2i-1}\} \,, \tag{15.33}$$

4 faces with dimension 2:

$$\begin{aligned} &\mathbf{\Gamma}^{(2)}_{2i-1,1} \supset \{Q_1^{2i-1}, Q_2^{2i-1}, Q_3^{2i-1}\} \,, \quad \mathbf{\Gamma}^{(2)}_{2i-1,2} \supset \{Q_1^{2i-1}, Q_2^{2i-1}, Q_4^{2i-1}\} \,, \\ &\mathbf{\Gamma}^{(2)}_{2i-1,3} \supset \{Q_1^{2i-1}, Q_3^{2i-1}, Q_4^{2i-1}\} \,, \quad \mathbf{\Gamma}^{(2)}_{2i-1,4} \supset \{Q_2^{2i-1}, Q_3^{2i-1}, Q_4^{2i-1}\} \,, \end{aligned} \tag{15.34}$$

and 6 faces with dimension 1:

$$\begin{aligned} &\mathbf{\Gamma}^{(1)}_{2i-1,1} \supset \{Q_1^{2i-1}, Q_2^{2i-1}\} \,, \quad \mathbf{\Gamma}^{(1)}_{2i-1,2} \supset \{Q_1^{2i-1}, Q_3^{2i-1}\} \,, \\ &\mathbf{\Gamma}^{(1)}_{2i-1,3} \supset \{Q_1^{2i-1}, Q_4^{2i-1}\} \,, \quad \mathbf{\Gamma}^{(1)}_{2i-1,4} \supset \{Q_2^{2i-1}, Q_3^{2i-1}\} \,, \\ &\mathbf{\Gamma}^{(1)}_{2i-1,5} \supset \{Q_2^{2i-1}, Q_4^{2i-1}\} \,, \quad \mathbf{\Gamma}^{(1)}_{2i-1,6} \supset \{Q_3^{2i-1}, Q_4^{2i-1}\} \,. \end{aligned} \tag{15.35}$$

The truncations are easily derived. The normal cones are

$$\begin{aligned} &\mathrm{a}: \quad \mathbf{U}^{(3)}_{2i-1,1} = \{P\colon\, p_2 = p_{2i} = p_{2i+1} = p_{2i+2}\} \,, \\ &\mathrm{d}: \quad \mathbf{U}^{(2)}_{2i-1,1} = \{P\colon\, p_2 = p_{2i} = p_{2i+1} \,, \quad p_2 > p_{2i+2}\} \,, \\ &\mathrm{c}: \quad \mathbf{U}^{(2)}_{2i-1,2} = \{P\colon\, p_2 = p_{2i} = p_{2i+2} \,, \quad p_2 > p_{2i+1}\} \,, \\ &\mathrm{e}: \quad \mathbf{U}^{(2)}_{2i-1,3} = \{P\colon\, p_2 = p_{2i+1} = p_{2i+2} \,, \quad p_2 > p_{2i}\} \,, \\ &\mathrm{b}: \quad \mathbf{U}^{(2)}_{2i-1,4} = \{P\colon\, p_{2i} = p_{2i+1} = p_{2i+2} \,, \quad p_{2i} > p_2\} \,, \\ &\mathrm{h}: \quad \mathbf{U}^{(1)}_{2i-1,1} = \{P\colon\, p_2 = p_{2i} \,, \quad p_2 > p_{2i+1} \,, \quad p_2 > p_{2i+2}\} \,, \\ &\mathrm{k}: \quad \mathbf{U}^{(1)}_{2i-1,2} = \{P\colon\, p_2 = p_{2i+1} \,, \quad p_2 > p_{2i} \,, \quad p_2 > p_{2i+2}\} \,, \\ &\mathrm{j}: \quad \mathbf{U}^{(1)}_{2i-1,3} = \{P\colon\, p_2 = p_{2i+2} \,, \quad p_2 > p_{2i} \,, \quad p_2 > p_{2i+1}\} \,, \\ &\mathrm{g}: \quad \mathbf{U}^{(1)}_{2i-1,4} = \{P\colon\, p_{2i} = p_{2i+1} \,, \quad p_{2i} > p_2 \,, \quad p_{2i} > p_{2i+2}\} \,, \end{aligned}$$

$$
\begin{aligned}
\mathrm{f}: \quad & \mathbf{U}^{(1)}_{2i-1,5} = \{P\colon\ p_{2i} = p_{2i+2}\,, \quad p_{2i} > p_2\,, \quad p_{2i} > p_{2i+1}\}\,, \\
\mathrm{i}: \quad & \mathbf{U}^{(1)}_{2i-1,6} = \{P\colon\ p_{2i+1} = p_{2i+2}\,, \quad p_{2i+1} > p_2\,, \quad p_{2i+1} > p_{2i}\}\,.
\end{aligned}
\tag{15.36}
$$

15.3.3 Arc Equations: Initial Arc

We consider now the particular case of the first equation $f_1 = 0$ in (15.4), corresponding to the first basic arc.

In that equation, the term $O(x_1) = O(\mu)$ is negligible compared to the term $x_4 = y_1$.

The point Q_2^{2i-1} of the general case has disappeared. The minimal dominant subset $\mathbf{S}'_1$ contains only 3 points:

$$
\begin{aligned}
Q_1^1 &= (0,1,0,\ldots,0) = E_2\,, \\
Q_3^1 &= (0,0,1,0,\ldots,0) = E_3\,, \\
Q_4^1 &= (0,0,0,1,0,\ldots,0) = E_4\,.
\end{aligned}
\tag{15.37}
$$

This is equivalent to omitting the $O()$ terms in the equation, which reduces to

$$
f'_1 = x_4 - G_2x_2 + G_3(1-K)x_3 = 0\,. \tag{15.38}
$$

The Newton polyhedron is a triangle. We obtain the faces simply by eliminating from the lists (15.33) to (15.35) of the general case the faces which contain Q_2^{2i-1}. There remains:

$$
\mathbf{\Gamma}^{(2)}_{13} \supset \{Q_1^1, Q_3^1, Q_4^1\}\,, \tag{15.39}
$$

$$
\mathbf{\Gamma}^{(1)}_{12} \supset \{Q_1^1, Q_3^1\}\,, \quad \mathbf{\Gamma}^{(1)}_{13} \supset \{Q_1^1, Q_4^1\}\,, \quad \mathbf{\Gamma}^{(1)}_{16} \supset \{Q_3^1, Q_4^1\}\,. \tag{15.40}
$$

The corresponding truncations are

$$
\begin{aligned}
\widehat{f}^{(2)}_{13} &= x_4 - G_2x_2 + G_3(1-K)x_3\,, \qquad \widehat{f}^{(1)}_{12} = -G_2x_2 + G_3(1-K)x_3\,, \\
\widehat{f}^{(1)}_{13} &= x_4 - G_2x_2\,, \qquad \widehat{f}^{(1)}_{16} = x_4 + G_3(1-K)x_3\,.
\end{aligned}
\tag{15.41}
$$

The normal cones are

$$
\begin{aligned}
\mathrm{a}: \quad & \mathbf{U}^{(2)}_{13} = \{P\colon\ p_2 = p_3 = p_4\}\,, \\
\mathrm{d}: \quad & \mathbf{U}^{(1)}_{12} = \{P\colon\ p_2 = p_3\,, \quad p_2 > p_4\}\,, \\
\mathrm{c}: \quad & \mathbf{U}^{(1)}_{13} = \{P\colon\ p_2 = p_4\,, \quad p_2 > p_3\}\,, \\
\mathrm{b}: \quad & \mathbf{U}^{(1)}_{16} = \{P\colon\ p_3 = p_4\,, \quad p_3 > p_2\}\,.
\end{aligned}
\tag{15.42}
$$

15.3.4 Arc Equations: Final Arc

We consider now the particular case of the last equation $f_{2n-1} = 0$ in (15.4), corresponding to the last basic arc. This case is symmetrical of the previous one.

In that equation, the term $O(x_1) = O(\mu)$ is negligible compared to the term $x_{2n} = y_{n-1}$.

The point Q_4^{2i-1} of the general case has disappeared. The minimal dominant subset $\mathbf{S}'_{2n-1}$ contains only 3 points:

$$\begin{aligned} Q_1^{2n-1} &= (0,1,0,\dots,0) = E_2, \ , \\ Q_2^{2n-1} &= (0,\dots,0,1,0) = E_{2n} \ , \\ Q_3^{2n-1} &= (0,\dots,0,1) = E_{2n+1} \ . \end{aligned} \tag{15.43}$$

This is equivalent to omitting the $O()$ terms in the equation, which reduces to

$$f'_{2n-1} = -x_{2n} + G_2 s_n x_2 + G_3(1+Ks_n)x_{2n+1} = 0 \ . \tag{15.44}$$

The Newton polyhedron is a triangle. We obtain the faces by eliminating from the lists (15.33) to (15.35) the faces which contain Q_4^{2i-1}. There remains:

$$\begin{aligned} \mathbf{\Gamma}^{(2)}_{2n-1,1} &\supset \{Q_1^{2n-1}, Q_2^{2n-1}, Q_3^{2n-1}\} \ , \\ \mathbf{\Gamma}^{(1)}_{2n-1,1} &\supset \{Q_1^{2n-1}, Q_2^{2n-1}\} \ , \\ \mathbf{\Gamma}^{(1)}_{2n-1,2} &\supset \{Q_1^{2n-1}, Q_3^{2n-1}\} \ , \\ \mathbf{\Gamma}^{(1)}_{2n-1,4} &\supset \{Q_2^{2n-1}, Q_3^{2n-1}\} \ , \end{aligned} \tag{15.45}$$

The corresponding truncations are

$$\begin{aligned} \widehat{f}^{(2)}_{2n-1,1} &= -x_{2n} + G_2 s_n x_2 + G_3(1+Ks_n)x_{2n+1} \ , \\ \widehat{f}^{(1)}_{2n-1,1} &= -x_{2n} + G_2 s_n x_2 \ , \\ \widehat{f}^{(1)}_{2n-1,2} &= G_2 s_n x_2 + G_3(1+Ks_n)x_{2n+1} \ , \\ \widehat{f}^{(1)}_{2n-1,4} &= -x_{2n} + G_3(1+Ks_n)x_{2n+1} \ . \end{aligned} \tag{15.46}$$

The normal cones are

$$\begin{aligned} \text{a}: &\quad \mathbf{U}^{(2)}_{2n-1,1} = \{P\colon\ p_2 = p_{2n} = p_{2n+1}\} \ , \\ \text{c}: &\quad \mathbf{U}^{(1)}_{2n-1,1} = \{P\colon\ p_2 = p_{2n} \ , \quad p_2 > p_{2n+1}\} \ , \\ \text{d}: &\quad \mathbf{U}^{(1)}_{2n-1,2} = \{P\colon\ p_2 = p_{2n+1} \ , \quad p_2 > p_{2n}\} \ , \\ \text{b}: &\quad \mathbf{U}^{(1)}_{2n-1,4} = \{P\colon\ p_{2n} = p_{2n+1} \ , \quad p_{2n} > p_2\} \ . \end{aligned} \tag{15.47}$$

15.3.5 Additional Relations: General Case

We must also consider the additional relations (15.7). We begin with the general case, i.e. the central equation $f_{2n-1+i} = 0$ in (15.7)), which exists for $n \geqq 4$ with $i = 2$ to $n - 2$.

The term $O(x_1) = O(\mu)$ is negligible with respect to the terms $x_{2i} = y_{i-1}$ and $x_{2i+4} = y_{i+1}$.

The minimal dominant subset $\mathbf{S}'_{2n-1+i}$ contains the 4 points

$$\begin{aligned} &Q_1^{2n-1+i} = E_{2i} \,, \quad Q_2^{2n-1+i} = E_{2i+1} \,, \\ &Q_3^{2n-1+i} = E_{2i+3} \,, \quad Q_4^{2n-1+i} = E_{2i+4} \,. \end{aligned} \tag{15.48}$$

This is equivalent to omitting the $O()$ terms in the equation, which reduces to

$$f'_{2n-1+i} = -x_{2i} + x_{2i+4} + G_3(1 + Ks_i)x_{2i+1} + G_3(1 - Ks_i)x_{2i+3} = 0 \,. \tag{15.49}$$

The 4 points are not coplanar. The Newton polyhedron is a tetrahedron. We obtain 1 face with dimension 3, 4 faces with dimension 2, and 6 faces with dimension 1:

$$\begin{aligned} &\mathbf{\Gamma}^{(3)}_{2n-1+i,1} \supset \{Q_1^{2n-1+i}, Q_2^{2n-1+i}, Q_3^{2n-1+i}, Q_4^{2n-1+i}\} \,, \\ &\mathbf{\Gamma}^{(2)}_{2n-1+i,1} \supset \{Q_1^{2n-1+i}, Q_2^{2n-1+i}, Q_3^{2n-1+i}\} \,, \\ &\mathbf{\Gamma}^{(2)}_{2n-1+i,2} \supset \{Q_1^{2n-1+i}, Q_2^{2n-1+i}, Q_4^{2n-1+i}\} \,, \\ &\mathbf{\Gamma}^{(2)}_{2n-1+i,3} \supset \{Q_1^{2n-1+i}, Q_3^{2n-1+i}, Q_4^{2n-1+i}\} \,, \\ &\mathbf{\Gamma}^{(2)}_{2n-1+i,4} \supset \{Q_2^{2n-1+i}, Q_3^{2n-1+i}, Q_4^{2n-1+i}\} \,, \\ &\mathbf{\Gamma}^{(1)}_{2n-1+i,1} \supset \{Q_1^{2n-1+i}, Q_2^{2n-1+i}\} \,, \quad \mathbf{\Gamma}^{(1)}_{2n-1+i,2} \supset \{Q_1^{2n-1+i}, Q_3^{2n-1+i}\} \,, \\ &\mathbf{\Gamma}^{(1)}_{2n-1+i,3} \supset \{Q_1^{2n-1+i}, Q_4^{2n-1+i}\} \,, \quad \mathbf{\Gamma}^{(1)}_{2n-1+i,4} \supset \{Q_2^{2n-1+i}, Q_3^{2n-1+i}\} \,, \\ &\mathbf{\Gamma}^{(1)}_{2n-1+i,5} \supset \{Q_2^{2n-1+i}, Q_4^{2n-1+i}\} \,, \quad \mathbf{\Gamma}^{(1)}_{2n-1+i,6} \supset \{Q_3^{2n-1+i}, Q_4^{2n-1+i}\} \,. \end{aligned} \tag{15.50}$$

The truncations are easily derived. The normal cones are

$$\begin{aligned} &\mathrm{A}: \quad \mathbf{U}^{(3)}_{2n-1+i,1} = \{P\colon\ p_{2i} = p_{2i+1} = p_{2i+3} = p_{2i+4}\} \,, \\ &\mathrm{D}: \quad \mathbf{U}^{(2)}_{2n-1+i,1} = \{P\colon\ p_{2i} = p_{2i+1} = p_{2i+3} \,, \quad p_{2i} > p_{2i+4}\} \,, \\ &\mathrm{B}: \quad \mathbf{U}^{(2)}_{2n-1+i,2} = \{P\colon\ p_{2i} = p_{2i+1} = p_{2i+4} \,, \quad p_{2i} > p_{2i+3}\} \,, \\ &\mathrm{C}: \quad \mathbf{U}^{(2)}_{2n-1+i,3} = \{P\colon\ p_{2i} = p_{2i+3} = p_{2i+4} \,, \quad p_{2i} > p_{2i+1}\} \,, \\ &\mathrm{E}: \quad \mathbf{U}^{(2)}_{2n-1+i,4} = \{P\colon\ p_{2i+1} = p_{2i+3} = p_{2i+4} \,, \quad p_{2i+1} > p_{2i}\} \,, \\ &\mathrm{G}: \quad \mathbf{U}^{(1)}_{2n-1+i,1} = \{P\colon\ p_{2i} = p_{2i+1} \,, \quad p_{2i} > p_{2i+3} \,, \quad p_{2i} > p_{2i+4}\} \,, \\ &\mathrm{H}: \quad \mathbf{U}^{(1)}_{2n-1+i,2} = \{P\colon\ p_{2i} = p_{2i+3} \,, \quad p_{2i} > p_{2i+1} \,, \quad p_{2i} > p_{2i+4}\} \,, \end{aligned}$$

$$
\begin{aligned}
&\mathrm{F}: && \mathbf{U}^{(1)}_{2n-1+i,3} = \{P\colon\ p_{2i} = p_{2i+4}\ ,\quad p_{2i} > p_{2i+1}\ ,\quad p_{2i} > p_{2i+3}\}\ ,\\
&\mathrm{K}: && \mathbf{U}^{(1)}_{2n-1+i,4} = \{P\colon\ p_{2i+1} = p_{2i+3}\ ,\quad p_{2i+1} > p_{2i}\ ,\quad p_{2i+1} > p_{2i+4}\}\ ,\\
&\mathrm{I}: && \mathbf{U}^{(1)}_{2n-1+i,5} = \{P\colon\ p_{2i+1} = p_{2i+4}\ ,\quad p_{2i+1} > p_{2i}\ ,\quad p_{2i+1} > p_{2i+3}\}\ ,\\
&\mathrm{J}: && \mathbf{U}^{(1)}_{2n-1+i,6} = \{P\colon\ p_{2i+3} = p_{2i+4}\ ,\quad p_{2i+3} > p_{2i}\ ,\quad p_{2i+3} > p_{2i+1}\}\ .
\end{aligned}
\tag{15.51}
$$

Here we use the letters A to K to label the cases.

15.3.6 Additional Relations: First Relation

We consider now the particular case of the first additional relation $f_{2n} = 0$ in (15.7), corresponding to the first two basic arcs.

The term $O(x_1) = O(\mu)$ is negligible with respect to the term $x_6 = y_2$.

The point Q_1^{2n-1+i} of the general case has disappeared. The minimal dominant subset $\mathbf{S}'_{2n}$ contains only 3 points:

$$
Q_2^{2n} = E_3\ ,\qquad Q_3^{2n} = E_5\ ,\qquad Q_4^{2n} = E_6\ . \tag{15.52}
$$

This is equivalent to omitting the $O()$ terms in the relation, which reduces to

$$
f'_{2n} = x_6 + G_3(1-K)x_3 + G_3(1+K)x_5 = 0\ . \tag{15.53}
$$

The Newton polyhedron is a triangle. We obtain the faces by eliminating from the list (15.50) of the general case the faces which contain Q_1^{2n-1+i}. There remains:

$$
\begin{aligned}
&\mathbf{\Gamma}^{(2)}_{2n,1} \supset \{Q_2^{2n}, Q_3^{2n}, Q_4^{2n}\}\ , \qquad \mathbf{\Gamma}^{(1)}_{2n,1} \supset \{Q_2^{2n}, Q_3^{2n}\}\ ,\\
&\mathbf{\Gamma}^{(1)}_{2n,2} \supset \{Q_2^{2n}, Q_4^{2n}\}\ , \qquad \mathbf{\Gamma}^{(1)}_{2n,3} \supset \{Q_3^{2n}, Q_4^{2n}\}\ .
\end{aligned}
\tag{15.54}
$$

The truncations are

$$
\begin{aligned}
\widehat{f}^{(2)}_{2n,1} &= x_6 + G_3(1-K)x_3 + G_3(1+K)x_5\ ,\\
\widehat{f}^{(1)}_{2n,1} &= G_3(1-K)x_3 + G_3(1+K)x_5\ ,\\
\widehat{f}^{(1)}_{2n,2} &= x_6 + G_3(1-K)x_3\ ,\\
\widehat{f}^{(1)}_{2n,3} &= x_6 + G_3(1+K)x_5\ .
\end{aligned}
\tag{15.55}
$$

The normal cones are

$$
\begin{aligned}
&\mathrm{A}: && \mathbf{U}^{(2)}_{2n,1} = \{P\colon\ p_3 = p_5 = p_6\}\ ,\\
&\mathrm{D}: && \mathbf{U}^{(1)}_{2n,1} = \{P\colon\ p_3 = p_5\ , p_3 > p_6\}\ ,\\
&\mathrm{B}: && \mathbf{U}^{(1)}_{2n,2} = \{P\colon\ p_3 = p_6\ , p_3 > p_5\}\ ,\\
&\mathrm{C}: && \mathbf{U}^{(1)}_{2n,3} = \{P\colon\ p_5 = p_6\ , p_5 > p_3\}\ .
\end{aligned}
\tag{15.56}
$$

15.3.7 Additional Relations: Last Relation

We consider now the particular case of the last additional relation $f_{3n-2} = 0$ in (15.7), corresponding to the last two basic arcs. This case is symmetrical of the previous one.

The term $O(x_1) = O(\mu)$ is negligible with respect to the term $x_{2n-2} = y_{n-2}$.

The point Q_4^{2n-1+i} of the general case has disappeared. The minimal dominant subset $\mathbf{S}'_{3n-2}$ contains only 3 points:

$$Q_1^{3n-2} = E_{2n-2}\,, \qquad Q_2^{3n-2} = E_{2n-1}\,, \qquad Q_3^{3n-2} = E_{2n+1}\,. \tag{15.57}$$

This is equivalent to omitting the $O()$ terms in the relation, which reduces to

$$f'_{3n-2} = -x_{2n-2}+G_3(1+Ks_{n-1})x_{2n-1}+G_3(1-Ks_{n-1})x_{2n+1} = 0\,. \tag{15.58}$$

The Newton polyhedron is a triangle. We obtain the faces by eliminating from (15.50) the faces which contain Q_4^{2n-1+i}. There remains:

$$\begin{aligned}&\mathbf{\Gamma}^{(2)}_{3n-2,1} \supset \{Q_1^{3n-2}, Q_2^{3n-2}, Q_3^{3n-2}\}\,, \quad \mathbf{\Gamma}^{(1)}_{3n-2,1} \supset \{Q_1^{3n-2}, Q_2^{3n-2}\}\,,\\ &\mathbf{\Gamma}^{(1)}_{3n-2,2} \supset \{Q_1^{3n-2}, Q_3^{3n-2}\}\,, \quad \mathbf{\Gamma}^{(1)}_{3n-2,3} \supset \{Q_2^{3n-2}, Q_3^{3n-2}\}\,.\end{aligned} \tag{15.59}$$

The truncations are

$$\begin{aligned}\hat{f}^{(2)}_{3n-2,1} &= -x_{2n-2} + G_3(1 + Ks_{n-1})x_{2n-1} + G_3(1 - Ks_{n-1})x_{2n+1}\,,\\ \hat{f}^{(1)}_{3n-2,1} &= -x_{2n-2} + G_3(1 + Ks_{n-1})x_{2n-1}\,,\\ \hat{f}^{(1)}_{3n-2,2} &= -x_{2n-2} + G_3(1 - Ks_{n-1})x_{2n+1}\,,\\ \hat{f}^{(1)}_{3n-2,3} &= G_3(1 + Ks_{n-1})x_{2n-1} + G_3(1 - Ks_{n-1})x_{2n+1}\,.\end{aligned} \tag{15.60}$$

The normal cones are

$$\begin{aligned}\mathrm{A}: &\quad \mathbf{U}^{(2)}_{3n-2,1} = \{P\colon\ p_{2n-2} = p_{2n-1} = p_{2n+1}\}\,,\\ \mathrm{B}: &\quad \mathbf{U}^{(1)}_{3n-2,1} = \{P\colon\ p_{2n-2} = p_{2n-1}\,, p_{2n-2} > p_{2n+1}\}\,,\\ \mathrm{C}: &\quad \mathbf{U}^{(1)}_{3n-2,2} = \{P\colon\ p_{2n-2} = p_{2n+1}\,, p_{2n-2} > p_{2n-1}\}\,,\\ \mathrm{D}: &\quad \mathbf{U}^{(1)}_{3n-2,3} = \{P\colon\ p_{2n-1} = p_{2n+1}\,, p_{2n-1} > p_{2n-2}\}\,.\end{aligned} \tag{15.61}$$

15.3.8 Additional Relations: Case $n = 2$

In that particular case, we have the single relation (15.8).

The points Q_1^{2n-1+i} and Q_4^{2n-1+i} of the general case have disappeared. The minimal dominant subset $\mathbf{S}'_4$ contains only 2 points:

$$Q_2^4 = E_3\,, \qquad Q_3^4 = E_5\,. \tag{15.62}$$

This is equivalent to omitting the $O()$ terms in the relation, which reduces to

$$f_4' = G_3(1-K)x_3 + G_3(1+K)x_5 = 0\,. \tag{15.63}$$

The Newton polyhedron is a segment. There is a single face

$$\mathbf{\Gamma}_{4,1}^{(1)} \supset \{Q_2^4, Q_3^4\}\,. \tag{15.64}$$

There is one truncation, which is identical with (15.63):

$$\widehat{f}_{4,1}^{(1)} = G_3(1-K)x_3 + G_3(1+K)x_5\,. \tag{15.65}$$

The corresponding normal cone is

$$\text{A}: \qquad \mathbf{U}_{4,1}^{(1)} = \{P\colon\ p_3 = p_5\}\,. \tag{15.66}$$

15.4 Intersections with the Cone of the Problem

The next step in the analysis (Bruno 2.8) consists in retaining only the faces for which the intersection of the normal cone with the cone of the problem $\mathbf{K}$ (see (15.16)) is non-empty. This turns out to be the case for all faces found above. This is easily shown by exhibiting in each case a vector which belongs to the normal cone and to $\mathbf{K}$. For instance, in the case of the normal cones (15.30), we can take

$$\begin{aligned}
&\text{for } \mathbf{U}_{2i,1}^{(2)}: \quad p_1 = -2\,, \quad p_{2i+1} = p_{2i+2} = p_{2i+3} = -1\,;\\
&\text{for } \mathbf{U}_{2i,1}^{(1)}: \quad p_1 = p_{2i+3} = -2\,, \quad p_{2i+1} = p_{2i+2} = -1\,;\\
&\text{for } \mathbf{U}_{2i,2}^{(1)}: \quad p_1 = p_{2i+1} = -2\,, \quad p_{2i+2} = p_{2i+3} = -1\,;\\
&\text{for } \mathbf{U}_{2i,3}^{(1)}: \quad p_1 = -3\,, \quad p_{2i+1} = p_{2i+2} = p_{2i+3} = -1\,.
\end{aligned} \tag{15.67}$$

Similar vectors can be found in the other cases.

15.5 Coherent Boundary Subsets

Up to now we have analysed the conditions under which a solution of the form (15.9) can satisfy one equation in (15.4) considered separately. For each equation, we have found that the vector P must belong to one of the normal cones which we have identified, while X satisfies the corresponding truncated equation. For instance, in order to satisfy the equation $f_1 = 0$, the vector P must belong to one of the 4 normal cones $U_{1j}^{(d)}$ listed in (15.42), while X satisfies the corresponding truncated equation $\widehat{f}_{1j}^{(d)} = 0$, with $\widehat{f}_{1j}^{(d)}$ given by (15.41).

Now we must find the solutions (15.9) which satisfy simultaneously all fundamental equations (15.4). Bruno (1.9) sketches a procedure for doing this. Essentially, we consider all possible combinations formed by: the cone of the problem (15.16); one normal cone for f_1; one normal cone for f_2; ...; one normal cone for f_{m_B}. For each combination, we determine the intersection of the $m_B + 1$ cones. If that intersection is empty, the combination can be eliminated. If the intersection is non-empty, we have a *valid combination.* The intersection is called *cone of truncation* (Bruno 2.6) and designated by $\mathbf{\Pi}$. The corresponding combination of boundary subsets is called by Bruno (1.9) a *coherent aggregate of boundary subsets.*

Unfortunately, the number of combinations, and thus the amount of work needed, grow exponentially with the number of equations m_B. In the case of the system (15.4), there are 4 possible normal cones for each of the equations $f_2 = 0, \ldots, f_{2n-2} = 0$; 11 possible normal cones for each of the equations $f_3 = 0, \ldots, f_{2n-3} = 0$; 4 normal cones for the first equation $f_1 = 1$ and 4 normal cones for the last equation $f_{2n-1} = 0$. Therefore the number of possible combinations is

$$4^{n+1} \times 11^{n-2} . \tag{15.68}$$

For $n = 2, 3, 4, \ldots$, this equals 64, 2816, 123904, ... So the method quickly becomes impracticable.

We consider now how to find the intersection $\mathbf{\Pi}$ for a given combination of normal cones. We remark that the definition of each normal cone is a set of linear homogeneous equalities and strict inequalities on the coordinates p_j of P. This is also the case for the cone of the problem. Therefore, for any given combination, we need to solve a system of linear homogeneous equalities and strict inequalities for the p_j.

As an example, we consider the case 1P2, and we select the normal cone d for the initial arc equation (equ. (15.42b)), the normal cone a for the encounter equation (equ. (15.30a)), and the normal cone d for the final arc equation (equ. (15.47c)). This case will be called "dad" below(see Sect. 15.5.4). The vector P must belong to the intersection of the cone of the problem (15.16) and these 3 normal cones: its 5 coordinates must satisfy the following set of conditions

$$\begin{aligned}
&p_1 < 0\,, \quad p_2 < 0\,, \quad p_3 < 0\,, \quad p_4 < 0\,, \quad p_5 < 0\,,\\
&-p_2 + p_3 = 0\,, \quad -p_2 + p_4 < 0\,,\\
&-p_1 + p_3 + p_4 = 0\,, \quad -p_1 + p_4 + p_5 = 0\,,\\
&-p_2 + p_5 = 0\,, \quad -p_2 + p_4 < 0\,.
\end{aligned} \tag{15.69}$$

There are 4 equalities and 7 strict inequalities. We need to find all solutions of this system.

15.5.1 The Motzkin-Burger Algorithm

An algorithm exists for the solution of a somewhat similar problem. Consider a system of *non-strict* inequalities

$$a_{j1}p_1 + a_{j2}p_2 + \ldots + a_{jn_B}p_{n_B} \leqq 0, \quad j = 1, \ldots, m\,, \tag{15.70}$$

The set of its solutions forms a *polyhedral cone* **C** (Bruno 1.1). It can be written in the form (Bruno 1.4)

$$P = \sum_{i=1}^{t} \lambda_i N^i + \sum_{j=1}^{s} \mu_j B^j\,, \tag{15.71}$$

where the N^i and the B^j are fixed vectors; each λ_i can take any positive or zero value; and each μ_j can take any real value. The set

$$\mathbf{N}(\mathbf{C}) = \{N^1, \ldots, N^t\} \tag{15.72}$$

is called a *fundamental system of solutions*, while the set

$$\mathbf{B}(\mathbf{C}) = \{B^1, \ldots, B^s\} \tag{15.73}$$

defines a *maximal linear subspace* (Bruno 1.4).

The *Motzkin-Burger algorithm* allows the computation of the N^i and the B^j, thus giving the solution of the system. It works by adding the inequalities one by one, starting from the trivial inequality

$$0p_1 + \ldots + 0p_{n_B} \leqq 0 \tag{15.74}$$

for which the solution is $t = 0$ (the set $\mathbf{N}(\mathbf{C})$ is empty), $s = n_B$, and $B^j = E_j$, $j = 1, \ldots, n_B$ (any P is a solution).

Our case differs from (15.70) in that it does not have non-strict inequalities, but instead a mixture of equalities and strict inequalities. We need to adapt our equations so as to be able to use the algorithm. First, we replace each equality by two non-strict inequalities, in which the '=' sign is replaced by '$\leqq$' and '$\geqq$' respectively. This does not change the set of solutions. Second, we replace each strict inequality by a non-strict inequality, i.e. we replace '$<$' by '$\leqq$'. In so doing, we enlarge the set of solutions, i.e. we obtain all the solutions of our problem, plus some parasitic solutions. Later, we will eliminate these parasitic solutions (Sect. 15.5.2).

We modify slightly the procedure described by Bruno: we start, not from the inequality (15.74), but from the set of n_B inequalities (15.16) which define the cone of the problem. (These strict inequalities are replaced by non-strict inequalities $p_j \leqq 0$, as explained above.) Then, initially or at later stages of the algorithm, a solution (15.71) never includes B vectors, i.e. the set $\mathbf{B}(\mathbf{C})$ is empty. **C** is a *polyhedral forward cone* (Bruno 1.1). This is immediately seen from the fact that the cone of the problem does not include any straight line. The set of solutions can then be described simply as

$$P = \sum_{i=1}^{t} \lambda_i N^i\,; \qquad \lambda_i \geqq 0\,, \quad i = 1, \ldots, t\,. \tag{15.75}$$

The set of the N^i forms a *skeleton* of the polyhedral forward cone (Bruno 1.1).

In particular, in the starting state, we have $t = n_B$ and $N_i = -E_i$, $i = 1, \ldots, n_B$.

Next we add the other inequalities, coming from the normal cones, one by one as follows. Assume that at some point in the procedure we have a system of inequalities (15.70), and its solution (15.75), defined by a set (15.72). We add one inequality

$$l(P) = a_1 p_1 + \ldots + a_{n_B} p_{n_B} \leqq 0 \,. \tag{15.76}$$

We call $\mathbf{C}'$ the new cone of solutions, and we want to deduce a new fundamental system of solutions $\mathbf{N}(\mathbf{C}')$. This is achieved by Theorem 4.1 in Bruno 1.4. This theorem is considerably simplified by the absence of the B^j and reads here:

Theorem 15.5.1. *The system $\mathbf{N}(\mathbf{C}')$ is composed of those vectors N^i of the set (15.72) for which $l(N^i) \leqq 0$, and in the case $t > 1$ also of the vectors*

$$N(k, k') = N^k |l(N^{k'})| + N^{k'} |l(N^k)| \tag{15.77}$$

for each pair of elements N^k, $N^{k'}$ of the system $\mathbf{N}(\mathbf{C})$ which satisfies the following conditions:

1. $l(N^k) l(N^{k'}) < 0$.

2. We consider the inequalities of the system (15.70) which reduce to equalities for the two elements N^k and $N^{k'}$. Then no other element of the system $\mathbf{N}(\mathbf{C})$ reduces all these inequalities to equalities.

Another simplification can be made when we add an equality:

$$l(P) = a_1 p_1 + \ldots + a_{n_B} p_{n_B} = 0 \,. \tag{15.78}$$

As explained above, in principle we do this by introducing successively the two inequalities $l(P) \leqq 0$ and $l(P) \geqq 0$. In other words: assume that we already have a system of inequalities (15.70) and its solution (15.75). We add first the inequality $l(P) \leqq 0$. We obtain a new cone of solutions $\mathbf{C}'$, and the system of solutions $\mathbf{N}(\mathbf{C}')$ given by Theorem 15.5.1. Next we add the inequality $-l(P) \leqq 0$. We obtain a new cone of solutions $\mathbf{C}''$ and a new system of solutions $\mathbf{N}(\mathbf{C}'')$, by applying again Theorem 15.5.1. It is easily shown that the two operations can be fused into a single one, described by

Theorem 15.5.2. *The system $\mathbf{N}(\mathbf{C}'')$ is composed of those vectors N^i of the set (15.72) for which $l(N^i) = 0$, and in the case $t > 1$ also of the vectors*

$$N(k, k') = N^k |l(N^{k'})| + N^{k'} |l(N^k)| \tag{15.79}$$

for each pair of elements N^k, $N^{k'}$ of the system $\mathbf{N}(\mathbf{C})$ which satisfies the following conditions:

1. $l(N^k) l(N^{k'}) < 0$.

2. We consider the inequalities of the system (15.70) which reduce to equalities for the two elements N^k and $N^{k'}$. Then no other element of the system $\mathbf{N}(\mathbf{C})$ reduces all these inequalities to equalities.

This theorem is very similar to Theorem 15.5.1, the only difference being the replacement of $l(N^i) \leqq 0$ by $l(N^i) = 0$ in the second line.

15.5.2 Elimination of Parasitic Solutions

We go back now to our original system with strict inequalities. Its solutions are a subset of the set (15.75). Consider one of the strict inequalities, of the form

$$l_j(P) = a_{j1}p_1 + a_{j2}p_2 + \ldots + a_{jn_B}p_{n_B} < 0\,. \tag{15.80}$$

If we substitute the general solution (15.75), we obtain

$$l_j(P) = \sum_{i=1}^{t} \lambda_i l_j(N^i)\,. \tag{15.81}$$

Since N^i is one particular solution of the form (15.75), there is

$$l_j(N^i) \leqq 0\,. \tag{15.82}$$

We call I_j the set of the i such that

$$l_j(N^i) < 0\,. \tag{15.83}$$

(15.81) becomes

$$l_j(P) = \sum_{i \in I_j} \lambda_i l_j(N^i)\,. \tag{15.84}$$

It follows that the strict inequality $l_j(P) < 0$ is satisfied if and only if at least one λ_i with $i \in I_j$ is positive. This condition can be written

$$\sum_{i \in I_j} \lambda_i > 0\,. \tag{15.85}$$

Each strict inequality in the original system gives rise to a condition of the form (15.85), which eliminates some of the solutions (15.75).

As a particular case, it may happen that $l_j(N^i)$ vanishes for all i; the set I_j is empty. In that case $l_j(P) = 0$ for any solution (15.75), and the strict inequality cannot be satisfied. The combination of normal cones is not valid and can be eliminated.

As an example, we consider again the case dad, corresponding to the set of conditions (15.69). From the Motzkin-Burger algorithm it is found that the skeleton consists of the two vectors $N^1 = (-1, 0, 0, -1, 0)$ and $N^2 = (-2, -1, -1, -1, -1)$ (see Table 15.2). The first strict inequality is $l_1(P) = p_1 < 0$. So $l_1(N^1) = -1$, $l_1(N^2) = -2$, $I_1 = \{1, 2\}$, and the condition (15.85) is

$$\lambda_1 + \lambda_2 > 0\,. \tag{15.86}$$

In the same way, the 6 other strict inequalities in (15.69) give respectively the conditions

$$\begin{aligned} &\lambda_2 > 0\,, \quad \lambda_2 > 0\,, \quad \lambda_1 + \lambda_2 > 0\,, \\ &\lambda_2 > 0\,, \quad \lambda_1 > 0\,, \quad \lambda_1 > 0\,. \end{aligned} \tag{15.87}$$

The whole set of 7 conditions reduces to

$$\lambda_1 > 0\,, \quad \lambda_2 > 0\,. \tag{15.88}$$

Thus the general solution of the system (15.69) is

$$P = \sum_{i=1}^{2} \lambda_i N^i\,; \qquad \lambda_i > 0\,, \quad i = 1, 2\,. \tag{15.89}$$

15.5.3 Program

The algorithm was implemented in a computer program.
This program finds all valid combinations of normal cones for a bifurcation 1Pn, for an arbitrary value of n, on the basis of the normal cone definitions (15.30), (15.36), (15.42), (15.47). For each valid combination, the set of vectors **N** is printed.

Table 15.1 shows the number of valid combinations found and the computing time in seconds (on a HP 720 workstation) for $n = 2$ to 6. It can be seen that the number of valid combinations increases less steeply than the number of possible combinations, given by (15.68); however, it still appears to grow exponentially with n.

Table 15.1. Number of valid combinations for a bifurcation 1Pn.

n	number of valid combinations	computing time (seconds)
2	12	0
3	39	2
4	138	24
5	505	1143
6	1920	69600

15.5.4 The Case 1P2

From now on we restrict our attention to the simplest case 1P2. Larger values of n could in principle be also treated by the methods about to be described; however, the amount of work grows exponentially.

In the case 1P2, there are $n_B = 5$ variables and $m_B = 3$ equations. These equations are (15.38), (15.26) for $i = 1$, (15.44) for $n = 2$. We regroup them here:

$$\begin{aligned} f_1' &= x_4 - G_2 x_2 + G_3(1-K)x_3 = 0\,, \\ f_2' &= x_4(x_5 - x_3) + G_1 x_1 = 0\,, \\ f_3' &= -x_4 + G_2 x_2 + G_3(1+K)x_5 = 0\,. \end{aligned} \tag{15.90}$$

The number of possible combinations of normal cones is $4 \times 4 \times 4 = 64$ (see (15.68). The program finds 12 valid combinations, which are listed in Table 15.2. Each combination is identified in column 1 by a set of 3 letters, which are the labels of the selected normal cones in the first, second and third equation respectively, as defined in (15.42), (15.30) for $i = 1$, and (15.47) for $n = 2$. Column 2 is the index i of a vector N^i; the components of N^i are listed in column 3. The number t of the vectors N^i varies from 1 to 3.

Table 15.2. Valid combinations for the bifurcation 1P2.

Case	i	N^i	dim Π	d
aaa	1	-2 -1 -1 -1 -1	1	4
acc	1	0 0 0 0 -1	2	3
	2	-2 -1 -1 -1 -1		
aba	1	-1 0 0 0 0	2	3
	2	-2 -1 -1 -1 -1		
dad	1	-1 0 0 -1 0	2	3
	2	-2 -1 -1 -1 -1		
dbd	1	-1 0 0 0 0	3	2
	2	-1 0 0 -1 0		
	3	-2 -1 -1 -1 -1		
cac	1	-1 0 -1 0 -1	2	3
	2	-2 -1 -1 -1 -1		
ccc	1	0 0 0 0 -1	3	2
	2	-1 0 -1 0 -1		
	3	-2 -1 -1 -1 -1		
cda	1	0 0 -1 0 0	2	3
	2	-2 -1 -1 -1 -1		
cdc	1	0 0 -1 0 0	3	2
	2	-1 0 -1 0 -1		
	3	-2 -1 -1 -1 -1		
cbc	1	-1 0 0 0 0	3	2
	2	-1 0 -1 0 -1		
	3	-2 -1 -1 -1 -1		
bab	1	0 -1 0 0 0	2	3
	2	-2 -1 -1 -1 -1		
bbb	1	-1 0 0 0 0	3	2
	2	0 -1 0 0 0		
	3	-2 -1 -1 -1 -1		

Column 4 gives the dimension of the cone of truncation Π. This dimension is the rank of the $t \times n$ matrix formed by the components of the N^i. In the present case, $n = 2$, the table shows that $\dim \mathbf{\Pi}$ is always equal to t. This is not generally true, however; for $n > 2$, cases exist with $\dim \mathbf{\Pi} < t$.

We remark that the vector $(-2, -1, -1, -1, -1)$ is always present.

15.6 Truncated Systems of Equations

To each normal cone is associated a truncated equation. Therefore, to each combination of normal cones is associated a *truncated system of equations* (Bruno 2.6). For each of the 12 valid combinations, we must now consider this system and try to solve it.

The *dimension of the truncated system*, which we designate by d, is the codimension of the cone of truncation (Bruno 2.6):

$$d = n_B - \dim \mathbf{\Pi} \,. \tag{15.91}$$

This dimension appears in the last column of Table 15.2. It will play an important role below.

The truncated system is made of 3 equations $\widehat{f_i} = 0$, $i = 1, 2, 3$. The expressions of the $\widehat{f_i}$, read from (15.41), (15.29) for $i = 1$, (15.46) for $n = 2$, are given by Table 15.3 for each valid combination.

Table 15.3. Truncated systems of equations.

Case	$\widehat{f_{1P}}$	$\widehat{f_{2P}}$	$\widehat{f_{3P}}$
aaa	$x_4 - G_2x_2 + G_3(1-K)x_3$	$x_4(x_5 - x_3) + G_1x_1$	$-x_4 + G_2x_2 + G_3(1+K)x_5$
acc	$x_4 - G_2x_2 + G_3(1-K)x_3$	$x_4(-x_3) + G_1x_1$	$-x_4 + G_2x_2$
aba	$x_4 - G_2x_2 + G_3(1-K)x_3$	$x_4(x_5 - x_3)$	$-x_4 + G_2x_2 + G_3(1+K)x_5$
dad	$-G_2x_2 + G_3(1-K)x_3$	$x_4(x_5 - x_3) + G_1x_1$	$G_2x_2 + G_3(1+K)x_5$
dbd	$-G_2x_2 + G_3(1-K)x_3$	$x_4(x_5 - x_3)$	$G_2x_2 + G_3(1+K)x_5$
cac	$x_4 - G_2x_2$	$x_4(x_5 - x_3) + G_1x_1$	$-x_4 + G_2x_2$
ccc	$x_4 - G_2x_2$	$x_4(-x_3) + G_1x_1$	$-x_4 + G_2x_2$
cda	$x_4 - G_2x_2$	$x_4(x_5) + G_1x_1$	$-x_4 + G_2x_2 + G_3(1+K)x_5$
cdc	$x_4 - G_2x_2$	$x_4(x_5) + G_1x_1$	$-x_4 + G_2x_2$
cbc	$x_4 - G_2x_2$	$x_4(x_5 - x_3)$	$-x_4 + G_2x_2$
bab	$x_4 + G_3(1-K)x_3$	$x_4(x_5 - x_3) + G_1x_1$	$-x_4 + G_3(1+K)x_5$
bbb	$x_4 + G_3(1-K)x_3$	$x_4(x_5 - x_3)$	$-x_4 + G_3(1+K)x_5$

15.6.1 Degeneracy

It can happen that the truncated system is degenerate (Bruno 2.6, Remark 6.2). This is indeed the case here in the 4 cases cac, ccc, cdc, cbc: the first and third equations are identical.

The remedy is indicated by Bruno (ibid.): one of the original equations $f_i = 0$ should be replaced by an appropriate combination, which provides in a sense a finer description of the system. Here the nature of the degeneracy indicates that we should replace f_3 by the combination $f_1 + f_3$. We have already considered this combination: it is the additional relation (15.8).

In each of the 4 degenerate cases, we must replace the last normal cone c, defined in (15.47), by the normal cone A, defined by (15.66). We label the 4 cases thus redefined: caA, ccA, cdA, cbA. We apply the Motzkin-Burger algorithm to these new cases. We find that only the two cases caA and cbA are valid combinations.

In the truncated system of equations, the third equation should be replaced by (15.63). Thus the entries cac, ccc, cdc, cbc should be crossed out in Tables 15.2 and 15.3, and new entries should be inserted as indicated by Tables 15.4 and 15.5.

Table 15.4. Additional valid combinations for the bifurcation 1P2.

Case	i	N^i	dim Π	d
caA	1	-1 0 -1 0 -1	2	3
	2	-2 -1 -1 -1 -1		
cbA	1	-1 0 0 0 0	3	2
	2	-1 0 -1 0 -1		
	3	-2 -1 -1 -1 -1		

Table 15.5. Additional truncated systems of equations.

Case	$\widehat{f}_{1P}$	$\widehat{f}_{2P}$	$\widehat{f}_{3P}$
caA	$x_4 - G_2 x_2$	$x_4(x_5 - x_3) + G_1 x_1$	$(1-K)x_3 + (1+K)x_5$
cbA	$x_4 - G_2 x_2$	$x_4(x_5 - x_3)$	$(1-K)x_3 + (1+K)x_5$

15.7 Power Transformations

In each of the 10 cases, we must now study the truncated system of equations. It will be convenient to make in each case a change of variables, which Bruno (1.11, 2.3, 2.7) calls a *power transformation*, defined by

$$x_i = w_1^{\beta_{i1}} \dots w_{n_B}^{\beta_{in_B}}, \quad i = 1, \dots, n_B . \tag{15.92}$$

I use w_i for the new variables, instead of y_i as in Bruno, in order to avoid confusion with my own quantities y_i. Note that according to (15.9) and (15.10), there is

$$w_i \neq 0 . \tag{15.93}$$

Theorem 7.1 in Bruno 2.7 states that for a truncated system of m_B equations, with dimension d, it is possible to find a change of variables (15.92) in such a way that the new system again has m_B equations, but depends only on d of the new variables w_j.

The simplest method for finding this change of variables in the present case appears to be Method 2 described in Bruno 1.11:

1. We select ($\dim \mathbf{\Pi}$) linearly independent vectors N^i (in the case $n = 2$, this amounts to selecting all vectors N^i since in that case $\dim \mathbf{\Pi}$ always equals t).
2. We write these vectors as the last ($\dim \mathbf{\Pi}$) columns of the matrix $\beta = (\beta_{ij})$. Since all components of the N^i are negative or zero, it will be convenient to use $-N^i$ rather than N^i.
3. We fill the remaining columns, i.e. the first d columns (see (15.91) in such a way that the matrix β is regular, i.e. $\det \beta \neq 0$. If possible, it is convenient to do this in such a way that the matrix is unimodular, i.e. all β_{ij} are integers and $|\det \beta| = 1$; then all exponents are integers in (15.92) and also in the inverse change of variables $w(x)$.

Rearranging the w_i variables if necessary, we can write the new truncated system as (Bruno 2.8, (8.5))

$$\widehat{g}_i(w_1, \ldots, w_d) = 0 \,, \quad i = 1, \ldots, m_B \,. \tag{15.94}$$

If $d < m_B$, there are more equations than variables, and in general the system has no solutions. If $d \geqq m_B$, we must first compute the Jacobian

$$\frac{\partial(\widehat{g}_1, \ldots, \widehat{g}_{m_B})}{\partial(w_1, \ldots, w_d)} \tag{15.95}$$

(Bruno 2.8, (8.7)). Any point $(w_1, \ldots, w_d)$ where this Jacobian has a rank less than m_B is a *critical point*. Such critical points require a special treatment (Bruno 2.8). Fortunately, no critical points will be found below. Some solutions may also have to be rejected because they contradict (15.93).

We apply now this method to each of the 10 cases in turn.

15.7.1 Case aaa

We consider first the case aaa, where the dimension of the truncated system is $d = 4$. We write N^1, given by Table 15.2, in the last column of β, and we fill the first 4 columns with the vectors E_2, E_3, E_4, E_5:

$$\beta = (E_2, E_3, E_4, E_5, -N^1) = \begin{pmatrix} 0 & 0 & 0 & 0 & 2 \\ 1 & 0 & 0 & 0 & 1 \\ 0 & 1 & 0 & 0 & 1 \\ 0 & 0 & 1 & 0 & 1 \\ 0 & 0 & 0 & 1 & 1 \end{pmatrix} . \tag{15.96}$$

There is $\det \beta = 2$: the matrix is regular but not unimodular.

The equations of the change of variables are then

$$x_1 = w_5^2 \,, \quad x_2 = w_1 w_5 \,, \quad x_3 = w_2 w_5 \,, \quad x_4 = w_3 w_5 \,, \quad x_5 = w_4 w_5 \,. \tag{15.97}$$

Substituting in the truncated equations given by Table 15.3, we obtain

$$\begin{aligned} &w_3 w_5 - G_2 w_1 w_5 + G_3(1-K) w_2 w_5 = 0 \,, \\ &w_3 w_5 (w_4 w_5 - w_2 w_5) + G_1 w_5^2 = 0 \,, \\ &-w_3 w_5 + G_2 w_1 w_5 + G_3(1+K) w_4 w_5 = 0 \,, \end{aligned} \tag{15.98}$$

or, dividing by w_5, w_5^2, w_5 respectively (we recall that the w_i never vanish),

$$\begin{aligned} &\widehat{g}_1 = w_3 - G_2 w_1 + G_3(1-K) w_2 = 0 \,, \\ &\widehat{g}_2 = w_3 (w_4 - w_2) + G_1 = 0 \,, \\ &\widehat{g}_3 = -w_3 + G_2 w_1 + G_3(1+K) w_4 = 0 \,. \end{aligned} \tag{15.99}$$

We have 3 equations for 4 variables, as expected. The Jacobian (15.95) is

$$\frac{\partial g}{\partial w} = \begin{pmatrix} -G_2 & G_3(1-K) & 1 & 0 \\ 0 & -w_3 & w_4 - w_2 & w_3 \\ G_2 & 0 & -1 & G_3(1+K) \end{pmatrix} \,. \tag{15.100}$$

The minor formed by the columns 1, 2, 4 has a determinant

$$\begin{vmatrix} -G_2 & G_3(1-K) & 0 \\ 0 & -w_3 & w_3 \\ G_2 & 0 & G_3(1+K) \end{vmatrix} = 2 G_2 G_3 w_3 \neq 0 \,. \tag{15.101}$$

Thus the matrix always has rank 3: there is no critical point.

Solutions of the system (15.99) form a one-parameter family. It is convenient to take w_3 as parameter; we have then a system of 3 linear equations for w_1, w_2, w_4, which is easily solved:

$$w_1 = \frac{w_3}{G_2} + \frac{(1-K^2) G_1 G_3}{2 G_2 w_3} \,, \quad w_2 = \frac{(1+K) G_1}{2 w_3} \,, \quad w_4 = \frac{(K-1) G_1}{2 w_3} \,. \tag{15.102}$$

The value $w_3 = 0$ is excluded according to (15.93). Thus w_1, w_2 and w_4 are always defined. Also, w_2 and w_4 are always non-zero. However, w_1 vanishes for the two values

$$w_3 = \pm \sqrt{\frac{(K^2-1) G_1 G_3}{2}} \tag{15.103}$$

when $|K| > 1$. These two values of w_3 must then also be excluded.

Going back to the original variables x_i with (15.97), we obtain

$$\begin{aligned} &x_1 = w_5^2 \,, \quad x_2 = w_5 \left(\frac{w_3}{G_2} + \frac{(1-K^2) G_1 G_3}{2 G_2 w_3} \right) \,, \\ &x_3 = w_5 \frac{(1+K) G_1}{2 w_3} \,, \quad x_4 = w_3 w_5 \,, \quad x_5 = w_5 \frac{(K-1) G_1}{2 w_3} \,. \end{aligned} \tag{15.104}$$

This is now a system with 2 parameters w_3 and w_5. Going back to my notations with (15.2), we obtain

$$\begin{aligned} &w_5 = \mu^{1/2}\,, \quad w_3 = y_1\mu^{-1/2}\,, \\ &\Delta C = \mu^{1/2}\left(\frac{y_1\mu^{-1/2}}{G_2} + \frac{(1-K^2)G_1G_3}{2G_2y_1\mu^{-1/2}}\right)\,, \\ &\Delta a_1 = \mu^{1/2}\frac{(1+K)G_1}{2y_1\mu^{-1/2}}\,, \quad \Delta a_2 = \mu^{1/2}\frac{(K-1)G_1}{2y_1\mu^{-1/2}}\,. \end{aligned} \tag{15.105}$$

There is a single vector $N^i = (-2,-1,-1,-1,-1)$ and we have from (15.75): $p_1 = -2\lambda_1$, $p_2 = -\lambda_1$, with $\lambda_1 \geqq 0$. The elimination of parasitic solutions (Sect. 15.5.2) gives $\lambda_1 > 0$. Thus from (15.14) we have

$$\nu = \frac{1}{2}\,. \tag{15.106}$$

Therefore the present case aaa should correspond to the case $\nu = 1/2$, the equations of which were established in Sect. 12.5 and studied in detail in Chaps. 13 and 14. Indeed, applying the changes of variables (12.111) and (12.112), we obtain

$$W = Y_1 + \frac{1-K^2}{2Y_1}\,, \quad X_1 = -\frac{1+K}{2Y_1}\,, \quad X_2 = \frac{K-1}{2Y_1}\,. \tag{15.107}$$

We recover the equation of the characteristic (13.41) for the case 1P2, and the equations for X_1 and X_2 deduced from (13.1a), with one proviso: in the case $|K| > 1$, the two intersections of the characteristic with the axis $W = 0$ (see Fig. 13.1) are not covered by the present treatment. This is because our fundamental assumption (15.9), (15.10) requires all variables to be non-zero. These points require a different change of variables; this will be covered in case bab (Sect. 15.7.7).

Additional computations would give the error terms (see Bruno 2.8). An example of these computations will be given in the case dad (Sect. 15.7.4).

15.7.2 Case acc

We consider next the 6 cases where the dimension of the truncated system is $d = 3$. We write N^1 and N^2, given by Tables 15.2 and 15.4, in the last two columns of β, and we fill the first three columns with appropriate unit vectors. We try to obtain relations as simple as possible for x_1 et x_2, which correspond respectively to the fundamental parameters μ et ΔC.

We obtain in the new variables a system of 3 equations for 3 variables, for which we expect a finite number of isolated solutions.

In the case acc, we take

$$\beta = (E_2, E_3, E_4, -N^1, -N^2)\,. \tag{15.108}$$

This is the same change of variables as in the case aaa (see (15.97)). The truncated equations become

$$\begin{aligned}\widehat{g}_1 &= w_3 - G_2 w_1 + G_3(1-K)w_2 = 0\,,\\ \widehat{g}_2 &= -w_3 w_2 + G_1 = 0\,,\\ \widehat{g}_3 &= -w_3 + G_2 w_1 = 0\,.\end{aligned} \tag{15.109}$$

This system has no solution: the first and last equations give $w_2 = 0$, which contradicts (15.93). In addition, the second equation cannot be satisfied since $G_1 \neq 0$.

15.7.3 Case aba

We take

$$\beta = (E_3, E_4, E_5, -N^1, -N^2)\,. \tag{15.110}$$

Here $\det\beta = 1$: the matrix is unimodular. The equations of the change of variables are

$$x_1 = w_4 w_5^2\,,\quad x_2 = w_5\,,\quad x_3 = w_1 w_5\,,\quad x_4 = w_2 w_5\,,\quad x_5 = w_3 w_5\,. \tag{15.111}$$

The truncated equations become

$$\begin{aligned}\widehat{g}_1 &= w_2 - G_2 + G_3(1-K)w_1 = 0\,,\\ \widehat{g}_2 &= w_3 - w_1 = 0\,,\\ \widehat{g}_3 &= -w_2 + G_2 + G_3(1+K)w_3 = 0\,.\end{aligned} \tag{15.112}$$

This system has a unique solution

$$w_1 = w_3 = 0\,,\quad w_2 = G_2\,. \tag{15.113}$$

However, this solution must be rejected because it contradicts (15.93).

15.7.4 Case dad

We take

$$\beta = (E_3, E_4, E_5, -N^1, -N^2)\,. \tag{15.114}$$

The equations of the change of variables are

$$\begin{aligned}&x_1 = w_4 w_5^2\,,\quad x_2 = w_5\,,\quad x_3 = w_1 w_5\,,\\ &x_4 = w_2 w_4 w_5\,,\quad x_5 = w_3 w_5\,.\end{aligned} \tag{15.115}$$

The truncated equations become

$$\begin{aligned}\widehat{g}_1 &= -G_2 + G_3(1-K)w_1 = 0\,,\\ \widehat{g}_2 &= w_2(w_3 - w_1) + G_1 = 0\,,\\ \widehat{g}_3 &= G_2 + G_3(1+K)w_3 = 0\,.\end{aligned} \tag{15.116}$$

This system has a unique solution

$$w_i = \psi_i, \quad i = 1, 2, 3 \, , \tag{15.117}$$

with

$$\psi_1 = \frac{G_2}{G_3(1-K)} \, , \qquad \psi_3 = -\frac{G_2}{G_3(1+K)} \, , \qquad \psi_2 = \frac{G_1 G_3 (1-K^2)}{2G_2} \, . \tag{15.118}$$

The Jacobian is

$$\frac{\partial g}{\partial w} = \begin{pmatrix} G_3(1-K) & 0 & 0 \\ -w_2 & w_3 - w_1 & w_2 \\ 0 & 0 & G_3(1+K) \end{pmatrix} . \tag{15.119}$$

Its determinant is

$$G_3^2(1-K^2)(w_3 - w_1) = -2G_2G_3 \neq 0 \, . \tag{15.120}$$

Thus the solution (15.117) is not a critical point. It is acceptable since $\psi_i \neq 0$ for $i = 1, 2, 3$.

Going back to the variables x_i, we obtain

$$x_1 = w_4 w_5^2 \, , \quad x_2 = w_5 \, , \quad x_3 = \frac{G_2}{G_3(1-K)} w_5 \, ,$$
$$x_4 = \frac{G_1 G_3 (1-K^2)}{2G_2} w_4 w_5 \, , \quad x_5 = -\frac{G_2}{G_3(1+K)} w_5 \, . \tag{15.121}$$

This is now a system with 2 parameters w_4 et w_5. Going back to my notations with (15.2), we obtain the relations between these two parameters and the fundamental parameters μ, ΔC:

$$w_5 = \Delta C \, , \quad w_4 = \mu \Delta C^{-2} \, , \tag{15.122}$$

and

$$\Delta a_1 = \frac{G_2}{G_3(1-K)} \Delta C \, , \quad y_1 = \frac{G_1 G_3 (1-K^2)}{2G_2} \mu \Delta C^{-1} \, ,$$
$$\Delta a_2 = -\frac{G_2}{G_3(1+K)} \Delta C \, . \tag{15.123}$$

Here there are two vectors

$$N^1 = (-1, 0, 0, -1, 0) \, , \qquad N^2 = (-2, -1, -1, -1, -1) \, , \tag{15.124}$$

and therefore from (15.75):

$$p_1 = -\lambda_1 - 2\lambda_2 \, , \quad p_2 = -\lambda_2 \, , \quad p_3 = -\lambda_2 \, ,$$
$$p_4 = -\lambda_1 - \lambda_2 \, , \quad p_5 = -\lambda_2 \, . \tag{15.125}$$

The elimination of parasitic solutions (Sect. 15.5.2) gives

$$\lambda_1 > 0 \, , \qquad \lambda_2 > 0 \, . \tag{15.126}$$

From (15.14) we have

$$\nu = \frac{\lambda_2}{\lambda_1 + 2\lambda_2} \tag{15.127}$$

and using (15.126):

$$0 < \nu < \frac{1}{2} \,. \tag{15.128}$$

Conversely, for any value of ν in that interval, we can find positive values of λ_1 and λ_2 which satisfy (15.127).

Thus the present case dad should correspond to one of the cases $0 < \nu < 1/2$, considered in Sect. 12.4.2. Indeed, the equations (15.123) are identical (except for the error terms) with the equations (12.110a), (12.108) corresponding to the case of two S-arcs.

It is possible to recover also the error terms in these equations. We describe the computations in detail in the present case dad; similar computations can be done in the other cases.

We go back to the whole equations: (15.38), (15.26) for $i = 1$, (15.44) for $n = 2$. We apply the change of variables (15.115). We obtain

$$\begin{aligned} g_1(w_1, \ldots, w_5) &= w_2 w_4 - G_2 + G_3(1-K)w_1 + O(w_5) + O(w_1^2 w_5) = 0 \,, \\ g_2(w_1, \ldots, w_5) &= w_2(w_3 - w_1) + G_1 \\ &\quad + O(w_2 w_4 w_5) + O[w_5(w_3 - w_1)] = 0 \,, \\ g_3(w_1, \ldots, w_5) &= -w_2 w_4 + G_2 + G_3(1+K)w_3 + O(w_5) + O(w_3^2 w_5) = 0 \,. \end{aligned} \tag{15.129}$$

(cf. Bruno 2.8, (8.4)).

We effect now the new change of variables (Bruno 2.8, (8.8))

$$w_i = \psi_i + z_i, \qquad i = 1, 2, 3 \,, \tag{15.130}$$

with the ψ_i given by (15.118). There is $\psi_i = \Theta(1)$, $i = 1, 2, 3$, and also $\psi_3 - \psi_1 = \Theta(1)$. On the other hand the z_i are small quantities. Terms of order 0 (with respect to the z_i) disappear and we obtain

$$\begin{aligned} h_1 &= \psi_2 w_4 + z_2 w_4 + G_3(1-K)z_1 + O(w_5) = 0 \,, \\ h_2 &= \psi_2(z_3 - z_1) + z_2(\psi_3 - \psi_1) + z_2(z_3 - z_1) + O(w_4 w_5) + O(w_5) = 0 \,, \\ h_3 &= -\psi_2 w_4 - z_2 w_4 + G_3(1+K)z_3 + O(w_5) = 0 \end{aligned} \tag{15.131}$$

(cf. Bruno 2.8, between (8.8) and (8.9)).

w_4, w_5 are small quantities. Solving these equations for the z_i, we obtain

$$z_1, z_2, z_3 = O(w_4) + O(w_5) \,. \tag{15.132}$$

Therefore we can write

$$w_i = \psi_i[1 + O(w_4) + O(w_5)] \,. \tag{15.133}$$

Coming back to the x_i, we have

$$x_1 = w_4 w_5^2 \, , \quad x_2 = w_5 \, , \quad x_3 = \frac{G_2}{G_3(1-K)} w_5 [1 + O(w_4) + O(w_5)] \, ,$$
$$x_4 = \frac{G_1 G_3 (1-K^2)}{2G_2} w_4 w_5 [1 + O(w_4) + O(w_5)] \, ,$$
$$x_5 = -\frac{G_2}{G_3(1+K)} w_5 [1 + O(w_4) + O(w_5)] \, . \tag{15.134}$$

Going back to my notations, we obtain as before

$$w_5 = \Delta C \, , \quad w_4 = \mu \Delta C^{-2} \, , \tag{15.135}$$

and

$$\Delta a_1 = \frac{G_2}{G_3(1-K)} \Delta C \left[1 + O(\Delta C) + O(\mu \Delta C^{-2})\right] \, ,$$
$$y_1 = \frac{G_1 G_3 (1-K^2)}{2G_2} \mu \Delta C^{-1} \left[1 + O(\Delta C) + O(\mu \Delta C^{-2})\right] \, ,$$
$$\Delta a_2 = -\frac{G_2}{G_3(1+K)} \Delta C \left[1 + O(\Delta C) + O(\mu \Delta C^{-2})\right] \, . \tag{15.136}$$

We have now recovered exactly the results (12.110a), (12.108), including the error terms.

15.7.5 Case caA

We take

$$\beta = (E_3, E_4, E_5, -N^1, -N^2) \, . \tag{15.137}$$

The equations of the change of variables are

$$x_1 = w_4 w_5^2 \, , \quad x_2 = w_5 \, , \quad x_3 = w_1 w_4 w_5 \, , \quad x_4 = w_2 w_5 \, , \quad x_5 = w_3 w_4 w_5 \, . \tag{15.138}$$

The truncated equations become

$$\widehat{g}_1 = w_2 - G_2 = 0 \, ,$$
$$\widehat{g}_2 = w_2 (w_3 - w_1) + G_1 = 0 \, ,$$
$$\widehat{g}_3 = (1-K) w_1 + (1+K) w_3 = 0 \, . \tag{15.139}$$

This system has a unique solution

$$w_2 = G_2 \, , \qquad w_1 = \frac{(K+1)G_1}{2G_2} \, , \qquad w_3 = \frac{(K-1)G_1}{2G_2} \, . \tag{15.140}$$

The determinant of the Jacobian is

$$\begin{vmatrix} 0 & 1 & 0 \\ -w_2 & w_3 - w_1 & w_2 \\ 1-K & 0 & 1+K \end{vmatrix} = 2w_2 = 2G_2 \neq 0 \, . \tag{15.141}$$

The solution is not a critical point. It is acceptable since $w_i \neq 0$, $i = 1, 2, 3$.

Going back to the variables x_i, we obtain

$$x_1 = w_4 w_5^2 \,, \quad x_2 = w_5 \,, \quad x_3 = \frac{(K+1)G_1}{2G_2} w_4 w_5 \,,$$
$$x_4 = G_2 w_5 \,, \quad x_5 = \frac{(K-1)G_1}{2G_2} w_4 w_5 \,. \tag{15.142}$$

This is a system with 2 parameters w_4 et w_5. Going back to my notations with (15.2), we obtain the relations between these two parameters and the fundamental parameters μ, ΔC:

$$w_5 = \Delta C \,, \quad w_4 = \mu \Delta C^{-2} \,. \tag{15.143}$$

These relations are the same as in the previous case dad. We have also

$$\Delta a_1 = \frac{(K+1)G_1}{2G_2} \mu \Delta C^{-1} \,, \quad y_1 = G_2 \Delta C \,,$$
$$\Delta a_2 = \frac{(K-1)G_1}{2G_2} \mu \Delta C^{-1} \,. \tag{15.144}$$

There are two vectors

$$N^1 = (-1, 0, -1, 0, -1) \,, \qquad N^2 = (-2, -1, -1, -1, -1) \,. \tag{15.145}$$

Therefore

$$p_1 = -\lambda_1 - 2\lambda_2 \,, \quad p_2 = -\lambda_2 \,, \quad p_3 = -\lambda_1 - \lambda_2 \,,$$
$$p_4 = -\lambda_2 \,, \quad p_5 = -\lambda_1 - \lambda_2 \,. \tag{15.146}$$

The elimination of parasitic solutions gives

$$\lambda_1 > 0 \,, \qquad \lambda_2 > 0 \,. \tag{15.147}$$

From (15.14)

$$\nu = \frac{\lambda_2}{\lambda_1 + 2\lambda_2} \tag{15.148}$$

and as in the case dad we obtain

$$0 < \nu < \frac{1}{2} \,. \tag{15.149}$$

Thus the present case caA should also correspond to one of the cases considered in Sect. 12.4.2. Indeed, the equations (15.144) are identical (except for the error terms) with the equations (12.109) corresponding to the case of a T-arc.

15.7.6 Case cda

We take

$$\beta = (E_2, E_4, E_5, -N^1, -N^2) \,. \tag{15.150}$$

The equations of the change of variables are

$$x_1 = w_5^2\,,\quad x_2 = w_1 w_5\,,\quad x_3 = w_4 w_5\,,\quad x_4 = w_2 w_5\,,\quad x_5 = w_3 w_5\,. \tag{15.151}$$

The truncated equations become

$$\begin{aligned}
\widehat{g}_1 &= w_2 - G_2 w_1 = 0\,,\\
\widehat{g}_2 &= w_2 w_3 + G_1 = 0\,,\\
\widehat{g}_3 &= -w_2 + G_2 w_1 + G_3(1+K) w_3 = 0\,.
\end{aligned} \tag{15.152}$$

This case is similar to acc (Sect. 15.7.2): the system has no solution.

15.7.7 Case bab

We take

$$\beta = (E_3, E_4, E_5, -N^1, -N^2)\,. \tag{15.153}$$

The equations of the change of variables are

$$x_1 = w_5^2\,,\quad x_2 = w_4 w_5\,,\quad x_3 = w_1 w_5\,,\quad x_4 = w_2 w_5\,,\quad x_5 = w_3 w_5\,. \tag{15.154}$$

The truncated equations become

$$\begin{aligned}
\widehat{g}_1 &= w_2 + G_3(1-K) w_1 = 0\,,\\
\widehat{g}_2 &= w_2(w_3 - w_1) + G_1 = 0\,,\\
\widehat{g}_3 &= -w_2 + G_3(1+K) w_3 = 0\,.
\end{aligned} \tag{15.155}$$

This system has two solutions if $|K| > 1$:

$$w_2 = \pm\sqrt{\frac{G_1 G_3 (K^2-1)}{2}}\,,\quad w_1 = \pm\frac{\sqrt{\frac{G_1(K^2-1)}{2G_3}}}{K-1}\,,\quad w_3 = \pm\frac{\sqrt{\frac{G_1(K^2-1)}{2G_3}}}{K+1}\,. \tag{15.156}$$

The determinant of the Jacobian is

$$\begin{vmatrix} G_3(1-K) & 1 & 0\\ -w_2 & w_3 - w_1 & w_2\\ 0 & -1 & G_3(1+K) \end{vmatrix} = 4G_3 w_2 \neq 0\,. \tag{15.157}$$

The solutions are not critical points. They are acceptable since $w_i \neq 0$, $i = 1, 2, 3$.

Going back to the variables x_i, we obtain

$$\begin{aligned}
&x_1 = w_5^2\,,\quad x_2 = w_4 w_5\,,\quad x_3 = \pm\frac{\sqrt{\frac{G_1(K^2-1)}{2G_3}}}{K-1} w_5\,,\\
&x_4 = \pm\sqrt{\frac{G_1 G_3(K^2-1)}{2}}\, w_5\,,\quad x_5 = \pm\frac{\sqrt{\frac{G_1(K^2-1)}{2G_3}}}{K+1} w_5\,.
\end{aligned} \tag{15.158}$$

This is a system with 2 parameters w_4 et w_5. Going back to my notations with (15.2), we obtain the relations between these two parameters and the fundamental parameters μ, ΔC:

$$w_5 = \mu^{1/2} \,, \quad w_4 = \mu^{-1/2}\Delta C \,. \tag{15.159}$$

We have also

$$\Delta a_1 = \pm\frac{\sqrt{\frac{G_1(K^2-1)}{2G_3}}}{K-1}\mu^{1/2} \,, \quad y_1 = \pm\sqrt{\frac{G_1G_3(K^2-1)}{2}}\mu^{1/2} \,,$$
$$\Delta a_2 = \pm\frac{\sqrt{\frac{G_1(K^2-1)}{2G_3}}}{K+1}\mu^{1/2} \,. \tag{15.160}$$

There are two vectors:

$$N^1 = (0,-1,0,0,0) \,, \qquad N^2 = (-2,-1,-1,-1,-1) \,. \tag{15.161}$$

Therefore

$$p_1 = -2\lambda_2 \,, \quad p_2 = -\lambda_1 - \lambda_2 \,, \quad p_3 = -\lambda_2 \,, \quad p_4 = -\lambda_2 \,, \quad p_5 = -\lambda_2 \,. \tag{15.162}$$

The elimination of parasitic solutions gives

$$\lambda_1 > 0 \,, \qquad \lambda_2 > 0 \,. \tag{15.163}$$

From (15.14):

$$\nu = \frac{\lambda_1 + \lambda_2}{2\lambda_2} \tag{15.164}$$

and we obtain

$$\nu > \frac{1}{2} \,. \tag{15.165}$$

Thus the present case bab corresponds to the case $\nu > 1/2$ considered in Sect. 12.6.

Finally, applying the changes of variables (12.111) and (12.112), we have

$$X_1 = \pm'\frac{\sqrt{(K^2-1)/2}}{K-1} \,, \quad X_2 = \mp'\frac{\sqrt{(K^2-1)/2}}{K+1} \,,$$
$$Y_1 = \mp'\sqrt{(K^2-1)/2} \,, \tag{15.166}$$

with

$$\pm' = \pm\mathrm{sign}(\cos\theta) \,. \tag{15.167}$$

15.7.8 Case dbd

We consider finally the 3 cases where the dimension of the truncated system is $d = 2$. We write N^1, N^2 and N^3, given by Tables 15.2 and 15.4, in the last three columns of β, and we fill the first two columns with appropriate unit vectors. We obtain in the new variables a system of 3 equations for 2 variables, which in general should not have solutions; this is indeed what we will find.

In the case dbd, we take

$$\beta = (E_3, E_4, -N^1, -N^2, -N^3) . \tag{15.168}$$

The equations of the change of variables are

$$x_1 = w_3 w_4 w_5^2 , \quad x_2 = w_5 , \quad x_3 = w_1 w_5 , \quad x_4 = w_2 w_4 w_5 , \quad x_5 = w_3 w_5 . \tag{15.169}$$

The truncated equations become

$$\begin{aligned} \widehat{g}_1 &= -G_2 + G_3(1-K)w_1 = 0 , \\ \widehat{g}_2 &= w_3 - w_1 = 0 , \\ \widehat{g}_3 &= G_2 + G_3(1+K)w_3 = 0 . \end{aligned} \tag{15.170}$$

Combining the first and last equations, we obtain $w_1 + w_3 = 0$, and with the second equation we have $w_1 = w_3 = 0$. But this contradicts (15.93). It also contradicts the first and last equations. Thus this system has no solution.

15.7.9 Case cbA

We take

$$\beta = (E_3, E_4, -N^1, -N^2, -N^3) . \tag{15.171}$$

The equations of the change of variables are

$$x_1 = w_3 w_4 w_5^2 , \quad x_2 = w_5 , \quad x_3 = w_1 w_4 w_5 , \quad x_4 = w_2 w_5 , \quad x_5 = w_4 w_5 . \tag{15.172}$$

The truncated equations become

$$\begin{aligned} \widehat{g}_1 &= w_2 - G_2 = 0 , \\ \widehat{g}_2 &= 1 - w_1 = 0 , \\ \widehat{g}_3 &= (1-K)w_1 + (1+K) = 0 . \end{aligned} \tag{15.173}$$

The last two equations give two different values for w_1. The system has no solution.

15.7.10 Case bbb

We take

$$\beta = (E_3, E_5, -N^1, -N^2, -N^3)\,. \tag{15.174}$$

The equations of the change of variables are

$$x_1 = w_3 w_5^2\,, \quad x_2 = w_4 w_5\,, \quad x_3 = w_1 w_5\,, \quad x_4 = w_5\,, \quad x_5 = w_2 w_5\,. \tag{15.175}$$

The truncated equations become

$$\begin{aligned} \widehat{g}_1 &= 1 + G_3(1-K)w_1 = 0\,, \\ \widehat{g}_2 &= w_2 - w_1 = 0\,, \\ \widehat{g}_3 &= -1 + G_3(1+K)w_2 = 0\,. \end{aligned} \tag{15.176}$$

This case is similar to dbd (Sect. 15.7.8). The system has no solution.

15.8 Total Bifurcation of Type 1

Up to now in the present chapter we have applied the Newton approach to the case of a partial bifurcation of type 1. We now show how it can just as well be applied to the case 1Tn, i.e. a total bifurcation of type 1.

The orbit is then made of n basic arcs (see Sect. 6.2.1). n must be even (Sect. 6.2.1.2).

There are $n_B = 2n + 2$ variables, which we write in Bruno's notations:

$$\begin{aligned} &x_1 = \mu\,, \quad x_2 = \Delta C\,, \quad x_3 = \Delta a_1\,, \quad x_4 = y_1\,, \quad x_5 = \Delta a_2\,, \\ &x_6 = y_2\,, \quad \ldots, \quad x_{2n+1} = \Delta a_n\,, \quad x_{2n+2} = y_n\,. \end{aligned} \tag{15.177}$$

The fundamental equations (12.32) become

$$\begin{aligned} f_1' &= x_4 - x_{2n+2} - G_2 x_2 + G_3(1-K)x_3 = 0\,, \\ f_{2i}' &= x_{2i+2}(x_{2i+3} - x_{2i+1}) + G_1 x_1 = 0\,, \qquad i = 1, \ldots, n-1\,, \\ f_{2i-1}' &= x_{2i+2} - x_{2i} + G_2 s_i x_2 + G_3(1+Ks_i)x_{2i+1} = 0\,, \quad i = 2, \ldots, n\,, \\ f_{2n}' &= x_{2n+2}(x_3 - x_{2n+1}) + G_1 x_1 = 0 \end{aligned} \tag{15.178}$$

with the cyclical boundary conditions

$$y_n = y_0\,, \qquad \Delta a_{n+1} = \Delta a_1\,. \tag{15.179}$$

Their number is $m_B = 2n$. We again have $m_B = n_B - 2$.

The additional relations (12.33) are still needed. Their number is n. We describe them with indices from $2n+1$ to $3n$:

$$f_{2n+i}' = -x_{2i} + x_{2i+4} + G_3(1+Ks_i)x_{2i+1} + G_3(1-Ks_i)x_{2i+3} = 0\,, \quad i = 1, \ldots, n\,. \tag{15.180}$$

In the particular case $n = 2$, there is a single additional relation

$$f'_5 = G_3(1 - K)x_3 + G_3(1 + K)x_5 = 0 . \tag{15.181}$$

The method of solution described in Sect. 15.2 is applicable without change. We can again use the study of the Newton polyhedra made in Sect. 15.3. There are 4 possible normal cones for each of the encounter equations $f_{2i} = 0$ in (15.178), and 11 possible normal cones for each of the arc equations $f_{2i-1} = 0$; thus the number of possible combinations of normal cones is

$$44^n . \tag{15.182}$$

The program implementing the Motzkin-Burger algorithm can again be used to find all valid combinations of normal cones. Table 15.6 shows the number of valid combinations and the computing time in seconds (on a HP 720 workstation) for $n = 2$ to 5.

Table 15.6. Number of valid combinations for a bifurcation 1Tn.

n	number of valid combinations	computing time (seconds)
2	32	0
3	85	10
4	300	487
5	1095	26079

15.8.1 The Case 1T2

We consider the case 1T2 as an example. There are $n_B = 6$ variables and $m_B = 4$ equations. These equations are

$$\begin{aligned} f'_1 &= x_4 - x_6 - G_2x_2 + G_3(1 - K)x_3 = 0 , \\ f'_2 &= x_4(x_5 - x_3) + G_1x_1 = 0 , \\ f'_3 &= x_6 - x_4 + G_2x_2 + G_3(1 + K)x_5 = 0 , \\ f'_4 &= x_6(x_3 - x_5) + G_1x_1 = 0 . \end{aligned} \tag{15.183}$$

The number of possible combinations of normal cones is $11 \times 4 \times 11 \times 4 = 1936$. The program finds 32 valid combinations, which are listed in Table 15.7. Each combination is identified in column 1 by a set of 4 letters, which are the labels of the selected cones (see (15.36) and (15.30)). Other columns have the same meaning as in Table 15.2.

Here, in the case kbkb, there is $t = 5$ but $\dim \mathbf{\Pi} = 4$ only: the matrix formed by the components of the N^i is singular. Indeed there is

$$N^1 - N^2 - N^4 + N^5 = 0 . \tag{15.184}$$

Table 15.7. Valid combinations for the bifurcation 1T2

Case	i	N^i	$\dim \Pi$	d
aaaa	1	-2 -1 -1 -1 -1 -1	1	5
accd	1	0 0 0 0 -1 0	2	4
	2	-2 -1 -1 -1 -1 -1		
abab	1	-1 0 0 0 0 0	2	4
	2	-2 -1 -1 -1 -1 -1		
daeb	1	-1 0 0 -1 0 0	2	4
	2	-2 -1 -1 -1 -1 -1		
dbeb	1	-1 0 0 0 0 0	3	3
	2	-1 0 0 -1 0 0		
	3	-2 -1 -1 -1 -1 -1		
caca	1	-1 0 -1 0 -1 0	2	4
	2	-2 -1 -1 -1 -1 -1		
cccd	1	0 0 0 0 -1 0	3	3
	2	-1 0 -1 0 -1 0		
	3	-2 -1 -1 -1 -1 -1		
cdac	1	0 0 -1 0 0 0	2	4
	2	-2 -1 -1 -1 -1 -1		
cdcc	1	0 0 -1 0 0 0	3	3
	2	-1 0 -1 0 -1 0		
	3	-2 -1 -1 -1 -1 -1		
cbcb	1	-1 0 0 0 0 0	3	3
	2	-1 0 -1 0 -1 0		
	3	-2 -1 -1 -1 -1 -1		
ebda	1	-2 -1 -1 -1 -1 -1	2	4
	2	-1 0 0 0 0 -1		
ebdb	1	-1 0 0 0 0 0	3	3
	2	-2 -1 -1 -1 -1 -1		
	3	-1 0 0 0 0 -1		
baba	1	0 -1 0 0 0 0	2	4
	2	-2 -1 -1 -1 -1 -1		
bcfd	1	0 -1 0 0 0 0	3	3
	2	0 0 0 0 -1 0		
	3	-2 -1 -1 -1 -1 -1		
bbbb	1	-1 0 0 0 0 0	3	3
	2	0 -1 0 0 0 0		
	3	-2 -1 -1 -1 -1 -1		
hajb	1	-1 0 0 -1 0 0	3	3
	2	-1 0 -1 0 -1 0		
	3	-2 -1 -1 -1 -1 -1		
hbjb	1	-1 0 0 0 0 0	4	2
	2	-1 0 0 -1 0 0		
	3	-1 0 -1 0 -1 0		
	4	-2 -1 -1 -1 -1 -1		
kaka	1	-2 -1 -1 -1 -1 -1	2	4
	2	-1 0 0 -1 0 -1		
kakb	1	-1 0 0 -1 0 0	3	3
	2	-2 -1 -1 -1 -1 -1		
	3	-1 0 0 -1 0 -1		

Case	i	N^i	$\dim \Pi$	d
kbka	1	-2 -1 -1 -1 -1 -1	3	3
	2	-1 0 0 0 0 -1		
	3	-1 0 0 -1 0 -1		
kbkb	1	-1 0 0 0 0 0	4	2
	2	-1 0 0 -1 0 0		
	3	-2 -1 -1 -1 -1 -1		
	4	-1 0 0 0 0 -1		
	5	-1 0 0 -1 0 -1		
jbha	1	-1 0 -1 0 -1 0	3	3
	2	-2 -1 -1 -1 -1 -1		
	3	-1 0 0 0 0 -1		
jbhb	1	-1 0 0 0 0 0	4	2
	2	-1 0 -1 0 -1 0		
	3	-2 -1 -1 -1 -1 -1		
	4	-1 0 0 0 0 -1		
gaib	1	0 -1 0 0 0 0	3	3
	2	-1 0 0 -1 0 0		
	3	-2 -1 -1 -1 -1 -1		
gbib	1	-1 0 0 0 0 0	4	2
	2	0 -1 0 0 0 0		
	3	-1 0 0 -1 0 0		
	4	-2 -1 -1 -1 -1 -1		
fafa	1	0 -1 0 0 0 0	3	3
	2	-1 0 -1 0 -1 0		
	3	-2 -1 -1 -1 -1 -1		
fcfd	1	0 -1 0 0 0 0	4	2
	2	0 0 0 0 -1 0		
	3	-1 0 -1 0 -1 0		
	4	-2 -1 -1 -1 -1 -1		
fdbc	1	0 -1 0 0 0 0	3	3
	2	0 0 -1 0 0 0		
	3	-2 -1 -1 -1 -1 -1		
fdfc	1	0 -1 0 0 0 0	4	2
	2	0 0 -1 0 0 0		
	3	-1 0 -1 0 -1 0		
	4	-2 -1 -1 -1 -1 -1		
fbfb	1	-1 0 0 0 0 0	4	2
	2	0 -1 0 0 0 0		
	3	-1 0 -1 0 -1 0		
	4	-2 -1 -1 -1 -1 -1		
ibga	1	0 -1 0 0 0 0	3	3
	2	-2 -1 -1 -1 -1 -1		
	3	-1 0 0 0 0 -1		
ibgb	1	-1 0 0 0 0 0	4	2
	2	0 -1 0 0 0 0		
	3	-2 -1 -1 -1 -1 -1		
	4	-1 0 0 0 0 -1		

We remark that the vector $(-2,-1,-1,-1,-1)$ is always present.

We do not pursue here the study of these 32 cases as it would take up too much space.

15.9 Conclusions

With the help of the Newton approach, we have thus recovered in a more rigorous way the results of the previous Chapters 12 to 14, at least in the simplest case `1P2`. In principle at least, bifurcations `1Pn` with larger values of n, and also bifurcations `1Tn`, could be treated in the same way.

However, the Newton approach has one fundamental shortcoming: it can be only applied for a specific value of n. Our objective is to obtain general results, valid for any n; this objective was indeed achieved in Chap. 12.

An additional practical problem with the Newton approach is that it proceeds by enumeration and study of all possible cases, one by one. Therefore, as has been seen, the amount of work grows exponentially with n, and very quickly becomes prohibitive. This is why we had to limit the use of this approach to the case `1P2`.

16. Proving General Results

We develop now a third approach, in which we retain part of the formalism of the Newton approach; but instead of considering the values of n and the valid combinations one by one, we try to prove general results, valid for any n.

16.1 Variables and Equations

We use again the variables x_j defined in (15.2) and (15.177). We recall that

$$x_1 = \mu \,, \qquad x_2 = \Delta C \,; \tag{16.1}$$

the variable Δa_i, corresponding to arc i, is represented by x_{2i+1}; and the variable y_i, corresponding to encounter i, is represented by x_{2i+2}.

We reproduce here for convenience the fundamental equations with which we will work, so as to make this chapter self-contained. We have first the encounter equation (15.26)

$$x_{2i+2}(x_{2i+3} - x_{2i+1}) + G_1 x_1 = 0 \,. \tag{16.2}$$

Next we have the arc equation in the general case (15.32)

$$x_{2i+2} - x_{2i} + G_2 s_i x_2 + G_3(1 + K s_i) x_{2i+1} = 0 \,. \tag{16.3}$$

For the initial basic arc in a partial bifurcation, with $i = 1$, the second term in (16.3) vanishes (see (15.38)). For the final arc, with $i = n$, the first term vanishes (see (15.44)).

Finally we have the additional relation in the general case (15.49)

$$-x_{2i} + x_{2i+4} + G_3(1 + K s_i) x_{2i+1} + G_3(1 - K s_i) x_{2i+3} = 0 \,. \tag{16.4}$$

For the first relation in a partial bifurcation, with $i = 1$, the first term in (16.4) vanishes (see (15.53)). For the last relation, with $i = n - 1$, the second term vanishes (see (15.58)).

If $n = 2$, the first and second terms vanish (see (15.63)).

16.2 Method of Solution

We again consider solutions of the form (15.9):

$$x_i = b_i \tau^{p_i}(1 + o(1)) \,, \qquad i = 1, \ldots, n_B \,, \tag{16.5}$$

and, guided by the results obtained in Chaps. 12 to 14, we try to derive general properties of the coefficients p_i and b_i directly from the equations. No reference is made here to Newton polyhedra: the present approach is algebraic rather than geometric. We will generally treat the cases of partial and total bifurcation together, although on occasion separate proofs are needed.

For the comparison with the results of Chap. 12, we recall the relation (15.14):

$$\nu = \frac{p_2}{p_1} \,. \tag{16.6}$$

Substituting (15.9) into (16.2), we find that the asymptotic values of the three terms are

$$b_{2i+2} b_{2i+3} \tau^{p_{2i+2}+p_{2i+3}} \,, \quad -b_{2i+2} b_{2i+1} \tau^{p_{2i+2}+p_{2i+1}} \,, \quad G_1 b_1 \tau^{p_1} \,. \tag{16.7}$$

We call c_P the maximum of the three exponents of τ (cf. Bruno 2.2). The *dominant terms* are those for which the exponent of τ equals c_P. *At least two terms must be dominant.* Deleting the non-dominant terms and dividing by τ^{c_P}, we obtain an equation for the b_i. Thus, depending on which terms are dominant, one of the following 4 cases must hold:

$$
\begin{aligned}
\text{a:} \quad & p_{2i+2} + p_{2i+1} = p_{2i+2} + p_{2i+3} = p_1 \\
& \qquad \text{and} \quad -b_{2i+2} b_{2i+1} + b_{2i+2} b_{2i+3} + G_1 b_1 = 0 \,; \\
\text{b:} \quad & p_{2i+2} + p_{2i+1} = p_{2i+2} + p_{2i+3} > p_1 \\
& \qquad \text{and} \quad -b_{2i+2} b_{2i+1} + b_{2i+2} b_{2i+3} = 0 \,; \\
\text{c:} \quad & p_{2i+2} + p_{2i+1} = p_1 > p_{2i+2} + p_{2i+3} \\
& \qquad \text{and} \quad -b_{2i+2} b_{2i+1} + G_1 b_1 = 0 \,; \\
\text{d:} \quad & p_{2i+2} + p_{2i+3} = 1 > p_{2i+2} + p_{2i+1} \\
& \qquad \text{and} \quad b_{2i+2} b_{2i+3} + G_1 b_1 = 0 \,.
\end{aligned} \tag{16.8}
$$

In each case, the first half corresponds to one of the normal cones (15.30), and the second half to the associated truncation (15.29).

A similar decomposition into cases takes place for the other equations (see Sect. 15.3).

16.3 Two General Propositions

We consider first an encounter i. We show first that

Proposition 16.3.1. *For any $i \in \mathcal{C}$,*

$$p_{2i+2} \leqq \max(p_2, p_1/2) . \tag{16.9}$$

Proof: we consider first the case of partial bifurcation. Then $0 < i < n$. Assume that (16.9) is violated at one or more places. Let i be the first such place, i.e. the smallest i such that

$$p_{2i+2} > \max(p_2, p_1/2) . \tag{16.10}$$

Consider the equation (16.3) for the arc preceding the encounter i. The terms 2 and 3 are negligible since by hypothesis $p_{2i+2} > p_{2i}$ and $p_{2i+2} > p_2$. Therefore the terms 1 and 4 must balance:

$$p_{2i+2} = p_{2i+1} , \qquad b_{2i+2} = -G_3(1 + Ks_i)b_{2i+1} . \tag{16.11}$$

In the particular case $i = 1$, the second term in (16.3) vanishes, and we obtain the same result.

We consider next the equation (16.2) for the encounter i. The term 3 is negligible since $p_{2i+2} + p_{2i+1} = 2p_{2i+2} > p_1$. Therefore the terms 1 and 2 must balance:

$$p_{2i+3} = p_{2i+1}, \qquad b_{2i+3} = b_{2i+1} . \tag{16.12}$$

We consider now (16.3) for $i + 1$:

$$x_{2i+4} - x_{2i+2} - G_2 s_i x_2 + G_3(1 - Ks_i)x_{2i+3} = 0 . \tag{16.13}$$

The terms 2 and 4 are of the same order; their sum is

$$[G_3(1 + Ks_i)b_{2i+1} + G_3(1 - Ks_i)b_{2i+1}]\tau^{p_{2i+1}} = 2G_3 b_{2i+1}\tau^{p_{2i+1}} . \tag{16.14}$$

This sum does not vanish. The term 3 is therefore negligible, and the sum (16.14) must be balanced by the term 1:

$$p_{2i+4} = p_{2i+1} , \qquad b_{2i+4} = -2G_3 b_{2i+1} . \tag{16.15}$$

We consider next (16.2) for $i + 1$:

$$x_{2i+4}(x_{2i+5} - x_{2i+3}) + G_1 x_1 = 0 . \tag{16.16}$$

The term 3 is negligible since $p_{2i+4} + p_{2i+3} = 2p_{2i+2} > p_1$, and we obtain

$$p_{2i+5} = p_{2i+3} = p_{2i+1} , \qquad b_{2i+5} = b_{2i+3} = b_{2i+1} . \tag{16.17}$$

Continuing in this way, we find that all successive exponents p_{2i+6}, p_{2i+7}, ..., are equal to p_{2i+1}, while the coefficients never vanish and are

$$b_{2i+2j+1} = b_{2i+1}, \quad j = 1, 2, \dots ,$$
$$b_{2i+2j+2} = \begin{cases} -G_3(j+1)b_{2i+1} & \text{for } j \text{ odd} = 1, 3, 5, \dots, \\ -G_3(j+1+Ks_i)b_{2i+1} & \text{for } j \text{ even} = 0, 2, 4, \dots . \end{cases} \tag{16.18}$$

However, sooner or later we reach the final arc. Then the first term in (16.3) vanishes, and the arc equation cannot be balanced. We have thus reached an impossibility.

We consider next the case of a total bifurcation. Assume that (16.9) is violated at one of more places. If the p_{2i+2} are not all equal, we can select an encounter i such that $p_{2i+2} > \max(p_2, p_1/2)$ and $p_{2i} < p_{2i+2}$, and by the same reasoning as above we arrive at an impossibility.

If all p_{2i+2} equal a common value p^* (with $p^* > \max(p_2, p_1/2)$), assume first that for some i there is $p_{2i+1} > p_1 - p^*$. Then the term 3 in (16.2) is negligible and we obtain $p_{2i+3} = p_{2i+1} > p_1 - p^*$, $b_{2i+3} = b_{2i+1}$. We can continue and we obtain $p_{2i+2j+1} = p_{2i+1}$, $b_{2i+2j+1} = b_{2i+1}$ for all j. If we sum the additional relation (16.4) over every other value of i, starting from an arbitrary value, the terms 1 and 2 are eliminated. The terms 3 and 4 are all of the same order and must balance:

$$G_3(1 + Ks_i)(b_{2i+1} + b_{2i+5} + \ldots) \\ +G_3(1 - Ks_i)(b_{2i+3} + b_{2i+7} + \ldots) = 0 , \tag{16.19}$$

which reduces to

$$G_3 n b_{2i+1} = 0 . \tag{16.20}$$

This contradicts the fundamental assumption (15.10).

Finally, we have the case where $p_{2i+1} \leqq p_1 - p^*$ for all i. In (16.3), there is $p_{2i+2} = p_{2i} = p^*$. Since $p^* > p_2$, the term 3 is negligible. Since $p^* > p_1/2$, we have $p_{2i+1} < p_1/2 < p^*$ and the term 4 is negligible. Therefore the terms 1 and 2 must balance. So all b_{2i+2} are equal to a common value b^*. We write

$$\gamma_{2i+1} = \lim_{\tau \to +\infty} \frac{x_{2i+1}}{\tau^{p_1 - p^*}} . \tag{16.21}$$

There is

$$\gamma_{2i+1} = \begin{cases} b_{2i+1} & \text{if } p_{2i+1} = p_1 - p^*, \\ 0 & \text{if } p_{2i+1} < p_1 - p^*. \end{cases} \tag{16.22}$$

Dividing (16.2) by τ^{p_1} and taking the limit $\tau \to +\infty$, we have then

$$b^*(\gamma_{2i+3} - \gamma_{2i+1}) + G_1 b_1 = 0 . \tag{16.23}$$

Summing over all encounters, we obtain

$$G_1 n b_1 = 0 , \tag{16.24}$$

which is impossible. ■

Next we have

Proposition 16.3.2. *For any $i \in \mathcal{A}$,*

$$p_{2i+1} \leqq \max(p_2, p_1/2) . \tag{16.25}$$

This follows immediately from (16.9): if a basic arc i existed for which $p_{2i+1} > \max(p_2, p_1/2)$, then the term 4 in (16.3) would be the only dominant term. ■

Propositions 16.3.1 and 16.3.2 suggest that the solutions of the system will depend critically on which of the two quantities p_2 and $p_1/2$ is largest. So we distinguish 3 cases.

16.4 The Case $p_2 = p_1/2$

As shown by (16.6), this corresponds to: $\nu = 1/2$.

Proposition 16.4.1. *If $p_2 \leqq p_1/2$, then $p_{2i+2} = p_1/2$ for any $i \in \mathcal{C}$.*

Proof: we have then $p_{2i+2} \leqq p_1/2, p_{2i+1} \leqq p_1/2$ from (16.9) and (16.25). If $p_{2i+2} < p_1/2$, then both terms 1 and 2 in (16.2) are negligible in comparison to term 3; this is impossible. ∎

In view of (16.25), we can write

$$\gamma_{2i+1} = \lim_{\tau \to +\infty} \frac{x_{2i+1}}{\tau^{p_1/2}} . \tag{16.26}$$

γ_{2i+1} is finite or zero. Dividing (16.2) by τ^{p_1} and taking the limit $\tau \to +\infty$, we have

$$b_{2i+2}(\gamma_{2i+3} - \gamma_{2i+1}) + G_1 = 0 . \tag{16.27}$$

Dividing (16.3) by $\tau^{p_1/2}$ and taking the limit $\tau \to +\infty$, we have

$$b_{2i+2} - b_{2i} + G_2 s_i b_2 + G_3(1 + K s_i)\gamma_{2i+1} = 0 . \tag{16.28}$$

(16.27) and (16.28) form a system of equations for the constants b_j and γ_j. We simplify this system by a change of variables similar to (12.112):

$$b_{2i+2} = s_i \sqrt{G_1 G_3}\, Y_i , \quad \gamma_{2i+1} = -s_i \sqrt{\frac{G_1}{G_3}}\, X_i , \quad b_2 = -\frac{1}{G_2}\sqrt{G_1 G_3}\, W . \tag{16.29}$$

The equations become

$$\begin{aligned} &Y_i(X_{i+1} + X_i) + 1 = 0 , \\ &Y_i + Y_{i-1} - W - (1 + K s_i)X_i = 0 . \end{aligned} \tag{16.30}$$

We have thus recovered the equations (12.114) which were obtained in Sect. 12.5 in the case $\nu = 1/2$.

16.5 The Case $p_2 < p_1/2$

This corresponds to $\nu > 1/2$. This case is similar to the previous one. We define again γ_{2i+1} by (16.26), and we obtain again (16.27). Instead of (16.28), we have now

$$b_{2i+2} - b_{2i} + G_3(1 + K s_i)\gamma_{2i+1} = 0 . \tag{16.31}$$

We apply again the change of variables (16.29) and we obtain the equations

$$\begin{aligned} &Y_i(X_{i+1} + X_i) + 1 = 0 , \\ &Y_i + Y_{i-1} - (1 + K s_i)X_i = 0 . \end{aligned} \tag{16.32}$$

We have thus recovered the equations (12.119) which were obtained in Sect. 12.6 in the case $\nu > 1/2$.

16.6 The Case $p_2 > p_1/2$

This corresponds to $\nu < 1/2$.

We know from (16.9) and (16.25) that $p_{2i+2} \leqq p_2$, $p_{2i+1} \leqq p_2$. We write

$$\beta_{2i+1} = \lim_{\tau \to +\infty} \frac{x_{2i+1}}{\tau^{p_2}}, \qquad \beta_{2i+2} = \lim_{\tau \to +\infty} \frac{x_{2i+2}}{\tau^{p_2}}. \tag{16.33}$$

There is

$$\beta_{2i+1} = \begin{cases} b_{2i+1} & \text{if } p_{2i+1} = p_2, \\ 0 & \text{if } p_{2i+1} < p_2, \end{cases} \tag{16.34}$$

and similarly for β_{2i+2}. Dividing (16.3) by τ^{p_2} and taking the limit $\tau \to +\infty$, we have

$$\beta_{2i+2} - \beta_{2i} + G_2 s_i b_2 + G_3(1 + K s_i)\beta_{2i+1} = 0. \tag{16.35}$$

Dividing (16.2) by τ^{2p_2} and taking the limit $\tau \to +\infty$, we have

$$\beta_{2i+2}(\beta_{2i+3} - \beta_{2i+1}) = 0\,. \tag{16.36}$$

We will call *node** an encounter i for which $p_{2i+2} < p_2$, and *antinode** an encounter for which $p_{2i+2} = p_2$. There is $\beta_{2i+2} = 0$ for a node*, $\beta_{2i+2} = b_{2i+2}$ for an antinode*. (It will be shown below that these definitions are natural generalizations for $\mu > 0$ of the definitions of nodes and antinodes introduced in Sect. 8.1 in the case of generating orbits. We use asterisks to distinguish them.)

We call also *arc** the part of the bifurcation orbit or bifurcating arc extending from one node* (or end of the bifurcating arc) to the next. (Again this is a natural generalization of the definition of an arc introduced in Sect. 4.1 in the case of generating orbits.)

16.6.1 No Arcs*

We consider first the special case where there are no arcs*. This can only happen in the case of a total bifurcation with no nodes*. Then

$$p_{2i+2} = p_2 \tag{16.37}$$

for all encounters.

Assume first that for some i there is $p_{2i+1} > p_1 - p_2$. Then the term 3 in (16.2) is negligible and we obtain $p_{2i+3} = p_{2i+1} > p_1 - p_2$, $b_{2i+3} = b_{2i+1}$. We can continue and we obtain $p_{2i+2j+1} = p_{2i+1}$, $b_{2i+2j+1} = b_{2i+1}$ for all j. If we sum the additional relation (16.4) over every other value of i, starting from an arbitrary value, the terms 1 and 2 are eliminated. The terms 3 and 4 are all of the same order and must balance:

$$\begin{aligned} &G_3(1 + K s_i)(b_{2i+1} + b_{2i+5} + \ldots) \\ &\quad + G_3(1 - K s_i)(b_{2i+3} + b_{2i+7} + \ldots) = 0\,, \end{aligned} \tag{16.38}$$

which reduces to

$$G_3 n b_{2i+1} = 0 \,. \tag{16.39}$$

This contradicts the fundamental assumption (15.10).

Therefore $p_{2i+1} \leqq p_1 - p_2$ for all i. Since $p_2 > p_1/2$, there is $p_{2i+1} < p_1/2 < p_2$ and the term 4 in (16.3) is negligible. The terms 1, 2, and 3 are of the same order. Therefore

$$b_{2i+2} - b_{2i} + G_2 s_i b_2 = 0 \,. \tag{16.40}$$

We can use this relation to compute the successive values of the b_{2i+2}, starting from $i = 2$. We obtain

$$b_{2i+2} = \begin{cases} b_4 & \text{for } i \text{ odd} = 3,\, 5,\, \ldots, \\ b_4 - G_2 b_2 & \text{for } i \text{ even} = 2,\, 4,\, \ldots . \end{cases} \tag{16.41}$$

We write

$$\gamma_{2i+1} = \lim_{\tau \to +\infty} \frac{x_{2i+1}}{\tau^{p_1 - p_2}} \,. \tag{16.42}$$

Dividing (16.2) by τ^{p_1} and taking the limit $\tau \to +\infty$, we have then

$$b_{2i+2}(\gamma_{2i+3} - \gamma_{2i+1}) + G_1 b_1 = 0 \,. \tag{16.43}$$

Dividing this by b_{2i+2} given by (16.41), and summing over all values of i, we obtain

$$\frac{G_1 b_1 n}{2b_4} + \frac{G_1 b_1 n}{2(b_4 - G_2 b_2)} = 0 \,, \tag{16.44}$$

from which

$$b_4 = \frac{G_2 b_2}{2} \tag{16.45}$$

and more generally, using (16.41) again:

$$b_{2i+2} = -(-1)^i \frac{G_2 b_2}{2} \,. \tag{16.46}$$

We can then compute the successive values of the γ_{2i+1} from (16.43), obtaining

$$\gamma_{2i+1} = \begin{cases} \gamma_3 & \text{for } i \text{ odd,} \\ \gamma_3 - \dfrac{2G_1 b_1}{G_2 b_2} & \text{for } i \text{ even.} \end{cases} \tag{16.47}$$

Summing once more the additional relation (16.4) over every other value of i, dividing by $\tau^{p_1 - p_2}$, taking the limit $\tau \to +\infty$, and using (16.47), we obtain

$$\frac{G_3(1+K)n}{2}\left(\gamma_3 - \frac{2G_1 b_1}{G_2 b_2}\right) + \frac{G_3(1-K)n}{2}\gamma_3 = 0 \,, \tag{16.48}$$

from which

$$\gamma_3 = (1+K)\frac{G_1 b_1}{G_2 b_2} \tag{16.49}$$

and more generally, using (16.47) again:

$$\gamma_{2i+1} = \left[K - (-1)^i\right] \frac{G_1 b_1}{G_2 b_2} . \tag{16.50}$$

These values are non-zero; therefore

$$p_{2i+1} = p_1 - p_2 \qquad \text{and} \qquad b_{2i+1} = \gamma_{2i+1} \tag{16.51}$$

for all i.

All asymptotic coefficients have been determined. It can be remarked that they repeat with a period 2. Therefore we have here a total bifurcation 1T2.

We go back to the physical variables. From (15.2), (15.9), (16.37), (16.46) we obtain

$$y_i = -(-1)^i \frac{G_2}{2} \Delta C . \tag{16.52}$$

From (15.2), (15.9), (16.51), (16.50) we obtain

$$\Delta a_i = \left[K - (-1)^i\right] \frac{G_1}{G_2} \mu \Delta C^{-1} . \tag{16.53}$$

We have thus recovered the main terms of the expressions (12.78) for a first species orbit in the case $0 < \nu < 1/2$.

16.6.2 Arcs*

We consider now the case where there are arcs*. The whole bifurcation orbit or bifurcating arc can them be decomposed into a sequence of arcs*, which we analyse individually. We consider an arc* made of m basic arcs, extending from encounter i to encounter $i+m$. All encounters inside the arc* are antinodes*; therefore

$$\beta_{2i+2j+2} = b_{2i+2j+2} \neq 0 , \qquad j = 1, \ldots, m-1 \tag{16.54}$$

where the β are defined by (16.33). Dividing (16.3) by τ^{p_2} and taking the limit, we obtain

$$\beta_{2i+2j+2} - \beta_{2i+2j} + G_2 s_{i+j} b_2 + G_3(1 + K s_{i+j})\beta_{2i+2j+1} = 0 , \qquad j = 1, \ldots, m . \tag{16.55}$$

From (16.36) we have

$$\beta_{2i+2j+3} - \beta_{2i+2j+1} = 0 , \qquad j = 1, \ldots, m-1 . \tag{16.56}$$

Finally, the two ends of the arc* are nodes*, and

$$\beta_{2i+2} = 0 , \qquad \beta_{2i+2m+2} = 0 . \tag{16.57}$$

In the case of a partial bifurcation, these equations can be made to cover also the case where one end of the arc* is also an end of the bifurcating arc (i.e. $i = 0$ or/and $i + m = n$) by defining

$$\beta_2 = 0\,, \quad \beta_{2n+2} = 0\,. \tag{16.58}$$

The equations (16.55), (16.56), and (16.57) form a system of $2m+1$ linear equations for the $2m+1$ variables $\beta_{2i+2}, \ldots, \beta_{2i+2m+2}$. This system is easily solved by elimination.

16.6.2.1 *S*-Arcs*. For m odd, we obtain

$$\beta_{2i+2j+1} = \frac{G_2 s_i b_2}{G_3(m - K s_i)}\,, \qquad j = 1, \ldots, m\,,$$

$$\beta_{2i+2j+2} = \begin{cases} \dfrac{-jG_2 s_i b_2}{m - K s_i} & \text{for } j \text{ even} = 2, 4, \ldots, m-1; \\ \dfrac{(m-j)G_2 s_i b_2}{m - K s_i} & \text{for } j \text{ odd} = 1, 3, 5, \ldots, m-2. \end{cases} \tag{16.59}$$

We call this a *S-arc**. All values in the right-hand sides are non-zero; therefore the asymptotic exponents and coefficients are determined:

$$\begin{aligned} p_{2i+2j+1} &= p_2\,, & b_{2i+2j+1} &= \beta_{2i+2j+1}\,, & j &= 1, \ldots, m\,, \\ p_{2i+2j+2} &= p_2\,, & b_{2i+2j+2} &= \beta_{2i+2j+2}\,, & j &= 1, \ldots, m-1\,. \end{aligned} \tag{16.60}$$

We go back to the physical variables. From (15.2), (15.9), (16.60), (16.59) we obtain

$$\Delta a_{i+j} = \frac{G_2 s_i}{G_3(m - K s_i)} \Delta C\,,$$

$$y_{i+j} = \begin{cases} \dfrac{-jG_2 s_i}{m - K s_i} \Delta C & \text{for } j \text{ even,} \\ \dfrac{(m-j)G_2 s_i}{m - K s_i} \Delta C & \text{for } j \text{ odd.} \end{cases} \tag{16.61}$$

We have thus recovered the main terms of the expressions (12.110) for a S-arc in the case $0 < \nu < 1/2$.

16.6.2.2 *T*-Arcs*. For $m = 2$, we obtain

$$\beta_{2i+3} = 0\,, \qquad \beta_{2i+5} = 0\,, \qquad \beta_{2i+4} = G_2 s_i b_2\,. \tag{16.62}$$

We call this a *T-arc**. We have

$$p_{2i+4} = p_2\,, \qquad b_{2i+4} = \beta_{2i+4} = G_2 s_i b_2\,. \tag{16.63}$$

The asymptotic coefficients of the two basic arcs are not determined at this point; we know only that

$$p_{2i+3} < p_2\,, \qquad p_{2i+5} < p_2\,. \tag{16.64}$$

A better characterization of the T-arcs* will be obtained in Sect. 16.6.2.4.

For m even and larger than 2, the resolution of the system gives $\beta_{2i+2j+2} = 0$ for $j = 2, 4, \ldots, m-2$; this contradicts (16.54). Therefore this case is not possible.

16.6.2.3 Nodes*. We consider now a node* i. If the previous arc* is a S-arc* made of m_b basic arcs, we have from (16.59a) with $j = m$

$$\beta_{2i+1} = -\frac{G_2 s_i b_2}{G_3(m_b + K s_i)} . \tag{16.65}$$

This value is non-zero; therefore $p_{2i+1} = p_2$ and $b_{2i+1} = \beta_{2i+1}$.

If the previous arc* is a T-arc, we have from (16.62)

$$p_{2i+1} < p_2 , \qquad \beta_{2i+1} = 0 . \tag{16.66}$$

Similarly, il the following arc* is a S-arc* made of m_a basic arcs, we have from (16.59a) with $j = 1$

$$\beta_{2i+3} = \frac{G_2 s_i b_2}{G_3(m_a - K s_i)} \tag{16.67}$$

from which $p_{2i+3} = p_2$ and $b_{2i+3} = \beta_{2i+3}$.

If the following arc* is a T-arc*, we have

$$p_{2i+3} < p_2 , \qquad \beta_{2i+3} = 0 . \tag{16.68}$$

Thus, for a SS node* (joining a S-arc* to a S-arc*), the terms 1 and 2 in (16.2) are of the same order and their sum has the asymptotic value

$$\frac{(m_b + m_a) G_2 s_i b_2 b_{2i+2} \tau^{p_2 + p_{2i+2}}}{G_3(m_b + K s_i)(m_a - K s_i)} . \tag{16.69}$$

This sum does not vanish. Therefore it must be balanced by the term 3, and we have

$$p_{2i+2} = p_1 - p_2 , \qquad b_{2i+2} = -\frac{G_1 G_3 b_1 (m_b + K s_i)(m_a - K s_i)}{(m_b + m_a) G_2 s_i b_2} . \tag{16.70}$$

For a ST node*, the term 1 in (16.2) is negligible compared to the term 2. The terms 2 and 3 must balance, and we have

$$p_{2i+2} = p_1 - p_2 , \qquad b_{2i+2} = -\frac{G_1 G_3 b_1 (m_b + K s_i)}{G_2 s_i b_2} . \tag{16.71}$$

For a TS node*, we obtain in the same way

$$p_{2i+2} = p_1 - p_2 , \qquad b_{2i+2} = -\frac{G_1 G_3 b_1 (m_a - K s_i)}{G_2 s_i b_2} . \tag{16.72}$$

We can go back to the physical variables, using (15.2), (15.9), (16.70) to (16.72). We recover the main terms of the expressions (12.108), (12.107), (12.106) for SS, ST, and TS nodes in the case $0 < \nu < 1/2$.

The last case would be a TT node*. It can be proved, however, that this case never arises:

Proposition 16.6.1. *There cannot be two T-arcs* in succession.*

The proof is somewhat lengthy and involved, and has been relegated to Sect. 16.8.

We remark that this proposition is similar to Proposition 4.3.2, which was proved in Sect. 12.2. There are differences, however. Proposition 4.3.2 was about ordinary generating orbits, and T-arcs defined by Definition 4.3.2. The present result is about periodic orbits in the vicinity of a generating orbit, and T-arcs* defined in Sect. 16.6.2.2. (It can be noticed also that the present proof is more complex.)

16.6.2.4 *T*-Arcs* Again. Finally we refine the estimates (16.64) for a T-arc*, running from node* i to node* $i+2$. We consider first the case where both ends of the T-arc* are junctions with a S-arc*. We have then from (16.71) and (16.72)

$$p_{2i+2} = p_1 - p_2 \,, \qquad p_{2i+6} = p_1 - p_2 \,. \tag{16.73}$$

Assume that $p_{2i+3} > p_1 - p_2$. Then the term 3 in (16.2) for $i+1$ is negligible, and the terms 1 and 2 must balance:

$$p_{2i+3} = p_{2i+5} \,, \qquad b_{2i+5} - b_{2i+3} = 0 \,. \tag{16.74}$$

The terms 1 and 2 in the additional relation (16.4) for $i+1$ are negligible as a consequence of our assumption, and the terms 3 and 4 must balance:

$$G_3(1 - Ks_i)b_{2i+3} + G_3(1 + Ks_i)b_{2i+5} = 0 \,. \tag{16.75}$$

But the two equations (16.74b) and (16.75) imply $b_{2i+3} = b_{2i+5} = 0$, which is impossible. Therefore

$$p_{2i+3} \leqq p_1 - p_2 \,. \tag{16.76}$$

In the same way it can be shown that

$$p_{2i+5} \leqq p_1 - p_2 \,. \tag{16.77}$$

We write

$$\gamma_{2i+3} = \lim_{\tau \to +\infty} \frac{x_{2i+3}}{\tau^{p_1 - p_2}} \,, \qquad \gamma_{2i+5} = \lim_{\tau \to +\infty} \frac{x_{2i+5}}{\tau^{p_1 - p_2}} \,. \tag{16.78}$$

Dividing (16.2) for $i+1$ by τ^{p_1}, letting $\tau \to +\infty$, and using (16.63), we obtain

$$G_2 s_i b_2 (\gamma_{2i+5} - \gamma_{2i+3}) + G_1 b_1 = 0 \,. \tag{16.79}$$

Dividing (16.4) for $i+1$ by $\tau^{p_1 - p_2}$, letting $\tau \to +\infty$, and using (16.71), (16.72), we obtain

$$-b_{2i+2} + b_{2i+6} + G_3(1 - Ks_i)\gamma_{2i+3} + G_3(1 + Ks_i)\gamma_{2i+5} = 0 \,. \tag{16.80}$$

We consider now the cases where one or both ends of the T-arc* coincide with ends of the bifurcating arc. Proceeding as above, we obtain an equation similar to (16.80), but with one or both of the first two terms absent. We define g' and g'' as in Sect. 12.4.2: $g' = 0$ if the encounter i is the beginning of a bifurcating arc; $g' = 1$ if it is a junction with a S-arc* made of m_b basic arcs; $g'' = 0$ if the encounter $i+2$ is the end of a bifurcating arc; $g'' = 1$ if it is a junction with a S-arc* made of m_a basic arcs.

Using (16.71) and (16.72), we can then generalize (16.80) into a single formula which covers all cases:

$$g'\frac{G_1 b_1(m_b + Ks_i)}{G_2 s_i b_2} - g''\frac{G_1 b_1(m_a - Ks_i)}{G_2 s_i b_2} \\ +(1 - Ks_i)\gamma_{2i+3} + (1 + Ks_i)\gamma_{2i+5} = 0\,. \tag{16.81}$$

Solving the two equations (16.79) and (16.81) for γ_{2i+3} and γ_{2i+5}, we obtain

$$\gamma_{2i+3} = \frac{G_1 b_1\left[(Ks_i + 1) - g'(m_b + Ks_i) + g''(m_a - Ks_i)\right]}{2G_2 s_i b_2}\,,$$
$$\gamma_{2i+5} = \frac{G_1 b_1\left[(Ks_i - 1) - g'(m_b + Ks_i) + g''(m_a - Ks_i)\right]}{2G_2 s_i b_2}\,. \tag{16.82}$$

In most cases, γ_{2i+3} and γ_{2i+5} are different from zero. Then the coefficients are

$$p_{2i+3} = p_{2i+5} = p_1 - p_2\,, \quad b_{2i+3} = \gamma_{2i+3}\,, \quad b_{2i+5} = \gamma_{2i+5}\,. \tag{16.83}$$

However, in the particular case $g' = 1$, $m_b = 1$, $g'' = 0$, γ_{2i+3} vanishes. In that case the coefficients p_{2i+3} and b_{2i+3} cannot be determined without further computation. Similarly, in the particular case $g' = 0$, $g'' = 1$, $m_a = 1$, γ_{2i+5} vanishes and the coefficients p_{2i+5} and b_{2i+5} are not known.

We can go back to the physical variables, using (15.2), (15.9), (16.63), (16.78), (16.82). We recover the main terms of the expressions (12.109) for a T-arc in the case $0 < \nu < 1/2$.

16.7 Conclusions

We have thus succeeded in rederiving all basic results for a bifurcation of type 1, previously obtained in Chap. 12, in a completely independent way, using only the fundamental equations. In particular, we have made no recourse to the qualitative analysis of bifurcations which was presented in Volume I; on the contrary, we have rederived here the existence and properties of nodes, antinodes, S-arcs, and T-arcs.

The present approach is strongly dependent on the particular form of the system of fundamental equations in our problem; it is an *ad hoc* method. It cannot be generalized to other problems in a straightforward manner. In other words, there is no guarantee that a similar method could be devised in another given problem.

16.8 Appendix: No TT Node*

We prove here Proposition 16.6.1. We recall that this takes place inside Section 16.6, i.e. we are considering the case

$$p_2 > \frac{p_1}{2} . \qquad (16.84)$$

Suppose that the bifurcation orbit or bifurcating arc contains a sequence of two or more T-arcs*. We call this a *T-sequence.*

16.8.1 Partial T-Sequence

We consider first the case of a *partial T-sequence,* made of l T-arcs*, extending from encounter i to encounter $i+2l$ (with $l \geqq 2$). Each T-arc* is made of two basic arcs, numbered $i+2j-1$ and $i+2j$.

The inside encounters of odd rank with respect to i, i.e. $i+1$, $i+3$, ..., $i+2l-1$, are antinodes*; from (16.63) we have

$$p_{2i+4j} = p_2 , \qquad b_{2i+4j} = G_2 s_i b_2 , \qquad j = 1, \ldots, l . \qquad (16.85)$$

The inside encounters of even rank, i.e. $i+2$, $i+4$, ..., $i+2l-2$, are nodes*; therefore

$$p_{2i+4j+2} < p_2, \qquad j = 1, \ldots, l-1 . \qquad (16.86)$$

For the arcs, we have from (16.64)

$$p_{2i+4j-1} < p_2 , \qquad p_{2i+4j+1} < p_2 , \qquad j = 1, \ldots, l . \qquad (16.87)$$

Finally, each end of the partial T-sequence is either a junction with a S-arc* or, in the case of a partial bifurcation, an end of the bifurcating arc. We will consider the case where both ends of the sequence are junctions with a S-arc*; the other cases can be treated in a similar way.

We have then from (16.71)and (16.72)

$$p_{2i+2} = p_1 - p_2 , \qquad p_{2i+4l+2} = p_1 - p_2 . \qquad (16.88)$$

We define

$$p^* = \max(p_{2i+3}, \ldots, p_{2i+4l+1}) . \qquad (16.89)$$

There is

$$p_2 > p^* > p_1 - p_2 . \qquad (16.90)$$

The first inequality results from (16.87); the second inequality holds because otherwise there would be $p_{2i+4j-1} \leqq p_1 - p_2$, $p_{2i+4j+1} \leqq p_1 - p_2$ for all j, and the terms 1 and 2 would be negligible in (16.2) for the inside encounters of even rank. We have also

$$p_{2i+4j+2} \geqq p_1 - p^*, \qquad j = 1, \ldots, l-1 , \qquad (16.91)$$

again because otherwise the terms 1 and 2 would be negligible in (16.2) for the inside encounters of even rank.

We prove next: *if* $p_{2i+4j-1} = p^*$, *then* $p_{2i+4j+1} = p^*$ *also; and conversely. Moreover,*

$$b_{2i+4j+1} - b_{2i+4j-1} = 0 . \tag{16.92}$$

The proof is immediate: from (16.85) and (16.90) we find that the term 3 in (16.2) for encounter $i+2j-1$ is negligible; therefore the terms 1 and 2 must balance.

The T-arcs* for which $p_{2i+4j-1} = p_{2i+4j+1} = p^*$ form one or more subsequences inside the T-sequence. We consider one such subsequence, extending from encounter $i+2j'$ to encounter $i+2j''$.

We consider the initial encounter $i+2j'$. If $j'=0$, it is the beginning of the T-sequence, in which case $p_{2i+4j'+2} = p_1 - p_2$ from (16.88). If $j'>0$, there is a preceding T-arc*, with $p_{2i+4j'+1} < p^*$; applying (16.2) at $i+2j'$, we find that the term 2 is negligible, and therefore $p_{2i+4j'+2} = p_1 - p^*$ and

$$b_{2i+4j'+2} b_{2i+4j'+3} + G_1 b_1 = 0 . \tag{16.93}$$

In view of (16.90), we have therefore in all cases

$$p_{2i+4j'+2} \leqq p_1 - p^* . \tag{16.94}$$

Similarly, the final encounter $i+2j''$ is the end of the T-sequence if $j''=l$, or there is a following T-arc* if $j''<l$. In the second case, we have $p_{2i+4j''+2} = p_1 - p^*$ and

$$-b_{2i+4j''+2} b_{2i+4j''+1} + G_1 b_1 = 0 \tag{16.95}$$

and in both cases we have

$$p_{2i+4j''+2} \leqq p_1 - p^* . \tag{16.96}$$

We consider three separate cases for p^*.

16.8.1.1 $p^* > p_1/2$. We have from (16.94)

$$p_{2i+4j'+2} < p^* . \tag{16.97}$$

From (16.4) applied to the first T-arc*, we have then

$$p_{2i+4j'+6} \leqq p^* \tag{16.98}$$

because otherwise the term 2 would be the single dominant term. Applying again (16.4) to the following T-arcs*, we obtain generally

$$p_{2i+4j+2} \leqq p^* , \qquad j = j', \ldots, j'' . \tag{16.99}$$

We write

$$\beta_{2i+4j+2} = \lim_{T \to +\infty} \frac{x_{2i+4j+2}}{T^{p^*}} , \qquad j = j', \ldots, j'' . \tag{16.100}$$

From (16.94) and (16.96) we have

$$\beta_{2i+4j'+2} = 0\,, \qquad \beta_{2i+4j''+2} = 0\,. \tag{16.101}$$

Dividing (16.2) by τ^{2p^*} and letting $\tau \to +\infty$, we obtain

$$\beta_{2i+4j+2}(b_{2i+4j+3} - b_{2i+4j+1}) = 0\,, \qquad j = j'+1, \ldots, j''-1\,. \tag{16.102}$$

The term 3 has vanished. Dividing (16.4) for $i + 2j - 1$ by τ^{p^*} and letting $\tau \to +\infty$, we obtain

$$\begin{aligned} &-\beta_{2i+4j-2} + \beta_{2i+4j+2} + G_3(1 - Ks_i)b_{2i+4j-1} \\ &\qquad +G_3(1 + Ks_i)b_{2i+4j+1} = 0\,, \qquad j = j'+1, \ldots, j''\,. \end{aligned} \tag{16.103}$$

Using (16.92), we can rewrite this as

$$-\beta_{2i+4j-2} + \beta_{2i+4j+2} + 2G_3 b_{2i+4j+1} = 0\,, \qquad j = j'+1, \ldots, j''\,. \tag{16.104}$$

We multiply now this equation by $b_{2i+4j+1}$, we sum it for $j = j'+1, \ldots, j''$, we add also (16.102) for $j = j'+1, \ldots, j''-1$, and we use (16.92) and (16.101). We obtain

$$2G_3 \sum_{j=j'+1}^{j''} b^2_{2i+4j+1} = 0\,. \tag{16.105}$$

But the $b_{2i+4j+1}$ are non-zero according to (15.10), and there is at least one term in the sum. We have thus reached an impossibility.

16.8.1.2 $p^* = p_1/2$. We have from (16.94)

$$p_{2i+4j'+2} \leqq p_1/2\,. \tag{16.106}$$

From (16.4) applied successively to the T-arcs*, we have then

$$p_{2i+4j+2} \leqq p_1/2\,, \qquad j = j', \ldots, j''\,. \tag{16.107}$$

We write now

$$\beta_{2i+4j+2} = \lim_{\tau \to +\infty} \frac{x_{2i+4j+2}}{\tau^{p_1/2}}\,, \qquad j = j', \ldots, j''\,. \tag{16.108}$$

We consider the initial encounter $i + 2j'$. If $j' = 0$, it is the beginning of the T-sequence. From (16.88) and (16.84, we have $p_{2i+2} < p_1/2$, and therefore $\beta_{2i+2} = 0$. If $j' > 0$, there is a preceding T-arc*, with $p_{2i+4j'+1} < p^*$; applying (16.2) at $i + 2j'$, we find that the term 2 is negligible, and therefore $p_{2i+4j'+2} = p_1/2$, $\beta_{2i+4j'+2} = b_{2i+4j'+2}$, and

$$\beta_{2i+4j'+2} b_{2i+4j'+3} + G_1 b_1 = 0\,. \tag{16.109}$$

We define

$$g' = H(j')\,, \tag{16.110}$$

i.e. $g' = 0$ if $j' = 0$, $g' = 1$ if $j' > 0$. Both cases can then be written as a single equation:

$$\beta_{2i+4j'+2}b_{2i+4j'+3} + g'G_1b_1 = 0\,. \tag{16.111}$$

Similarly, we consider the final encounter $i+2j''$ and we define $g'' = H(l-j'')$. We obtain

$$-\beta_{2i+4j''+2}b_{2i+4j''+1} + g''G_1b_1 = 0\,. \tag{16.112}$$

We consider now an inside encounter. Dividing (16.2) for $i+2j$ by τ^{p_1} and letting $\tau \to +\infty$, we obtain

$$\beta_{2i+4j+2}(b_{2i+4j+3} - b_{2i+4j+1}) + G_1b_1 = 0\,, \qquad j = j'+1,\ldots,j''-1\,. \tag{16.113}$$

Dividing (16.4) for $i+2j-1$ by $\tau^{p_1/2}$ and letting $\tau \to +\infty$, we obtain

$$\begin{aligned} &-\beta_{2i+4j-2} + \beta_{2i+4j+2} + G_3(1-Ks_i)b_{2i+4j-1} \\ &\quad +G_3(1+Ks_i)b_{2i+4j+1} = 0\,, \qquad j = j'+1,\ldots,j''\,. \end{aligned} \tag{16.114}$$

This equation is identical with (16.103). Using (16.92) we can rewrite this as

$$-\beta_{2i+4j-2} + \beta_{2i+4j+2} + 2G_3b_{2i+4j+1} = 0\,, \qquad j = j'+1,\ldots,j''\,. \tag{16.115}$$

This is identical with (16.104).

We multiply now this equation by $b_{2i+4j+1}$, we sum it for $j = j'+1,\ldots,j''$, we add also (16.113) for $j = j'+1,\ldots,j''-1$, and we use (16.92), (16.111), and (16.112). We obtain

$$2G_3 \sum_{j=j'+1}^{j''} b_{2i+4j+1}^2 + G_1b_1[g' + g'' + (j''-j'-1)] = 0\,. \tag{16.116}$$

The $b_{2i+4j+1}$ are non-zero, and there is at least one term in the sum; therefore the Σ is positive. Also there is $g' \geqq 0$, $g'' \geqq 0$, $j''-j'-1 \geqq 0$. We have thus reached an impossibility.

16.8.1.3 $p^* < p_1/2$. We have from (16.94)

$$p_{2i+4j'+2} \leqq p_1 - p^*\,. \tag{16.117}$$

Consider (16.4) applied to the first T-arc*. Assume that $p_{2i+4j'+6} > p_1 - p^*$. Then the first term is negligible compared to the second. Also we have $p_{2i+4j'+6} > p_1/2 > p^*$; therefore the terms 3 and 4 are also negligible. This is impossible; therefore

$$p_{2i+4j'+6} \leqq p_1 - p^*\,. \tag{16.118}$$

Applying again (16.4) to the following T-arcs*, we obtain generally

$$p_{2i+4j+2} \leqq p_1 - p^*\,, \qquad j = j',\ldots,j''\,. \tag{16.119}$$

We write

$$\beta_{2i+4j+2} = \lim_{\tau\to+\infty} \frac{x_{2i+4j+2}}{\tau^{p_1-p^*}}\,, \qquad j = j',\ldots,j''\,. \tag{16.120}$$

As in the previous case, we obtain (16.111) and (16.112). Dividing (16.2) by τ^{p_1} and letting $\tau \to +\infty$, we obtain

$$\beta_{2i+4j+2}(b_{2i+4j+3} - b_{2i+4j+1}) + G_1 b_1 = 0 \,, \qquad j = j' + 1, \ldots, j'' - 1 \,. \tag{16.121}$$

This is identical with (16.113). Dividing (16.4) by $\tau^{p_1 - p^*}$ and letting $\tau \to +\infty$, we obtain

$$-\beta_{2i+4j-2} + \beta_{2i+4j+2} = 0 \,, \qquad j = j' + 1, \ldots, j'' \,. \tag{16.122}$$

We multiply now this equation by $b_{2i+4j+1}$, we sum it for $j = j' + 1, \ldots, j''$, we add also (16.121) for $j = j' + 1, \ldots, j'' - 1$, and we use (16.92), (16.111), and (16.112). We obtain

$$G_1 b_1 [g' + g'' + (j'' - j' - 1)] = 0 \,. \tag{16.123}$$

There is $g' \geqq 0$, $g'' \geqq 0$, $j'' - j' - 1 \geqq 0$. Thus the only way to satisfy (16.123) is by taking $g' = 0$, $g'' = 0$, $j'' - j' = 1$. But this means that the subsequence contains only one T-arc*, and that its ends coincide with the ends of the partial T-sequence. Therefore the partial T-sequence itself contains only one T-arc*.

16.8.2 Total *T*-Sequence

In the case of a total bifurcation, we consider now the case of a *total T-sequence*, making up the whole bifurcation orbit. The proof follows the same lines as in the case of a partial T-sequence. We consider one period of the bifurcation orbit, from encounter i to encounter $i + n$, with i the beginning of a T-arc*. We define p^* as in (16.89). There is $p_{2i+4j-1} = p_{2i+4j+1} = p^*$ either for all T-arcs* or for a non-empty subset of them. In the second case, they form one or more subsequences. We consider one such subsequence. The proof proceeds then as above, with the simplification that there is always a preceding T-arc* and a following T-arc*, and $g' = 1$, $g'' = 1$.

In the first case (i.e. $p_{2i+4j-1} = p_{2i+4j+1} = p^*$ for all T-arcs*), we consider again the three separate cases for p^*.

16.8.2.1 $p^* > p_1/2$. Assume that there is $p_{2i+4j+2} > p^*$ for some j. Applying (16.4) to successive T-arcs*, we find that $p_{2i+4j+2} > p^*$ for all j. The term 3 in (16.2) is then negligible for encounters of even rank, and the terms 1 and 2 must balance:

$$b_{2i+4j+3} - b_{2i+4j+1} = 0 \qquad \text{for all } j \,. \tag{16.124}$$

Using also (16.92), we obtain

$$b_{2i+4j+1} = b_{2i+1} \,, \qquad b_{2i+4j+3} = b_{2i+1} \qquad \text{for all } j \,. \tag{16.125}$$

Summing (16.4) over every other value of i, we obtain the equation (16.19), and $G_3 n b_{2i+1} = 0$, which is impossible. Therefore

$$p_{2i+4j+2} \leqq p^* \qquad \text{for all } j \, . \tag{16.126}$$

We define then $\beta_{2i+4j+2}$ as in (16.100) and we continue as in Sect. 16.8.1.1; we obtain

$$2G_3 \sum_{j=1}^{n/2} b_{2i+4j+1}^2 = 0 \, , \tag{16.127}$$

which is impossible.

16.8.2.2 $p^* = p_1/2$. Reasoning as in the previous case, we can show that

$$p_{2i+4j+2} \leqq p_1/2 \qquad \text{for all } j \, . \tag{16.128}$$

We define then $\beta_{2i+4j+2}$ as in (16.108) and we continue as in Sect. 16.8.1.2; we obtain

$$2G_3 \sum_{1}^{n/2} b_{2i+4j+1}^2 + G_1 b_1 \frac{n}{2} = 0 \, , \tag{16.129}$$

which is impossible.

16.8.2.3 $p^* < p_1/2$. Reasoning as in the previous case, we can show that

$$p_{2i+4j+2} \leqq p_1 - p^* \qquad \text{for all } j \, . \tag{16.130}$$

We define then $\beta_{2i+4j+2}$ as in (16.120) and we continue as in Sect. 16.8.1.3; we obtain

$$G_1 b_1 \frac{n}{2} = 0 \, , \tag{16.131}$$

which is impossible.

This completes the proof of Proposition 16.6.1. ■

17. Quantitative Study of Type 2

We begin now the quantitative analysis of bifurcations of type 2, to which the remainder of the present volume will be devoted.

17.1 New Notations T^f, T^g

We introduce a new notation for the T-arcs, which will simplify the presentation for type 2.

Definition 17.1.1. *A T-arc will be designated by T^f if it is a T^i arc and $a > 1$, or if it is a T^e arc and $a < 1$.*

A T-arc will be designated by T^g if it is a T^e arc and $a > 1$, or if it is a T^i arc and $a < 1$.

We define symbols **f** and **g** for the arcs T^f and T^g, in the same way that we defined symbols **i** and **e** for the arcs T^i and T^e in Sect. 6.2.1.3.

A geometrical interpretation can be given. The unit circle (the orbit of M_2) divides the (X, Y) plane in two regions, and the bifurcation orbit lies entirely in one of these two regions (see Fig. 4.1, type 2). Then *a T^f arc is one which moves out of the region occupied by the bifurcation orbit, while a T^g arc is one which moves into the region occupied by the bifurcation orbit.*

By analogy with the side of passage σ, defined by (8.1), we define also the *relative side of passage* as

$$\sigma' = \text{sign}(a-1)\sigma = \text{sign}(a-1)\text{sign}(x_0-1) . \qquad (17.1)$$

σ' is positive if the side of passage lies in the same region as the bifurcation orbit, negative if it lies in the other region.

With these new notations, it is possible to simplify somewhat the results of Chap. 8 for bifurcations of type 2 and to make them independent of A.

1. For a node, instead of (8.43) and Table 8.2, we have: if $\text{sign}(\epsilon'\Delta C) = +1$, only S-arcs are present, and

$$\sigma' = -1 . \qquad (17.2)$$

If $\text{sign}(\epsilon'\Delta C) = -1$, the value of σ' is given by Table 17.1.

2. For an antinode, instead of (8.75) and (8.77), we have: for a S-arc:

Table 17.1. Value of σ' for a type 2 node, for $\text{sign}(\epsilon'\Delta C) = -1$.

		second arc		
		f	**g**	1, 2, ...
first arc	**f**		−	−
	g	+		+
	1, 2, ...	+	−	+

$$\sigma' = -1\,, \tag{17.3}$$

and for a first species orbit E:

$$\sigma' = \text{sign}(\epsilon'\Delta C)\,. \tag{17.4}$$

3. In Sect. 8.4.2, it was noted that in a bifurcation of type 2, all sides of passage change their sign σ if the sign of $a-1$ is changed and **i** and **e** are exchanged. For that reason, only the case $a-1>0$ was listed in Tables 8.12 and 8.18. We can now transform these tables into universal tables, valid for all values of a, as follows: (i) we substitute the symbols **f** and **g** for **i** and **e** respectively; (ii) we make the convention that the signs in the heading are now the relative sides of passage.

17.2 Fundamental Equations

We will follow again the general analysis presented in Chap. 11. We consider a bifurcation orbit Ω of type 2, and we number consecutively the collisions and the basic arcs implied in the bifurcation as explained in Sect. 11.3.1. We consider also neighbouring periodic orbits Ω_μ of the restricted problem for μ small but not zero.

The time t_i, introduced in a general way in Sect. 11.3.1, was precisely defined in the case of a bifurcation of type 1 as the time of intersection of the true orbit with the unit circle (the orbit of M_2) (see Sect. 12.1.1). For type 2, another definition turns out to be preferable: we define t_i as the time of *conjunction*, i.e. the time when M_3 crosses the x axis at encounter i. The velocity of M_3 in moving axes at a collision is non-zero (see (11.22)) and is parallel to the y axis; this guarantees that the neighbouring orbit Ω_μ has a well-defined intersection with the x axis in the vicinity of M_2.

We define h_i as the oriented distance from M_2 to M_3 at the conjunction; in other words the abscissa of the intersection point is

$$x = 1 + h_i\,. \tag{17.5}$$

There is

$$\text{sign}(h_i) = \sigma_i \tag{17.6}$$

where σ_i is the side of passage, defined by (8.1).

We consider now the intermediate orbit i (see Sect. 11.3.2). It intersects the x axis at a time t''_i. We define the abscissa of the intersection as $r''_i = r_i(t''_i)$. We define also the radial velocity as $u''_i = \dot{r}_i(t''_i)$. From Proposition 11.3.1 we obtain

$$t''_i = t_i + O(\mu) \; ; \qquad r''_i = 1 + h_i + O(\mu) \, . \tag{17.7}$$

Similarly the intermediate orbit $i+1$ has a conjunction in the vicinity of t_i, at a time t'_{i+1}. The abscissa and the radial velocity are r'_{i+1}, u'_{i+1}, and there is

$$t'_{i+1} = t_i + O(\mu) \; ; \qquad r'_{i+1} = 1 + h_i + O(\mu) \, . \tag{17.8}$$

17.2.1 Encounter Relations

We consider now the matching relation (11.79), which we project on the x axis:

$$d(u'_{i+1} - u''_i) = -\frac{2\mu\cos\varphi}{v} + O(\mu d) + O[\mu(u'_{i+1} - u''_i)] \tag{17.9}$$

where φ is the angle between j_b and the x axis. The quantities v, d, φ are taken for instance on the intermediate orbit i, at the time t''_i. But there is

$$d = \rho''_i \cos\varphi = h_i \cos\varphi + O(\mu) \, . \tag{17.10}$$

Therefore

$$h_i(u'_{i+1} - u''_i) = -\frac{2\mu}{v} + O(\mu h_i) + O[\mu(u'_{i+1} - u''_i)] \, , \qquad i \in \mathcal{C} \, . \tag{17.11}$$

For a partial bifurcation, the equations (11.71) give

$$h_0 = O(\mu) \, , \qquad h_n = O(\mu) \, . \tag{17.12}$$

17.2.2 Arc Relations

We must now obtain relations between r'_i, u'_i, r''_i, u''_i, expressing the fact that these quantities correspond to the same keplerian orbit: the intermediate orbit i.

For $\mu = 0$, the intermediate orbit becomes identical with the bifurcation orbit between times t_{i-1} and t_i. In fixed axes with origin in M_1 and an appropriate orientation, and also an appropriate origin of time, the equations of motion are given in Hénon (1968, Equ. (26)):

$$\begin{aligned} X &= \epsilon a(\epsilon'' \cos E - e) \, , \\ Y &= \epsilon\epsilon'\epsilon'' a\sqrt{1-e^2} \sin E \, , \\ t &= a^{3/2}(E - \epsilon'' e \sin E) \, , \end{aligned} \tag{17.13}$$

from which we deduce

$$r = a(1 - \epsilon'' e \cos E) , \qquad u = \dot{r} = \frac{\epsilon'' e \sin E}{\sqrt{a}(1 - \epsilon'' e \cos E)} . \tag{17.14}$$

a, e are the elements of the bifurcation ellipse. The Jacobi constant is given by (3.11):

$$C = 2\epsilon' \sqrt{a(1 - e^2} + \frac{1}{a} . \tag{17.15}$$

The motion of M_2 is given by

$$X = \cos t , \qquad Y = \sin t . \tag{17.16}$$

The collisions $i - 1$ and i take place for $t = \pm\tau$, $E = \pm\eta$; therefore we have the relations

$$\begin{aligned} \cos\tau &= \epsilon a(\epsilon'' \cos\eta - e) , \\ \sin\tau &= \epsilon\epsilon'\epsilon'' a\sqrt{1 - e^2} \sin\eta , \\ \tau &= a^{3/2}(\eta - \epsilon'' e \sin\eta) , \\ 1 &= a(1 - \epsilon'' e \cos\eta) . \end{aligned} \tag{17.17}$$

For $\mu \neq 0$, we still use a system of fixed axes with origin in M_1, such that the X axis passes through the pericenter and the apocenter of the intermediate orbit. We take the origin of time when M_2 passes through the point $(1, 0)$. The motion of M_2 is then still given by (17.16). The motion on the intermediate orbit i is given by

$$\begin{aligned} X &= \epsilon a_i(\epsilon'' \cos E - e_i) , \\ Y &= \epsilon\epsilon'\epsilon'' a_i\sqrt{1 - e_i^2} \sin E , \\ t - t_i^* &= a_i^{3/2}(E - \epsilon'' e_i \sin E) . \end{aligned} \tag{17.18}$$

Note the appearance of t_i^*, defined as the time of passage at pericenter (or apocenter), now generally non-zero.

Collisions are now replaced by encounters. A conjunction happens at time $t_i' \simeq -\tau$, for $E = E_i'$; we have then

$$\begin{aligned} r_i' \cos t_i' &= \epsilon a_i(\epsilon'' \cos E_i' - e_i) , \\ r_i' \sin t_i' &= \epsilon\epsilon'\epsilon'' a_i\sqrt{1 - e_i^2} \sin E_i' , \\ t_i' - t_i^* &= a_i^{3/2}(E_i' - \epsilon'' e_i \sin E_i') , \end{aligned} \tag{17.19}$$

from which we deduce

$$\begin{aligned} r_i' &= a_i(1 - \epsilon'' e_i \cos E_i') , \\ u_i' &= \frac{\epsilon'' e_i \sin E_i'}{\sqrt{a_i}(1 - \epsilon'' e_i \cos E_i')} . \end{aligned} \tag{17.20}$$

We have similar equations for the conjunction at time $t_i'' \simeq +\tau$, for $E = E_i''$:

$$
\begin{aligned}
r_i'' \cos t_i'' &= \epsilon a_i(\epsilon'' \cos E_i'' - e_i) , \\
r_i'' \sin t_i'' &= \epsilon\epsilon'\epsilon'' a_i \sqrt{1 - e_i^2} \sin E_i'' , \\
t_i'' - t_i^* &= a_i^{3/2}(E_i'' - \epsilon'' e_i \sin E_i'') ,
\end{aligned} \tag{17.21}
$$

and

$$
\begin{aligned}
r_i'' &= a_i(1 - \epsilon'' e_i \cos E_i'') , \\
u_i'' &= \frac{\epsilon'' e_i \sin E_i''}{\sqrt{a_i}(1 - \epsilon'' e_i \cos E_i'')} .
\end{aligned} \tag{17.22}
$$

We expand now these equations in terms of the small displacements from the bifurcation orbit.

17.2.2.1 Bifurcation Orbit. We recall first some basic equations for the case $\mu = 0$, $\Delta C = 0$, i.e. the bifurcation orbit of type 2 (see Sects. 4.4 and 6.2.1.3). The bifurcation ellipse is characterized by two mutually prime integers I and J, not both equal to 1, and a direction of motion $\epsilon' = \pm 1$ (Sects. 4.4, 6.2.1.3). Its semi-major axis and eccentricity are

$$
a = \left(\frac{I}{J}\right)^{2/3} , \qquad e = \frac{|a-1|}{a} . \tag{17.23}
$$

In fixed axes, for a collision in $X = 1$, $Y = 0$, the components of the velocity are

$$
V_X = 0 , \qquad V_Y = \epsilon' \sqrt{a(1-e^2)} = \epsilon' \sqrt{\frac{2a-1}{a}} . \tag{17.24}
$$

In rotating axes, the components are

$$
v_x = 0 , \qquad v_y = V_Y - 1 = \epsilon' \sqrt{\frac{2a-1}{a}} - 1 . \tag{17.25}
$$

The modulus of the relative velocity is therefore

$$
v = \left| \epsilon' \sqrt{\frac{2a-1}{a}} - 1 \right| . \tag{17.26}
$$

The Jacobi constant is

$$
C = 2\epsilon' \sqrt{\frac{2a-1}{a}} + \frac{1}{a} . \tag{17.27}
$$

The bifurcation orbit or the bifurcating arc consist of a sequence of n identical basic arcs. Each basic arc corresponds to a time interval $2\pi I$ and to J revolutions of M_3 on its supporting ellipse. Therefore, for a given basic arc, we can choose the origin of time and the origin of the eccentric anomaly in such a way that the two end collisions take place at times

$$t'_{i0} = -\pi I \ , \qquad t''_{i0} = \pi I \ , \tag{17.28}$$

and correspond to values of the eccentric anomaly

$$E'_{i0} = -\pi J \ , \qquad E''_{i0} = \pi J \ . \tag{17.29}$$

On the other hand, all variables h_i, u'_i, u''_i, t^*_i vanish, and $r'_i = r''_i = 1$. Finally, from the definitions of ϵ and ϵ'' in Hénon (1968) it is easily shown that

$$\epsilon = (-1)^I \text{sign}(a-1) \ , \qquad \epsilon'' = (-1)^J \text{sign}(a-1) \ . \tag{17.30}$$

17.2.2.2 Expansions. We consider now the intermediate orbit i. We write

$$\begin{aligned} &a_i = a + \Delta a_i \ , \qquad e_i = e + \Delta e_i \ , \qquad t'_i = -\pi I + \Delta t'_i \ , \\ &t''_i = +\pi I + \Delta t''_i \ , \qquad E'_i = -\pi J + \Delta E'_i \ , \qquad E''_i = +\pi J + \Delta E''_i \ . \end{aligned} \tag{17.31}$$

We call ΔC the displacement of the Jacobi constant for the true orbit with respect to the bifurcation orbit, and ΔC_i the displacement for the intermediate orbit i. We have (see (12.15))

$$\Delta C_i = \Delta C + O(\mu) \ . \tag{17.32}$$

The Δ quantities are small. We recall that the quantities h_i, u'_i, u''_i, t^*_i are also small.

We simplify now the equations by expanding with respect to the small quantities, and eliminating some variables. We obtain first from (17.15) and (17.32)

$$\Delta C = \left(\frac{V_Y}{a} - \frac{1}{a^2} \right) \Delta a_i - \frac{2|a-1|}{V_Y} \Delta e_i + O(\Delta a_i^2, \Delta e_i^2, \mu) \ . \tag{17.33}$$

We can also express e_i as a function of C and a_i; we obtain then

$$\Delta e_i = -\frac{V_Y}{2|a-1|} \Delta C + \frac{V_Y}{2|a-1|} \left(\frac{V_Y}{a} - \frac{1}{a^2} \right) \Delta a_i + O(\Delta C^2, \Delta a_i^2, \mu) \ . \tag{17.34}$$

This relation will be used later to eliminate Δe_i.

Subtracting (17.19c) from (17.21c), we eliminate t^*_i:

$$t''_i - t'_i = a_i^{3/2} [E''_i - E'_i - \epsilon'' e_i (\sin E''_i - \sin E'_i)] \ . \tag{17.35}$$

This becomes upon substitution of (17.31):

$$\begin{aligned} 2\pi I + \Delta t''_i - \Delta t'_i = (a + \Delta a_i)^{3/2} \\ [2\pi J + \Delta E''_i - \Delta E'_i - \epsilon''(e + \Delta e_i)(-1)^J (\sin \Delta E''_i - \sin \Delta E'_i)] \ . \end{aligned} \tag{17.36}$$

The last parenthesis can be written

$$\sin \Delta E''_i - \sin \Delta E'_i = (\Delta E''_i - \Delta E'_i)[1 + O(\Delta E''^2_i, \Delta E'^2_i)] \ . \tag{17.37}$$

Using (17.30b) and (17.23), we obtain

$$\Delta t_i'' - \Delta t_i' = 3\pi J\sqrt{a}\Delta a_i$$
$$+\sqrt{a}(\Delta E_i'' - \Delta E_i')[1 + O(\Delta a_i, \Delta e_i, \Delta E_i''^2, \Delta E_i'^2)] + O(\Delta a_i^2) \; . \tag{17.38}$$

Using (17.20a), we can write (17.19b) as

$$\sin t_i' = \epsilon\epsilon'\epsilon''\sqrt{1-e_i^2}\frac{\sin E_i'}{1-\epsilon'' e_i \cos E_i'} \tag{17.39}$$

which can be expanded, using (17.30), (17.24), (17.31), into

$$\Delta t_i'[1 + O(\Delta t_i'^2)] = \sqrt{a}V_Y\Delta E_i'[1 + O(\Delta e_i, \Delta E_i'^2)] \tag{17.40}$$

or, using the equation itself to evaluate $O(\Delta t_i'^2)$,

$$\Delta t_i' = \sqrt{a}V_Y\Delta E_i'[1 + O(\Delta e_i, \Delta E_i'^2)] \; . \tag{17.41}$$

Similarly, (17.21b) gives

$$\Delta t_i'' = \sqrt{a}V_Y\Delta E_i''[1 + O(\Delta e_i, \Delta E_i''^2)] \; . \tag{17.42}$$

We will need a more accurate expression for the differences $\Delta t_i'' - \Delta t_i'$, $\Delta E_i'' - \Delta E_i'$, which appear in (17.38). We subtract (17.39) from the similar equation for $\sin t_i''$:

$$\sin t_i'' - \sin t_i' = \epsilon\epsilon'\epsilon''\sqrt{1-e_i^2}\left[\frac{\sin E_i''}{1-\epsilon'' e_i\cos E_i''} - \frac{\sin E_i'}{1-\epsilon'' e_i \cos E_i'}\right] \; . \tag{17.43}$$

The left-hand side is treated as above. The expression between brackets expands into

$$\Delta E_i''(q_0 + q_2\Delta E_i''^2 + q_4\Delta E_i''^4 + \ldots)$$
$$-\Delta E_i'(q_0 + q_2\Delta E_i'^2 + q_4\Delta E_i'^4 + \ldots) \tag{17.44}$$

where the coefficients q_i are functions of $\epsilon'' e_i$. This can also be written

$$(\Delta E_i'' - \Delta E_i')[q_0 + O(\Delta E_i'^2, \Delta E_i''^2)] \; . \tag{17.45}$$

There is

$$q_0 = \frac{(-1)^J}{1-\epsilon'' e_i(-1)^J} \tag{17.46}$$

and (17.43) can be expanded into

$$\Delta t_i'' - \Delta t_i' = \sqrt{a}V_Y(\Delta E_i'' - \Delta E_i')[1 + O(\Delta e_i, \Delta E_i'^2, \Delta E_i''^2)] \; . \tag{17.47}$$

Note that this equation cannot be deduced directly from (17.41) and (17.42).

We can now eliminate $\Delta t_i'' - \Delta t_i'$ between the last equation and (17.38):

$$3\pi J\Delta a_i = v_y(\Delta E_i'' - \Delta E_i')[1 + O(\Delta a_i, \Delta e_i, \Delta E_i'^2, \Delta E_i''^2)] + O(\Delta a_i^2) \; . \tag{17.48}$$

The last term $O(\Delta a_i^2)$ can be eliminated by moving it to the left side, which becomes $3\pi J\Delta a_i[1+O(\Delta a_i)]$, and dividing both sides by $1+O(\Delta a_i)$.

We consider now (17.20b). Comparing with (17.39), we obtain

$$u_i' = \frac{\epsilon\epsilon' e_i}{\sqrt{a_i(1-e_i^2)}}\sin t_i' \,, \tag{17.49}$$

from which

$$u_i' = \frac{a-1}{aV_Y}\Delta t_i'[1+O(\Delta a_i, \Delta e_i, \Delta t_i'^2)] \,. \tag{17.50}$$

In the same way, (17.22b) gives

$$u_i'' = \frac{a-1}{aV_Y}\Delta t_i''[1+O(\Delta a_i, \Delta e_i, \Delta t_i''^2)] \,. \tag{17.51}$$

Here again we need a more accurate expression for the difference. This is obtained by subtracting (17.49) from the similar equation for u_i'' and t_i'':

$$u_i'' - u_i' = \frac{\epsilon\epsilon' e_i}{\sqrt{a_i(1-e_i^2)}}(\sin t_i'' - \sin t_i') \,, \tag{17.52}$$

which expands into

$$u_i'' - u_i' = \frac{a-1}{aV_Y}(\Delta t_i'' - \Delta t_i')[1+O(\Delta a_i, \Delta e_i, \Delta t_i'^2, \Delta t_i''^2)] \,. \tag{17.53}$$

We eliminate $\Delta t_i'' - \Delta t_i'$ with (17.47) and then $\Delta E_i'' - \Delta E_i'$ with (17.48):

$$3\pi J\Delta a_i = \frac{\sqrt{a}v_y}{a-1}(u_i'' - u_i')[1+O(\Delta a_i, \Delta e_i, u_i'^2, u_i''^2)] \,. \tag{17.54}$$

Here we have replaced the quantities $O(\Delta t_i'^2)$, $O(\Delta t_i''^2)$, $O(\Delta E_i'^2)$, $O(\Delta E_i''^2)$ by $O(u_i'^2)$ and $O(u_i''^2)$, using (17.41), (17.42), (17.50), and (17.51).

Finally we eliminate Δe_i with the help of (17.34):

$$3\pi J\Delta a_i = \frac{\sqrt{a}v_y}{a-1}(u_i'' - u_i')[1+O(\Delta C, \Delta a_i, u_i'^2, u_i''^2, \mu)] \,. \tag{17.55}$$

We consider now (17.20a). Using (17.8b) and expanding, we obtain

$$\begin{aligned} h_{i-1} = {} & \frac{1}{a}\Delta a_i - a\,\mathrm{sign}(a-1)\Delta e_i + \frac{a-1}{2}\Delta E_i'^2 \\ & + O(\Delta E_i'^4, \Delta a_i\Delta e_i, \Delta a_i\Delta E_i'^2, \Delta e_i\Delta E_i'^2, \mu) \,. \end{aligned} \tag{17.56}$$

(The expansion to second order in $\Delta E_i'$ will be needed later.) We eliminate $\Delta E_i'$ with (17.41) and (17.50), Δe_i with (17.34), and we obtain

$$2(a-1)h_{i-1} = aV_Y\Delta C + \frac{v_y}{a}\Delta a_i + au_i'^2 + O(\Delta C^2, \Delta a_i^2, u_i'^4, \mu) \,. \tag{17.57}$$

In the same way, (17.22a) and (17.7b) give

$$2(a-1)h_i = aV_Y\Delta C + \frac{v_y}{a}\Delta a_i + au_i''^2 + O(\Delta C^2, \Delta a_i^2, u_i''^4, \mu) \,. \tag{17.58}$$

The equations (17.11), (17.55), (17.57), (17.58) form a system of 4 relations for the 4 indexed quantities Δa_i, u'_i, u''_i, h_i; the 8 other quantities have been eliminated.

Here again, we can obtain a more accurate relation by subtracting (17.20a) from (17.22a):

$$h_i - h_{i-1} = -a_i\epsilon'' e_i(\cos E''_i - \cos E'_i) + O(\mu)\ , \tag{17.59}$$

or, after expansion

$$h_i - h_{i-1} = \frac{a-1}{2}(\Delta E''^2_i - \Delta E'^2_i)[1 + O(\Delta a_i, \Delta e_i, \Delta E'^2_i, \Delta E''^2_i)] + O(\mu) \tag{17.60}$$

and after elimination of $\Delta E'_i$, $\Delta E''_i$, and Δe_i

$$h_i - h_{i-1} = \frac{a}{2(a-1)}(u''^2_i - u'^2_i)[1 + O(\Delta C, \Delta a_i, u'^2_i, u''^2_i)] + O(\mu)\ . \tag{17.61}$$

Finally we substitute $u''_i - u'_i$ from (17.55), obtaining

$$h_i - h_{i-1} = \frac{3\pi J\sqrt{a}}{2v_y}\Delta a_i(u'_i + u''_i)[1 + O(\Delta C, \Delta a_i, u'^2_i, u''^2_i)] + O(\mu)\ . \tag{17.62}$$

This relation can be substituted to one of the two relations (17.57) and (17.58).

We collect and rearrange (17.11), (17.55), (17.57), (17.62), thus obtaining the set of fundamental equations for bifurcations of type 2:

$$h_i(u'_{i+1} - u''_i) = -\frac{2}{v}\mu[1 + O(h_i, u'_{i+1} - u''_i)]\ , \qquad i \in \mathcal{C}\ , \tag{17.63}$$

$$u''_i - u'_i = \frac{3\pi I(a-1)}{a^2 v_y}\Delta a_i[1 + O(\Delta C, \Delta a_i, u'^2_i, u''^2_i, \mu)]\ , \qquad i \in \mathcal{A}\ , \tag{17.64}$$

$$2(a-1)h_{i-1} = aV_Y\Delta C + \frac{v_y}{a}\Delta a_i + au'^2_i + O(\Delta C^2, \Delta a_i^2, u'^4_i, \mu)\ , \qquad i \in \mathcal{A}\ , \tag{17.65}$$

$$h_i - h_{i-1} = \frac{3\pi I}{2av_y}\Delta a_i(u'_i + u''_i)[1 + O(\Delta C, \Delta a_i, u'^2_i, u''^2_i)] + O(\mu)\ , \qquad i \in \mathcal{A}\ , \tag{17.66}$$

and the boundary conditions (17.12) for a partial bifurcation

$$h_0 = O(\mu)\ , \qquad h_n = O(\mu)\ . \tag{17.67}$$

Note that for (17.65) we could just as well have used (17.58) instead of (17.57).

17.2.3 Separation of the Case $n = 1$

The cases with a single basic arc, $n = 1$, have very different properties from the general case $n > 1$, both for a partial bifurcation (case 2P1) and for a total bifurcation (case 2T1). Therefore these two cases will be studied separately in Chap. 23, and for now we assume that $n > 1$.

17.3 The Case $\nu = 0$

We apply now the general method described in Sect. 11.4. We consider first the case $\mu = 0$, $\Delta C \neq 0$, corresponding to $\nu = 0$. As for type 1 (Sect. 12.3), this case is simple and it is not necessary to use the machinery described in Sect. 11.4. The solutions are known: they correspond to a displacement from the bifurcation orbit along one of the branches which emanate from it. The first species family is excluded here because it corresponds to a total bifurcation with $n = 1$. Therefore we have only to consider second species families. The bifurcating arc or the bifurcation orbit is made of a sequence of S and T-arcs. (If the symbolic sign of the branch, $\epsilon' \text{sign}(\Delta C)$, is positive, then only S-arcs are present; see Sect. 6.2.1.3.)

We determine now the orders of magnitude in ΔC of the variables Δa_i, u'_i, u''_i, h_i. The first fundamental equation (17.63) reduces to

$$h_i(u'_{i+1} - u''_i) = 0 \,. \tag{17.68}$$

This equation can be satisfied in two different ways:

$$h_i = 0 \qquad \text{or} \qquad u'_{i+1} = u''_i \,. \tag{17.69}$$

The first solution corresponds to the case where the collision becomes a node as we move away from the bifurcation: the impact distance remains zero. The second solution corresponds to an antinode: the velocity does not change.

For the two ends of a bifurcating arc, we have also

$$h_i = 0 \,. \tag{17.70}$$

As a consequence, we obtain for each arc T or S an independent set of equations, which can be solved separately.

17.3.1 T-Arc

We consider first the case of a T arc, running from collision $i - 1$ to collision i. There is

$$h_{i-1} = h_i = 0 \,. \tag{17.71}$$

(17.66) gives

$$0 = \Delta a_i(u'_i + u''_i) \,. \tag{17.72}$$

The solution corresponding to a T-arc is

$$\Delta a_i = 0 . \tag{17.73}$$

This corresponds to the fact that the period is constant along a T-arc family. The other solution $u'_i + u''_i = 0$ corresponds to a S-arc of length 1; this case will be treated in the next section. (17.64) gives then

$$u'_i = u''_i . \tag{17.74}$$

The radial velocity is the same at both ends of a T-arc. Finally, (17.65) gives

$$0 = aV_Y \Delta C + au_i'^2 + O(\Delta C^2, u_i'^4) , \tag{17.75}$$

which reduces to

$$u'_i = u''_i = \pm\sqrt{-V_Y \Delta C}[1 + O(\Delta C)] . \tag{17.76}$$

The + sign corresponds to a T^e arc, the − sign to a T^i arc.

17.3.2 S-Arc

We consider now the case of a S-arc made of m basic arcs, running from i to $i + m$. For the intermediate antinodes we have from (17.69b)

$$u'_{i+j+1} = u''_{i+j} , \qquad j = 1, \ldots, m - 1 . \tag{17.77}$$

All basic arcs in the S-arc have the same supporting ellipse, therefore they all have the same semi-major axis; we write

$$\Delta a_{i+j} = \Delta a , \qquad j = 1, \ldots, m . \tag{17.78}$$

Summing (17.64) over the m basic arcs, we obtain

$$u''_{i+m} - u'_{i+1} = O(\Delta a) . \tag{17.79}$$

Because of the symmetry of a S-arc, we have also

$$u''_{i+m} = -u'_{i+1} . \tag{17.80}$$

Therefore

$$u'_{i+1}, u''_{i+m} = O(\Delta a) \tag{17.81}$$

and more generally, from (17.64)

$$u'_{i+j}, u''_{i+j} = O(\Delta a) , \qquad j = 1, \ldots, m . \tag{17.82}$$

From (17.66), using the boundary condition $h_i = 0$, we have then

$$h_{i+j} = O(\Delta a^2) , \qquad j = 0, \ldots, m . \tag{17.83}$$

Substituting in (17.65), we obtain

$$\Delta a = -\frac{a^2 V_Y}{v_y} \Delta C[1 + O(\Delta C)] . \tag{17.84}$$

Incidentally, we verify that all Δa_{i+j} have the same value. Substituting in (17.64) and summing again, we have

$$u''_{i+m} - u'_{i+1} = -\frac{3\pi m I(a-1)V_Y}{v_y^2}\Delta C[1+O(\Delta C)] \, . \tag{17.85}$$

Using the symmetry (17.80), we can compute u'_{i+1} and u''_{i+m} separately. Using (17.64) again and summing over the first j basic arcs, we obtain the general equations

$$u'_{i+j} = (m-2j+2)\frac{3\pi I(a-1)V_Y}{2v_y^2}\Delta C[1+O(\Delta C)] \, ,$$
$$u''_{i+j} = (m-2j)\frac{3\pi I(a-1)V_Y}{2v_y^2}\Delta C[1+O(\Delta C)] \, , \quad j=1,\dots,m \, . \tag{17.86}$$

Finally, using (17.66) and summing over the first j basic arcs, we obtain

$$h_{i+j} = -j(m-j)\frac{9\pi^2 I^2 a(a-1)V_Y^2}{2v_y^4}\Delta C^2[1+O(\Delta C)] \, , \qquad j=0,\dots,m \, . \tag{17.87}$$

17.4 The Case $0 < \nu < 1/3$

We consider now the case $\mu > 0$, $\Delta C \neq 0$, corresponding to $\nu > 0$. We will try to extend each solution obtained in the previous section for $\nu = 0$ to the present case. This corresponds to the asymptotic branches of the families of orbits for $\mu \neq 0$, which are close to the branches for $\mu = 0$ (see Fig. 1.1).

We use the general method described in Sect. 11.4. Reference is made below to the successive steps. We continue to call T-arc or S-arc, for $\nu > 0$, a sequence of intermediate arcs which reduce to a T-arc or S-arc for $\nu \to 0$.

Step 1. We estimate orders of magnitude by extrapolating from the previous case $\nu = 0$.

(17.87) suggests that h_i is of the order of ΔC^2 inside a S-arc. (17.84), (17.86) suggest that Δa_i, u'_i, u''_i are of order ΔC in a S-arc. (17.76) suggests that u'_i, u''_i are of order $\sqrt{\Delta C}$ in a T-arc.

On the other hand, some quantities vanish for $\nu = 0$: h_i at a node (see (17.69)), Δa_i in a T-arc (see (17.73)). Their order of magnitude for $\nu > 0$ will only be estimated later (see below, Steps 8 and 9).

Step 2. This suggests the following changes of variables:

$$h_i = \Delta C^2 y_i^* \, , \qquad \Delta a_i = \Delta C z_i^* \, ; \tag{17.88}$$

for a S-arc:

$$u'_i = \Delta C x_i'^* \, , \qquad u''_i = \Delta C x_i''^* \, ; \tag{17.89}$$

for a T-arc:

$$u'_i = (-\epsilon'\Delta C)^{1/2} X'^*_i , \qquad u''_i = (-\epsilon'\Delta C)^{1/2} X''^*_i . \tag{17.90}$$

We need different changes of variables for u'_i and u''_i in the cases of a S-arc and a T-arc.

Step 3. It will be convenient to make a second change of variables in order to simplify the constant coefficients in the equations:

$$\begin{aligned} X'^*_i &= \pm\sqrt{\epsilon' V_Y}\, X'_i , \qquad X''^*_i = \pm\sqrt{\epsilon' V_Y}\, X''_i , \\ z^*_i &= \frac{a^2 V_Y}{v_y} z_i , \qquad x'^*_i = \frac{3\pi I(a-1)V_Y}{v_y^2} x'_i , \\ x''^*_i &= \frac{3\pi I(a-1)V_Y}{v_y^2} x''_i , \qquad y^*_i = \frac{9\pi^2 I^2 a(a-1)V_Y^2}{2v_y^4} y_i . \end{aligned} \tag{17.91}$$

In the first two equations, the sign is + for a T^e arc, – for a T^i arc; in that way, X'_i and X''_i are always positive. Note also that, by virtue of (17.88a) and (17.91f), there is

$$\operatorname{sign}(y_i) = \sigma'_i \tag{17.92}$$

where σ'_i is the relative side of passage, defined by (17.1).

Step 4. Substituting in the fundamental equations (17.63) to (17.67), collecting the dominant terms in the left-hand members, and dividing by the appropriate powers of ΔC, we obtain:

- For a T-arc:

$$\begin{aligned} &X''_i - X'_i = O(\Delta C^{1/2}) , \\ &X'^2_i - 1 - z_i = O(\Delta C, \mu\Delta C^{-1}) , \\ &z_i(X'_i + X''_i) = O(\Delta C^{1/2}, \mu\Delta C^{-3/2}) . \end{aligned} \tag{17.93}$$

  Note: since ΔC can be positive or negative, in principle we should write $O(|\Delta C|^{1/2})$, $O(\mu|\Delta C|^{-3/2})$, etc. However we will omit the absolute value symbols for simplicity; the intended meaning is clear in any case.

- For a S-arc:

$$y_{i+j}(x'_{i+j+1} - x''_{i+j}) = O(\mu\Delta C^{-3}) , \qquad j = 1, \dots, m-1 , \tag{17.94}$$

$$x''_{i+j} - x'_{i+j} - z_{i+j} = O(\Delta C, \mu) , \qquad j = 1, \dots, m , \tag{17.95}$$

$$z_{i+j} + 1 = O(\Delta C, \mu\Delta C^{-1}) , \qquad j = 1, \dots, m , \tag{17.96}$$

$$y_{i+j} - y_{i+j-1} - z_{i+j}(x'_{i+j} + x''_{i+j}) = O(\Delta C, \mu\Delta C^{-2}) , \quad j = 1, \dots, m . \tag{17.97}$$

- For a TT node (joining a T-arc to a T-arc):

$$y_i(X'_{i+1} + X''_i) = O(\mu\Delta C^{-5/2}) ; \tag{17.98}$$

  there is a + sign because two T-arcs in succession must have opposite signs (Proposition 4.3.2).

– For a TS node:

$$y_i X_i'' = O(\Delta C^{1/2}, \mu \Delta C^{-5/2}) . \tag{17.99}$$

– For a ST node:

$$y_i X_{i+1} = O(\Delta C^{1/2}, \mu \Delta C^{-5/2}) . \tag{17.100}$$

– For a SS node:

$$y_i (x_{i+1}' - x_i'') = O(\mu \Delta C^{-3}) . \tag{17.101}$$

– For the two ends of a bifurcating arc:

$$y_i = O(\mu \Delta C^{-2}) . \tag{17.102}$$

Step 5. We want all terms in the right-hand members to vanish for $\mu \to 0$. This is the case if

$$0 < \nu < \frac{1}{3} . \tag{17.103}$$

The asymptotic equations are then obtained by equating the right-hand members to zero.

Step 6. We solve now these asymptotic equations. (17.98) to (17.102) show that for a node, in all cases,

$$y_i = 0 \tag{17.104}$$

As a consequence, here again we obtain for each arc T or S an independent set of equations, which can be solved separately.

For a T-arc, the system (17.93) has three solutions:

$$\begin{aligned} z_i &= 0 , \quad X_i' = X_i'' = +1 ; \\ z_i &= 0 , \quad X_i' = X_i'' = -1 ; \\ z_i &= -1 , \quad X_i' = X_i'' = 0 . \end{aligned} \tag{17.105}$$

By continuity with the previous case $\nu = 0$, we choose the first solution.

For a S-arc, (17.96) gives

$$z_{i+j} = -1 . \tag{17.106}$$

(17.94) gives for an antinode

$$x_{i+j+1}' - x_{i+j}'' = 0 . \tag{17.107}$$

(17.95), (17.97), (17.107), plus (17.104) for the two ends of the arc, form a system of linear equations for the x_{i+j}', x_{i+j}'', y_{i+j} which is easily solved:

$$x_{i+j+1}' = x_{i+j}'' = \frac{m}{2} - j , \qquad y_{i+j} = -j(m-j) . \tag{17.108}$$

Step 7. For $0 < \mu \ll 1$, the solution described by (17.93) to (17.102) continues to exist if the error terms in the right-hand side are small and if the Jacobian $|J|$ of the asymptotic equations is not zero. The first condition is satisfied if $\nu < 1/3$. We must now compute the Jacobian.

There are $4n$ variables and $4n$ equations for a total bifurcation; $4n+1$ variables and $4n+1$ equations for a partial bifurcation. We consider first the case of a TT node. The partial derivatives of the expression f appearing on the left-hand side of (17.98) are

$$\frac{\partial f}{\partial y_i} = X'_{i+1} + X''_i \,, \qquad \frac{\partial f}{\partial X''_i} = y_i \,, \qquad \frac{\partial f}{\partial X'_{i+1}} = y_i \,. \tag{17.109}$$

In the solution (17.104) to (17.108), only the first of these derivatives is non-zero. Therefore the line in $|J|$ corresponding to (17.98) has a single non-zero element. We can then factor out this element and eliminate the corresponding line and column.

Similarly, for the other kinds of nodes corresponding to (17.99) to (17.101), and for the ends of a bifurcating arc corresponding to (17.102), we find that only the partial derivative $\partial f/\partial y_i$ is non-zero and the determinant can be simplified.

After this has been done for all nodes, $|J|$ is decomposed into a product of smaller determinants, each of which corresponds to a T- or S-arc. For a T-arc, there are 3 variables X'_i, X''_i, z_i, and the 3 equations (17.93); the Jacobian is

$$\begin{vmatrix} -1 & 1 & 0 \\ 2X'_i & 0 & -1 \\ z_i & z_i & X'_i + X''_i \end{vmatrix} \tag{17.110}$$

which equals -4 for the solution (17.105).

For a S-arc, the determinant has $4m-1$ lines and columns, given by (17.95), (17.96), (17.97), (17.94) for the variables x'_{i+1}, x''_{i+1}, z_{i+1}, y_{i+1}, x'_{i+2}, $\ldots$, z_{i+m}. (17.96) gives only one non-zero derivative: $\partial f/\partial z_{i+j} = 1$, and we can eliminate the corresponding line and column. There remains a determinant of $3m-1$ lines and columns, defined by (17.95), (17.97), (17.94) for the variables x'_{i+j}, x''_{i+j}, y_{i+j}. For (17.94), the derivatives are

$$\frac{\partial f}{\partial x''_{i+j}} = -y_{i+j} \,, \qquad \frac{\partial f}{\partial y_{i+j}} = x'_{i+j+1} - x''_{i+j} = 0 \,, \qquad \frac{\partial f}{\partial x'_{i+j+1}} = y_{i+j} \,. \tag{17.111}$$

We can factorize y_{i+j} (which is non-zero according to (17.108b)). The determinant takes then the form

$$\begin{vmatrix} -1 & 1 & & & & & & & \\ 1 & 1 & 1 & & & & & & \\ & -1 & & 1 & & & & & \\ & & & -1 & 1 & & & & \\ & & -1 & 1 & 1 & 1 & & & \\ & & & & -1 & & 1 & & \\ & & & & & & -1 & 1 & \\ & & & & & & & & \ddots \end{vmatrix} . \tag{17.112}$$

This determinant is easily computed by successive eliminations; its value is $-2m$.

Thus $|J|$ is non-zero. *For $0 < \nu < 1/3$, there exists for $\mu \neq 0$ a solution of the fundamental equations (17.93) to (17.102) close to the solution (17.104) to (17.108) of the asymptotic equations.* The error is of the order of the largest term in the right-hand members of (17.93) to (17.102), i.e. $O(\Delta C^{1/2}, \mu\Delta C^{-3})$. Thus the solution has the form

$$X_i' = 1 + O(\Delta C^{1/2}, \mu\Delta C^{-3}) , \tag{17.113}$$

etc.

Steps 8 and 9. We refine now the error estimates for some variables. We also go back to the physical variables, using (17.88) to (17.91). (17.98) shows that for a TT node,

$$y_i = O(\mu\Delta C^{-5/2}) , \qquad h_i = O(\mu\Delta C^{-1/2}) . \tag{17.114}$$

(17.101) shows that for a SS node,

$$y_i = O(\mu\Delta C^{-3}) , \qquad h_i = O(\mu\Delta C^{-1}) . \tag{17.115}$$

For a TS node, we have $u_i'' = \Theta(\Delta C^{1/2})$ and $u_{i+1}' = O(\Delta C)$; (17.63) gives then

$$h_i = O(\mu\Delta C^{-1/2}) , \qquad y_i = O(\mu\Delta C^{-5/2}) . \tag{17.116}$$

For a ST node, the result is the same.

For a T arc, we have $X_i' + X_i'' = 2$; therefore $u_i' + u_i'' = \Theta(\Delta C^{1/2})$, and (17.66) gives

$$\Delta a_i = O(\mu\Delta C^{-1}) . \tag{17.117}$$

(17.65) under consideration of (17.103) reduces to

$$V_Y \Delta C + u_i'^2 = O(\Delta C^2) , \tag{17.118}$$

from which

$$u_i' = \pm\sqrt{-V_Y\Delta C}[1 + O(\Delta C)] . \tag{17.119}$$

From (17.64) we have

$$u_i'' - u_i' = O(\mu\Delta C^{-1}) , \tag{17.120}$$

and therefore also

$$u_i'' = \pm\sqrt{-V_Y \Delta C}[1 + O(\Delta C)] . \tag{17.121}$$

In (17.119) and (17.121), the sign is + for a T^e arc, – for a T^i arc. These two equations are identical with (17.76).

For a S-arc, we have first from (17.96)

$$\Delta a = -\frac{a^2 V_Y}{v_y} \Delta C[1 + O(\Delta C)] . \tag{17.122}$$

which is identical with (17.84). On the other hand, the error terms in (17.94) to (17.97), and also in (17.115), (17.116), (17.102) for the ends, are all included in $O(\Delta C, \mu\Delta C^{-3})$. Therefore, going back to physical variables, we have

$$\begin{aligned} u'_{i+j} &= (m - 2j + 2)\frac{3\pi I(a-1)V_Y}{2v_y^2}\Delta C[1 + O(\Delta C, \mu\Delta C^{-3})] , \\ u''_{i+j} &= (m - 2j)\frac{3\pi I(a-1)V_Y}{2v_y^2}\Delta C[1 + O(\Delta C, \mu\Delta C^{-3})] , \\ j &= 1, \ldots, m , \end{aligned} \tag{17.123}$$

$$\begin{aligned} h_{i+j} &= -j(m - j)\frac{9\pi^2 I^2 a(a-1)V_Y^2}{2v_y^4}\Delta C^2[1 + O(\Delta C, \mu\Delta C^{-3})] , \\ j &= 0, \ldots, m . \end{aligned} \tag{17.124}$$

These equations are similar to (17.86) and (17.87); they differ only by the presence of an additional error term.

Finally, we can compute exactly the dominant term of h_i in a node with the help of (17.63), (17.119), (17.121), (17.123). We obtain

- At a TT node:

$$h_i = \pm\frac{1}{v}\mu(-V_Y\Delta C)^{-1/2}[1 + O(\Delta C^{1/2})] , \tag{17.125}$$

  with a + sign for a T^eT^i node and a – sign for a T^iT^e node.
- For a TS or ST node:

$$h_i = \pm\frac{2}{v}\mu(-V_Y\Delta C)^{-1/2}[1 + O(\Delta C^{1/2})] , \tag{17.126}$$

  with a + sign for a T^eS or ST^i node, and a – sign for a T^iS or ST^e node.
- For a SS node:

$$h_i = -\frac{4v}{3\pi I(m_a + m_b)(a-1)V_Y}\mu\Delta C^{-1}[1 + O(\Delta C, \mu\Delta C^{-3})] , \tag{17.127}$$

  where m_b and m_a are the values of m for the two S-arcs.

Substituting into (17.66), we obtain also the dominant term of Δa_i for a T-arc:

$$\Delta a_i = -\frac{a(g' + g'')\text{sign}(v_y)}{3\pi I V_Y}\mu\Delta C^{-1}[1 + O(\Delta C^{1/2})] \tag{17.128}$$

where

$$g' = \begin{cases} 2 & \text{if the previous arc is a } S\text{-arc,} \\ 1 & \text{if the previous arc is a } T\text{-arc,} \\ 0 & \text{if the previous arc does not exist;} \end{cases} \tag{17.129}$$

$$g'' = \begin{cases} 2 & \text{if the following arc is a } S\text{-arc,} \\ 1 & \text{if the following arc is a } T\text{-arc,} \\ 0 & \text{if the following arc does not exist.} \end{cases} \tag{17.130}$$

The previous or following arc does not exist in the case of an end of a bifurcating arc. Note that $g' + g'' = 0$ only in the case 2P1, which has been set aside (Sect. 17.2.3) and will be considered in Sect. 23.2.

The variations in order of magnitude of the physical quantities h_i, Δa_i, u'_i and u''_i as functions of ΔC, given by (17.119) to (17.128), are shown below in Fig. 17.2. Only the right part of each panel, corresponding to the interval $\mu^{1/3} < \Delta C < 1$, should be considered at the present stage; the left parts, corresponding to $\Delta C \leqq \mu^{1/3}$, will be described in the following sections.

17.5 The Case $\nu = 1/3$: Transition 2.1

We study now what happens when we reach the value $\nu = 1/3$, corresponding to $\Delta C = \Theta(\mu^{1/3})$.

In the case of type 1, we saw in Sect. 12.5 that when the value $\nu = 1/2$ is reached, all branches are joined two by two, and the story ends there. In the case of type 2, the situation is more complex. As will be found below, some branches are joined at $\nu = 1/3$, while other branches traverse this value, go on to $\nu > 1/3$, and meet another branch only later, at $\nu = 1/2$.

To distinguish these two cases, we introduce the notion of *transition*. This is intended as a kind of subdivision in a bifurcation. Generally, when a transition is reached, the exponents may change (see below Fig. 17.2), but the bifurcation is not yet ended.

We call thus *transition 2.1* (type 2, first transition) the case $\nu = 1/3$, and *transition 2.2* the case $\nu = 1/2$.

We start now the study of transition 2.1. We use again the general method of Sect. 11.4.

Step 1. The results of the previous section show that when we arrive at $\nu = 1/3$, the orders of magnitude are as follows:

– For a T-arc:

$$\Delta a_i = O(\mu^{2/3}) \ ; \qquad u'_i, u''_i = O(\mu^{1/6}) \ ; \tag{17.131}$$

– For a S-arc:

$$\Delta a_{i+j}, u'_{i+j}, u''_{i+j} = O(\mu^{1/3}) , \qquad j = 1, \dots, m ;$$
$$h_{i+j} = O(\mu^{2/3}) , \qquad j = 1, \dots, m-1 ; \tag{17.132}$$

– TT, TS, ST nodes:

$$h_i = O(\mu^{5/6}) ; \tag{17.133}$$

– SS node:

$$h_i = O(\mu^{2/3}) . \tag{17.134}$$

We remark that the h_i are now of the same order inside a S-arc and at a SS node. A *fusion* of consecutive S-arcs takes place: they form a single region, which we will call a *R-region.* (An isolated S-arc will also be called a R-region.) This fusion is the essence of transition 2.1. Two cases must be distinguished.

(i) In the case of a total bifurcation, and if only S-arcs are present, we have a *total transition 2.1.* The whole orbit obtained for $\nu = 1/3$ is a R-region, which will be called a *R-orbit.*

(ii) In the other cases, i.e. for a partial bifurcation and/or if T-arcs are present, we have a *partial transition 2.1.* One or more finite sequences of S-arcs may be present. Each sequence terminates on both sides either at a junction with a T-arc, or in one end of a bifurcating arc. Each such sequence is a R-region, which will be called a *R-arc.*

These new concepts will play a fundamental role, and it is essential to understand them. We illustrate them with a few examples (Fig. 17.1). Full lines represent schematically the original bifurcation orbit or bifurcating arc, with their division into S- and T-arcs. Dashed lines represent the R-regions.

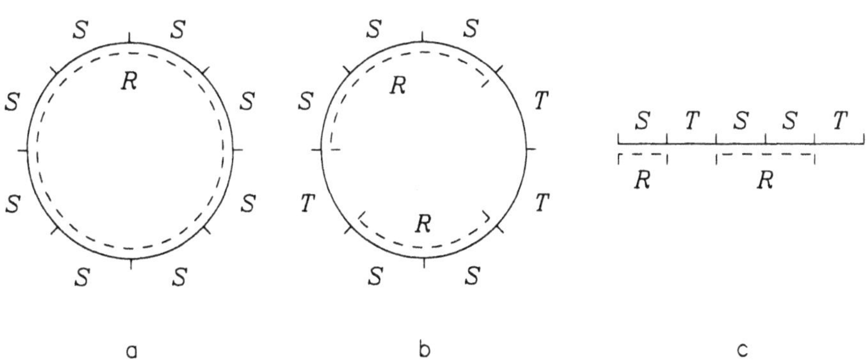

Fig. 17.1. Fusion of S-arcs into R-regions. a: total bifurcation, total transition 2.1. b: total bifurcation, partial transition 2.1. c: partial bifurcation, partial transition 2.1.

In case a, we have a total bifurcation, and only S-arcs are present; therefore this is a total transition 2.1. The R-region is a R-orbit, which coincides with the whole bifurcation orbit.

In case b, we still have a total bifurcation, but T-arcs are present; therefore we have a partial transition 2.1. The R-regions are R-arcs. In this case, it can happen that no S-arcs are present.

In case c, we have a partial bifurcation, and therefore a partial transition 2.1. The R-regions are again R-arcs. In this case, it can happen that no T-arcs are present, or no S-arcs.

In particular, it should be well understood that there is *no* one-to-one correspondence between total/partial bifurcation and total/partial transition. In case b, we have a total bifurcation and at the same time a partial transition 2.1.

A bifurcation orbit or a bifurcating arc is now formed of T-arcs and R-regions. The former division into nodes and antinodes is not appropriate any more. We define a *R-node* as an encounter lying at the junction between two T-arcs, or between a T-arc and a R-arc, and a *R-antinode* as an encounter lying inside a R-region.

We introduce a notation similar to the one introduced for second species bifurcations (Sects. 6.2.1 and 6.2.2). We use a letter `T` or `P` to designate a total or partial transition. We define the *order* of a R-region as the number $\tilde{n}$ of basic arcs which form the R-region. We use a short designation of the form `2.1`<`T` *or* `P`><*order*>; for instance: `2.1P3`.

By analogy with Sect. 11.3.1, we define $\tilde{\mathcal{C}}$, the *set of internal collisions* of a R-region, as

$$\tilde{\mathcal{C}} = \begin{cases} \{1,\ldots,\tilde{n}-1\} & \text{for a } R\text{-arc,} \\ \{0,\ldots,\tilde{n}-1\} & \text{for a } R\text{-orbit.} \end{cases} \tag{17.135}$$

We define also $\tilde{\mathcal{A}}$, the *set of basic arcs* of a R-region, as

$$\tilde{\mathcal{A}} = \{1,\ldots,\tilde{n}\} . \tag{17.136}$$

Step 2. The estimates in the previous step suggest the following changes of variables.

- For a T-arc:

$$u'_i = (-\epsilon'\Delta C)^{1/2} X'^{*}_i , \qquad u''_i = (-\epsilon'\Delta C)^{1/2} X''^{*}_i , \qquad \Delta a_i = \Delta C Z^*_i . \tag{17.137}$$

  We use here powers of ΔC rather than μ in order to show the continuity with the neighbouring cases $\nu < 1/3$ and $\nu > 1/3$. These relations are identical with (17.90) and (17.88b).
- For a R-region:

$$u'_i = \mu^{1/3} x'^{*}_i , \qquad u''_i = \mu^{1/3} x''^{*}_i , \qquad \Delta a_i = \mu^{1/3} z^*_i ; \tag{17.138}$$

– For all encounters:

$$h_i = \mu^{2/3} y_i^* ; \tag{17.139}$$

– For ΔC:

$$\Delta C = \mu^{1/3} w^* . \tag{17.140}$$

Note that according to the definition of ν, $w^* = \Theta(1)$.

Step 3. It will be convenient to make a second change of variables:

$$\begin{aligned}
&X_i'^* = \pm\sqrt{\epsilon' V_Y} X_i' , \qquad X_i''^* = \pm\sqrt{\epsilon' V_Y} X_i'' , \qquad Z_i^* = \frac{a^2 V_Y}{v_y} Z_i , \\
&x_i'^* = -L_1 x_i' , \qquad x_i''^* = -L_1 x_i'' , \qquad z_i^* = -\frac{a^2 v_y L_1}{3\pi I (a-1)} z_i , \\
&y_i^* = \frac{2}{v L_1} y_i , \qquad w^* = -\frac{v_y^2 L_1}{3\pi I V_Y (a-1)} w ,
\end{aligned} \tag{17.141}$$

with

$$L_1 = \left[\frac{4(a-1)}{av}\right]^{1/3} . \tag{17.142}$$

In the first two equations, the sign is + for a T^e arc, – for a T^i arc; in that way, X_i' and X_i'' are always positive. The first 3 relations are identical with the corresponding relations in (17.91). Here again, we have from (17.139), (17.141g), (17.142), and (17.1):

$$\operatorname{sign}(y_i) = \sigma_i' . \tag{17.143}$$

Note also that

$$\operatorname{sign}(w) = -\operatorname{sign}(\epsilon' \Delta C) . \tag{17.144}$$

Step 4. Substituting in the fundamental equations (17.63) to (17.67), we obtain:

– For a T-arc:

$$\begin{aligned}
&X_i'' - X_i' = O(\mu^{1/6}) , \qquad X_i'^2 - 1 - Z_i = O(\mu^{1/3}) , \\
&Z_i (X_i' + X_i'') = O(\mu^{1/6}) .
\end{aligned} \tag{17.145}$$

These equations are identical with (17.93), in the particular case $\Delta C = \Theta(\mu^{1/3})$. *The T-arcs are essentially unaffected by transition 2.1.*

– For a R-region:

$$\begin{aligned}
&y_{i+j}(x'_{i+j+1} - x''_{i+j}) - 1 = O(\mu^{1/3}) , \qquad j \in \tilde{C} , \\
&x''_{i+j} - x'_{i+j} - z_{i+j} = O(\mu^{1/3}) , \qquad j \in \tilde{A} , \\
&z_{i+j} + w = O(\mu^{1/3}) , \qquad j \in \tilde{A} , \\
&y_{i+j} - y_{i+j-1} - z_{i+j}(x'_{i+j} + x''_{i+j}) = O(\mu^{1/3}) , \qquad j \in \tilde{A} .
\end{aligned} \tag{17.146}$$

– For a TT R-node:

$$y_i(X'_{i+1} + X''_i) = O(\mu^{1/6}) ; \tag{17.147}$$

– For a TR R-node:

$$y_i X''_i = O(\mu^{1/6}) ; \tag{17.148}$$

– For a RT R-node:

$$y_i X'_{i+1} = O(\mu^{1/6}) ; \tag{17.149}$$

– For the end of a bifurcating arc:

$$y_i = O(\mu^{1/3}) . \tag{17.150}$$

Step 5. The right-hand members do tend to zero for $\mu \to 0$. The asymptotic equations are obtained by equating them to zero.

Step 6. We solve the asymptotic equations. For a T-arc, the system (17.145) becomes identical with the system (17.93). It has three solutions; by continuity with the previous case $\nu < 1/3$, we again choose the solution

$$Z_i = 0 , \qquad X'_i = X''_i = 1 . \tag{17.151}$$

For the R-nodes, each of the equations (17.147) to (17.149) has two solutions; by continuity we choose in all cases the solution

$$y_i = 0 . \tag{17.152}$$

For a R-region, we obtain from (17.146c)

$$z_{i+j} = -w . \tag{17.153}$$

We can then eliminate the z_{i+j} variables and we obtain the system

$$\begin{aligned} & y_{i+j}(x'_{i+j+1} - x''_{i+j}) - 1 = 0 , \qquad j \in \tilde{C} , \\ & x''_{i+j} - x'_{i+j} + w = 0 , \qquad j \in \tilde{A} , \\ & y_{i+j} - y_{i+j-1} + w(x'_{i+j} + x''_{i+j}) = 0 , \qquad j \in \tilde{A} . \end{aligned} \tag{17.154}$$

In addition, at the two ends of a R-arc there is from (17.150) and (17.152)

$$y_i = y_{i+\tilde{n}} = 0 . \tag{17.155}$$

The available evidence (see below) suggests that this system cannot be solved explicitly in general. It will be studied in detail in Chaps. 19 and 20.

Step 7. We must compute the Jacobian. We consider first the R-nodes. As in Sect. 17.4, Step 7, we find that only the partial derivative $\partial f/\partial y_i$ is non-zero in (17.147) to (17.149), and the determinant can be simplified. $|J|$ is decomposed into a product of smaller determinants, each of which corresponds to a T-arc or to a R-region. It is then sufficient to verify that each of these determinants is non-zero.

For a T-arc, the determinant is again given by (17.110) and equals -4. For a R-region, we can eliminate the lines corresponding to the equations

(17.146c) and the columns corresponding to the variables z_{i+j}; we are left with the determinant of the system (17.154), which we call *R-Jacobian*. It will only be possible to compute that determinant after the system has been solved (see Chaps. 19 and 20). We will find that it is generally non-zero; however it can vanish in isolated points on the characteristics.

Therefore there exists in general, for $\mu > 0$, a solution close to the asymptotic solution, the distance being of the order of $\mu^{1/6}$.

Steps 8 and 9. We refine now the error estimates for some variables. We also go back to the physical variables. We have for the R-nodes

$$h_i = O(\mu^{5/6}) \,. \tag{17.156}$$

For a T-arc, (17.66) gives then

$$\Delta a_i = O(\mu^{2/3}) \,. \tag{17.157}$$

(17.65) and (17.64) give

$$u_i', u_i'' = \pm\sqrt{-V_Y \Delta C}[1 + O(\mu^{1/3})] \,. \tag{17.158}$$

For a R-region, (17.65) gives

$$\Delta a_i = -\frac{a^2 V_Y}{v_y} \Delta C[1 + O(\mu^{1/3})] \,. \tag{17.159}$$

The Δa_i are asymptotically equal in a R-region. Guillaume (1971, pp. 134–135) already noted that the Δa_i are equal within $O(\mu^{2/3})$.

Since the system (17.154) is not exactly solvable, it is not possible to give explicit expressions for the other variables. If a particular solution of that system is known, then the physical variables can be computed with (17.137) to (17.141):

$$\begin{aligned}
u_{i+j}' &= -L_1 x_{i+j}' \mu^{1/3}[1 + O(\mu^{1/6})] \,, \qquad j \in \tilde{A} \,, \\
u_{i+j}'' &= -L_1 x_{i+j}'' \mu^{1/3}[1 + O(\mu^{1/6})] \,, \qquad j \in \tilde{A} \,, \\
h_{i+j} &= \frac{2}{vL_1} y_{i+j} \mu^{2/3}[1 + O(\mu^{1/6})] \,, \qquad j \in \tilde{A} \,, \\
\Delta C &= -\frac{v_y^2 L_1}{3\pi I V_Y (a-1)} w \mu^{1/3} \,.
\end{aligned} \tag{17.160}$$

17.6 The Case $1/3 < \nu < 1/2$

We continue to speak of T-arc or R-regions, by continuity, for $\nu > 1/3$, and for the same sequences of intermediate arcs.

We use once more the general method of Sect. 11.4.

Step 1. The results obtained for T-arcs in the two previous sections, for $0 < \nu < 1/3$ ((17.117), (17.119), (17.121)) and for $\nu = 1/3$ ((17.157),

(17.158), suggest that these arcs are not affected by transition 2.1, and that for $\nu > 1/3$ we continue to have $u'_i, u''_i = O(\Delta C^{1/2})$ and Δa_i small

For R-regions, on the other hand, the study of small values of $\tilde{n}$ in the case $\nu = 1/3$ (Sects. 19.2, 20.2) shows that the following two properties hold generally:

(i) w can tend towards 0 only on the positive side: $w \to 0^+$.

(ii) There is:

$$y_i = O(w^{1/2}) \; ; \qquad x'_i, x''_i = O(w^{-1/2}) \qquad \text{for } w \to 0^+ \; . \tag{17.161}$$

The only exceptions are:

- The case 2.1P1 (R-arc of order 1) (Sect. 19.2.1), for $w \to 0^-$. But this case corresponds to $\epsilon' \Delta C > 0$; therefore there are no T-arcs. The R-arc coincides with the bifurcating arc, and we have a bifurcation 2P1, which has been temporarily excluded (Sect. 17.2.3).
- The case 2.1T1 (R-orbit of order 1) (Sect. 20.2.1). But this corresponds to a bifurcation 2T1, also excluded.

These properties will be proved in Sect. 19.1.3.

Going back to the physical variables with the help of (17.137) to (17.141), we change (17.161) into

$$h_i = O(\mu^{1/2} \Delta C^{1/2}) \; ; \qquad u'_i, u''_i = O(\mu^{1/2} \Delta C^{-1/2}) \; . \tag{17.162}$$

The results for $\nu < 1/3$ (see (17.122)) and for $\nu = 1/3$ (see (17.159)) suggest that in a R-region, Δa_i is not affected by transition 2.1, and that for $\nu > 1/3$ we still have

$$\Delta a_i = O(\Delta C) \; . \tag{17.163}$$

Finally, for the R-nodes, the results for $\nu = 1/3$ (see (17.156)) suggest that the h_i are still small for $\nu > 1/3$.

Step 2. We make therefore the following changes of variables:

- For a T-arc:

$$u'_i = (-\epsilon' \Delta C)^{1/2} X'^*_i \; , \qquad u''_i = (-\epsilon' \Delta C)^{1/2} X''^*_i \; , \qquad \Delta a_i = \Delta C Z^*_i \; , \tag{17.164}$$

which is identical with (17.137).

- For a R-region:

$$u'_i = \mu^{1/2} (-\epsilon' \Delta C)^{-1/2} x'^*_i \; , \qquad u''_i = \mu^{1/2} (-\epsilon' \Delta C)^{-1/2} x''^*_i \; ,$$
$$\Delta a_i = \Delta C z^*_i \; ; \tag{17.165}$$

- For all encounters:

$$h_i = \mu^{1/2} (-\epsilon' \Delta C)^{1/2} y^*_i \; . \tag{17.166}$$

Step 3. It will be convenient to make a second change of variables:

$$X_i'^* = \pm\sqrt{\epsilon' V_Y} X_i' \,, \qquad X_i''^* = \pm\sqrt{\epsilon' V_Y} X_i'' \,, \qquad Z_i^* = \frac{a^2 V_Y}{v_y} Z_i \,,$$
$$x_i'^* = -L_2 x_i' \,, \qquad x_i''^* = -L_2 x_i'' \,, \qquad z_i^* = -\frac{a^2 V_Y}{v_y} z_i \,, \qquad y_i^* = \frac{2}{v L_2} y_i \,, \tag{17.167}$$

with

$$L_2 = \text{sign}(a-1) \left[\frac{2v}{3\pi I a V} \right]^{1/2} . \tag{17.168}$$

The factor sign$(a-1)$ has been introduced here so that x_i', x_i'', y_i have the same signs as the corresponding quantities defined in (17.141). In particular, we have from (17.166), (17.167g), (17.168), and (17.1):

$$\text{sign}(y_i) = \sigma_i' \,. \tag{17.169}$$

Step 4. Substituting in the fundamental equations (17.63) to (17.67), we obtain:

- For a T-arc:

$$X_i'' - X_i' = O(\Delta C^{1/2}) \,, \qquad X_i'^2 - 1 - Z_i = O(\mu^{1/2} \Delta C^{-1/2}) \,,$$
$$Z_i (X_i' + X_i'') = O(\mu^{1/2} \Delta C^{-1}) \,. \tag{17.170}$$

  These equations are identical with (17.145) if we substitute $\Delta C = \Theta(\mu^{1/3})$. This shows again that *the T-arcs are not affected by transition 2.1.*
- For a R-region:

$$y_{i+j}(x_{i+j+1}' - x_{i+j}'') - 1 = O(\mu^{1/2} \Delta C^{-1/2}) \,, \qquad j \in \tilde{C} \,,$$
$$x_{i+j}'' - x_{i+j}' = O(\mu^{-1/2} (\Delta C^{3/2}) \,, \qquad j \in \tilde{A} \,,$$
$$z_{i+j} + 1 = O(\mu \Delta C^{-2}) \,, \qquad j \in \tilde{A} \,,$$
$$2(y_{i+j} - y_{i+j-1}) - z_{i+j}(x_{i+j}' + x_{i+j}'') = O(\mu^{1/2} \Delta C^{-1/2}) \,, \qquad j \in \tilde{A} \,. \tag{17.171}$$

- For a TT R-node:

$$y_i (X_{i+1}' + X_i'') = O(\mu^{1/2} \Delta C^{-1}) \,; \tag{17.172}$$

- For a TR R-node:

$$y_i X_i'' = O(\mu^{1/2} \Delta C^{-1}) \,; \tag{17.173}$$

- For a RT R-node:

$$y_i X_{i+1}' = O(\mu^{1/2} \Delta C^{-1}) \,; \tag{17.174}$$

- For the end of a bifurcating arc:

$$y_i = O(\mu^{1/2} \Delta C^{-1/2}) \,. \tag{17.175}$$

Step 5. Examining the right-hand members, we find that they do tend to zero for $\mu \to 0$ if

$$1/3 < \nu < 1/2 . \tag{17.176}$$

The asymptotic equations are obtained by equating them to zero.

Step 6. We solve the asymptotic equations. For a T-arc, the system is once more identical with the previous cases (17.93) and (17.145), and by continuity with the previous cases, we again choose the solution

$$Z_i = 0 , \qquad X_i' = X_i'' = 1 . \tag{17.177}$$

Similarly, for the R-nodes, each of the equations (17.172) to (17.174) also has two solutions; by continuity we choose in all cases the solution

$$y_i = 0 . \tag{17.178}$$

For a R-region, (17.171c) gives

$$z_{i+j} = -1 . \tag{17.179}$$

(17.171b) gives

$$x_i'' = x_i' . \tag{17.180}$$

We call x_i this common value. (17.171a) and (17.171d) become

$$\begin{aligned} &y_{i+j}(x_{i+j+1} - x_{i+j}) - 1 = 0 , \qquad j \in \tilde{C} , \\ &y_{i+j} - y_{i+j-1} + x_{i+j} = 0 , \qquad j \in \tilde{A} . \end{aligned} \tag{17.181}$$

In addition, at the two ends of a R-arc there is from (17.175) and (17.178)

$$y_i = y_{i+\tilde{n}} = 0 . \tag{17.182}$$

The available evidence (see below) suggests that this system cannot be solved explicitly in general. It will be studied in detail in Chap. 18.

Step 7. We must compute the Jacobian. As in the previous cases (Step 7 in Sect 17.5), we find that the R-nodes can be eliminated; $|J|$ is decomposed into a product of smaller determinants, each of which corresponds to a T-arc or to a R-region; the determinant corresponding to a T-arc is non-zero. For a R-region, we can eliminate the lines corresponding to the equations (17.171c) and the columns corresponding to the variables z_{i+j}; we can also eliminate the lines corresponding to the equations (17.171b) and replace the two columns for x'_{i+j} and x''_{i+j} by a single column for x_{i+j}. We are left with the determinant of the system (17.181). It will only be possible to compute that determinant after the system has been solved. We will find that it never vanishes (Sect. 18.1.3).

Therefore in the interval (17.176), for $\mu > 0$, there exists a solution close to the asymptotic solution, the distance being of the order of

$$O(\mu^{-1/2}\Delta C^{3/2}, \mu^{1/2}\Delta C^{-1}) . \tag{17.183}$$

The first of these two error terms becomes $O(1)$ when ν approaches the value 1/3, while the second becomes $O(1)$ when ν approaches the value 1/2. For $1/3 < \nu < 1/2$, both error terms are $o(1)$.

Steps 8 and 9. We can now go back to the physical variables, and refine some error terms. (17.172) to (17.174) show that we have for all R-nodes

$$h_i = O(\mu\Delta C^{-1/2})\,. \tag{17.184}$$

For a T-arc, (17.66) gives then

$$\Delta a_i = O(\mu\Delta C^{-1})\,; \tag{17.185}$$

(17.65) reduces to

$$V_Y\Delta C + u_i'^2 = O(\mu\Delta C^{-1})\,, \tag{17.186}$$

from which

$$u_i' = \pm\sqrt{-V_Y\Delta C}[1 + O(\mu\Delta C^{-2})]\,. \tag{17.187}$$

(17.64) gives

$$u_i'' - u_i' = O(\mu\Delta C^{-1})\,, \tag{17.188}$$

and

$$u_i'' = \pm\sqrt{-V_Y\Delta C}[1 + O(\mu\Delta C^{-2})]\,. \tag{17.189}$$

For a R-region, we have from (17.171c)

$$\Delta a_{i+j} = -\frac{a^2V_Y}{v_y}\Delta C[1 + O(\mu\Delta C^{-2})]\,. \tag{17.190}$$

Since the system (17.181) is not exactly solvable, it is not possible to give explicit expressions for the other variables. If a particular solution of that system is known, then the physical variables can be computed by

$$\begin{aligned} u'_{i+j}, u''_{i+j} &= -L_2 x_{i+j}\mu^{1/2}(-\epsilon'\Delta C)^{-1/2} \\ &\qquad [1 + O(\mu^{-1/2}\Delta C^{3/2}, \mu^{1/2}\Delta C^{-1})]\,, \qquad j \in \tilde{A}\,, \\ h_{i+j} &= \frac{2}{vL_2} y_{i+j}\mu^{1/2}(-\epsilon'\Delta C)^{1/2} \\ &\qquad [1 + O(\mu^{-1/2}\Delta C^{3/2}, \mu^{1/2}\Delta C^{-1})]\,, \qquad j \in \tilde{A}\,. \end{aligned} \tag{17.191}$$

The main term in (17.187), (17.189), (17.190) is the same as in the previous cases $\nu < 1/3$ (see (17.119), (17.121), (17.122)) and $\nu = 1/3$ ((17.158), (17.159)).

Finally, we can compute exactly the dominant term of h_i in a R-node with the help of (17.63), (17.187), (17.189), (17.191). We obtain

– for a TT R-node:

$$h_i = \pm\frac{1}{v}\mu(-V_Y\Delta C)^{-1/2}[1 + O(\Delta C^{1/2}, \mu\Delta C^{-2})]\,, \tag{17.192}$$

with a + sign for a T^eT^i R-node and a − sign for a T^iT^e R-node.

– For a TR or RT R-node:

$$h_i = \pm\frac{2}{v}\mu(-V_Y\Delta C)^{-1/2}[1 + O(\Delta C^{1/2}, \mu^{1/2}\Delta C^{-1})] \,, \tag{17.193}$$

with a + sign for a T^eR or RT^i R-node, and a – sign for a T^iR or RT^e R-node.

Substituting into (17.66), we obtain also the dominant term of Δa_i for a T-arc:

$$\Delta a_i = -\frac{a(g' + g'')\text{sign}(v_y)}{3\pi I V_Y}\mu\Delta C^{-1}[1 + O(\Delta C^{1/2}, \mu^{1/2}\Delta C^{-1})] \tag{17.194}$$

where

$$g' = \begin{cases} 2 & \text{if the previous arc is a } R\text{-arc,} \\ 1 & \text{if the previous arc is a } T\text{-arc,} \\ 0 & \text{if the previous arc does not exist;} \end{cases} \tag{17.195}$$

$$g'' = \begin{cases} 2 & \text{if the following arc is a } R\text{-arc,} \\ 1 & \text{if the following arc is a } T\text{-arc,} \\ 0 & \text{if the following arc does not exist.} \end{cases} \tag{17.196}$$

These equations are identical with (17.125), (17.126), (17.128), with "S-arc" replaced by "R-arc" everywhere.

Fig. 17.2 shows in log-log plots the variations in order of magnitude of the quantities h_i, Δa_i, u_i' and u_i'' as functions of ΔC, given by (17.184) to (17.191) in the interval $\mu^{1/2} < \Delta C < \mu^{1/3}$. These curves join smoothly at $\Delta C = \mu^{1/3}$ with those found before in Sect. 17.4 for the interval $\mu^{1/3} < \Delta C < 1$.

We remark that this figure is more complex than the corresponding Fig. 12.3 for type 1.

17.7 The Case $\nu = 1/2$: Transition 2.2

We study now what happens when we reach the value $\nu = 1/2$, corresponding to $\Delta C = \Theta(\mu^{1/2})$.

Step 1. The results of the previous section show that when we arrive at $\nu = 1/2$, the orders of magnitude are in all cases (R-nodes, T-arcs, R-regions):

$$\Delta a_i = O(\mu^{1/2}) \,; \qquad u_i', u_i'' = O(\mu^{1/4}) \,; \qquad h_i = O(\mu^{3/4}) \,. \tag{17.197}$$

A general fusion takes place: all T-arcs and R-regions fuse into a single region. This is also apparent on Fig. 17.2. We will call this *transition 2.2* (type 2, second transition). This transition affects the totality of the bifurcation orbit or bifurcating arc, at the difference of transition 2.1 which did not affect the T arcs.

We distinguish a *total transition 2.2*, corresponding to a total bifurcation, and a *partial transition 2.2*, corresponding to a partial bifurcation. Here

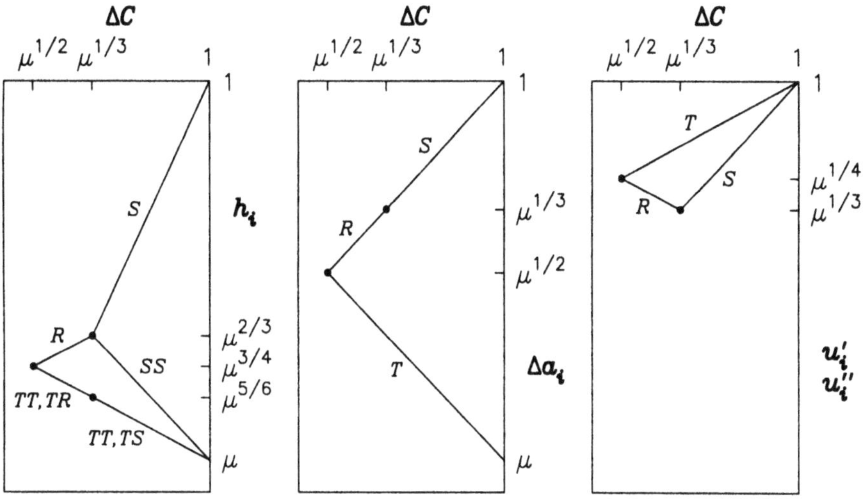

Fig. 17.2. Variations of h_i, Δa_i, u'_i and u''_i as functions of ΔC. Both scales are logarithmic. Dots correspond to transitions. In the left panel, TT, TS, SS, TR represent values of h_i at a node or a R-node, while S and R represent values at an antinode or a R-antinode, inside a S-arc or R-region respectively.

there is a one-to-one correspondence between total/partial bifurcation and total/partial transition; the situation is simpler than in the case of transition 2.1 (Sect. 17.5).

Step 2. We make changes of variables so as to obtain quantities $O(1)$:

$$\Delta a_i = \mu^{1/2} Z_i^* \,, \qquad u'_i = \mu^{1/4} X_i'^* \,, \qquad u''_i = \mu^{1/4} X_i''^* \,,$$
$$h_i = \mu^{3/4} Y_i^* \,, \qquad \Delta C = \mu^{1/2} W^* \,. \tag{17.198}$$

Step 3. It will be convenient to make a second change of variables:

$$X_i'^* = -L_3 X'_i \,, \qquad X_i''^* = -L_3 X''_i \,, \qquad Y_i^* = \frac{2}{v L_3} Y_i \,,$$
$$Z_i^* = -\frac{a^2 L_3^2}{v_y} Z_i \,, \qquad W^* = -\frac{L_3^2}{V_Y} W \,, \tag{17.199}$$

with

$$L_3 = \operatorname{sign}(a-1) \left[\frac{2v}{3\pi I a} \right]^{1/4} \,. \tag{17.200}$$

The factor $\operatorname{sign}(a-1)$ has been introduced so that X'_i, X''_i, Y_i have the same signs as x'_i, x''_i, y_i defined in (17.141) and also x'_i, x''_i, y_i defined in (17.167). In particular, we have from (17.198d), (17.199c), (17.200), and (17.1):

$$\operatorname{sign}(Y_i) = \sigma'_i \,. \tag{17.201}$$

Step 4. Substituting in the fundamental equations (17.63) to (17.67), we obtain

$$\begin{aligned}
&Y_i(X'_{i+1} - X''_i) - 1 = O(\mu^{1/4})\,, \qquad i \in \mathcal{C}\,,\\
&X''_i - X'_i = O(\mu^{-1/4})\,, \qquad i \in \mathcal{A}\,,\\
&X'^2_i - Z_i - W = O(\mu^{1/4})\,, \qquad i \in \mathcal{A}\,,\\
&2(Y_i - Y_{i-1}) - Z_i(X'_i + X''_i) = O(\mu^{1/4})\,, \qquad i \in \mathcal{A}\,. \qquad (17.202)
\end{aligned}$$

In addition, for the ends of a bifurcating arc,

$$Y_0, Y_n = O(\mu^{1/4})\,. \qquad (17.203)$$

Step 5. The right-hand members tend to zero for $\mu \to 0$. The asymptotic equations are obtained by equating them to zero.

Step 6. We solve the asymptotic equations. (17.202b) gives

$$X'_i = X''_i\,. \qquad (17.204)$$

We call X_i this common value. The system reduces to

$$\begin{aligned}
&Y_i(X_{i+1} - X_i) - 1 = 0\,, \qquad i \in \mathcal{C}\,,\\
&X_i{}^2 - Z_i - W = 0\,, \qquad i \in \mathcal{A}\,,\\
&Y_i - Y_{i-1} - Z_i X_i = 0\,, \qquad i \in \mathcal{A}\,, \qquad (17.205)
\end{aligned}$$

and at the two ends of a bifurcating arc,

$$Y_0 = Y_n = 0\,. \qquad (17.206)$$

We can eliminate Z_i between (17.205b) and (17.205c), obtaining

$$Y_i - Y_{i-1} - X_i(X_i{}^2 - W) = 0\,, \qquad i \in \mathcal{A}\,. \qquad (17.207)$$

However, the available evidence (see below) suggests that the system formed by (17.205a) and (17.207) cannot be solved explicitly in general. It will be studied in detail in Chaps. 21 and 22.

Step 7. We must compute the Jacobian. This will only be possible after the system has been solved. We will find that it is generally non-zero; however it can vanish in isolated points on the characteristics.

Therefore there exists in general, for $\mu > 0$, a solution close to the asymptotic solution, the distance being of the order of $\mu^{1/4}$.

Step 9. Since the system (17.205) is not exactly solvable, it is not possible to give explicit expressions for the physical variables. If a particular solution of that system is known, then the physical variables can be computed with (17.198) and (17.199):

$$\begin{aligned}
&\Delta a_i = -\frac{a^2 L_3^2}{v_y} Z_i \mu^{1/2}[1 + O(\mu^{1/4})]\,,\\
&u'_i, u''_i = -L_3 X_i \mu^{1/4}[1 + O(\mu^{1/4})]\,,\\
&h_i = \frac{2}{v L_3} Y_i \mu^{3/4}[1 + O(\mu^{1/4})]\,,\\
&\Delta C = -\frac{L_3^2}{V_Y} W \mu^{1/2}\,. \qquad (17.208)
\end{aligned}$$

17.8 The Case $\nu > 1/2$ Does Not Exist

The detailed study of the case $\nu = 1/2$ which will be made in Chaps. 21 and 22 shows that all branches are joined two by two. Thus the program outlined in Sect. 11.4 is completed.

At the difference of type 1, Sect. 12.6, there is no case $\nu > 1/2$, because W never vanishes (see (21.7), (22.5)). Thus, for type 2, $\nu = 1/2$ is the highest possible value of ν, corresponding to $\Delta C = \Theta(\mu^{1/2})$.

18. The Case $1/3 < \nu < 1/2$

In the previous chapter, three cases have been found for which the fundamental equations could not be explicitly solved:

1. For $\nu = 1/3$ (transition 2.1), the equations (17.154) of a R-region.
2. For $1/3 < \nu < 1/2$, the equations (17.181), again for a R-region.
3. For $\nu = 1/2$ (transition 2.2), the general equations (17.205).

We study now these cases in more detail. We begin with the second case, which is the simplest, and also because the results will be needed in the first and third cases.

18.1 R-Arc

To simplify the notations, we renumber the encounters and the basic arcs starting from the origin of the R-arc. The equations are then

$$\begin{aligned} &y_i - y_{i-1} + x_i = 0 , \qquad i = 1, \ldots, \tilde{n} , \\ &y_i(x_{i+1} - x_i) - 1 = 0 , \qquad i = 1, \ldots, \tilde{n} - 1 ; \\ &y_0 = 0 , \qquad y_{\tilde{n}} = 0 . \end{aligned} \tag{18.1}$$

This is a system of $2\tilde{n} + 1$ equations for $2\tilde{n} + 1$ variables. So we expect isolated solutions. The correspondence between the present variables (x_i, y_i) and the physical variables is given by (17.165), (17.166), and (17.167).These equations involve two parameters, μ and ΔC. Thus, for a given value of μ, an isolated solution of (18.1) generates a one-parameter family of orbits, obtained by varying ΔC.

18.1.1 Properties

1. The quantities y_i for $i = 1, \ldots, \tilde{n}-1$ can never vanish, because of (18.1b). Therefore each of them keeps a constant sign along a family. This is an illustration of the principle of the invariance of the side of passage (Chap. 8).

2. For any solution, by applying the fundamental symmetry Σ of the restricted problem (Sect. 2.7), we obtain another solution:
$$\Sigma: \qquad (y_i, x_i) \mapsto (y_{\tilde{n}-i}, -x_{\tilde{n}+1-i}) \,. \tag{18.2}$$
3. For any solution, there exists a symmetrical solution obtained by changing the signs of all variables y_i and x_i. We call this symmetry Σ':
$$\Sigma': \qquad (y_i, x_i) \mapsto (-y_i, -x_i) \,. \tag{18.3}$$
4. For $\tilde{n} \geqq 2$, we can eliminate the x_i and obtain relations involving three successive values of y_i:
$$y_{i-1} - 2y_i + y_{i+1} + \frac{1}{y_i} = 0 \,, \qquad i = 1, \ldots, \tilde{n} - 1 \,. \tag{18.4}$$
5. (18.4) can be rewritten as a system of $\tilde{n} - 1$ quadratic equations:
$$y_i(y_{i-1} - 2y_i + y_{i+1}) + 1 = 0 \,, \qquad i = 1, \ldots, \tilde{n} - 1 \,. \tag{18.5}$$
It follows that there are at most $2^{\tilde{n}-1}$ solutions.

18.1.2 Number of Solutions

We forget temporarily the last equation $y_{\tilde{n}} = 0$. We start from $y_0 = 0$ and a given value of x_1 and we compute successively the other variables y_1, x_2, y_2, ..., $x_{\tilde{n}}$, $y_{\tilde{n}}$, using alternately (18.1a) and (18.1b), rewritten as
$$y_i = y_{i-1} - x_i \,, \qquad x_{i+1} = x_i + \frac{1}{y_i} \,. \tag{18.6}$$
All variables are thus functions of x_1.

Proposition 18.1.1. *There is*
$$\frac{\mathrm{d}y_i}{\mathrm{d}x_1} < 0 \,, \qquad \frac{\mathrm{d}x_i}{\mathrm{d}x_1} > 0 \,, \qquad i = 1, \ldots, \tilde{n} \,. \tag{18.7}$$

Proof: this is obviously true for $i = 1$, since (18.6a) reduces then to $y_1 = -x_1$. Assume that it is true up to i. From (18.6b) we have
$$\frac{\mathrm{d}x_{i+1}}{\mathrm{d}x_1} = \frac{\mathrm{d}x_i}{\mathrm{d}x_1} - \frac{1}{y_i^2}\frac{\mathrm{d}y_i}{\mathrm{d}x_1} > 0 \,, \tag{18.8}$$
and from (18.6a)
$$\frac{\mathrm{d}y_{i+1}}{\mathrm{d}x_1} = \frac{\mathrm{d}y_i}{\mathrm{d}x_1} - \frac{\mathrm{d}x_{i+1}}{\mathrm{d}x_1} < 0 \,. \tag{18.9}$$

From (18.6) one can easily draw up the table of variations of the successive variables (Table 18.1).

This table shows that when x_1 takes all possible values by increasing from $-\infty$ to $+\infty$, y_1 vanishes once, y_2 vanishes twice, and generally y_i vanishes 2^{i-1} times. In particular $y_{\tilde{n}}$ vanishes $2^{\tilde{n}-1}$ times. Reintroducing now the last equation $y_{\tilde{n}} = 0$, we find that

Table 18.1. Variations of the successive variables when x_1 increases from $-\infty$ to $+\infty$.

<table>
<tr><td>x_1</td><td colspan="4">$-\infty$ ↗ $+\infty$</td></tr>
<tr><td>y_1</td><td colspan="4">$+\infty$ ↘ 0 ↘ $-\infty$</td></tr>
<tr><td>x_2</td><td colspan="2">$-\infty$ ↗ $+\infty$</td><td colspan="2">$-\infty$ ↗ $+\infty$</td></tr>
<tr><td>y_2</td><td colspan="2">$+\infty$ ↘ 0 ↘ $-\infty$</td><td colspan="2">$+\infty$ ↘ 0 ↘ $-\infty$</td></tr>
<tr><td>x_3</td><td>$-\infty$ ↗ $+\infty$</td><td>$-\infty$ ↗ $+\infty$</td><td>$-\infty$ ↗ $+\infty$</td><td>$-\infty$ ↗ $+\infty$</td></tr>
<tr><td>y_3</td><td>$+\infty$ ↘ 0 ↘ $-\infty$</td><td>$+\infty$ ↘ 0 ↘ $-\infty$</td><td>$+\infty$ ↘ 0 ↘ $-\infty$</td><td>$+\infty$ ↘ 0 ↘ $-\infty$</td></tr>
<tr><td>...</td><td colspan="4">...</td></tr>
</table>

Proposition 18.1.2. *The system (18.1) has* $2^{\tilde{n}-1}$ *distinct solutions.*

We remark that this is the maximal number found in the previous section (Property 5).

Moreover, as can be seen easily from the table,

Proposition 18.1.3. *There is exactly one solution for each choice of the signs of* $y_1, \ldots, y_{\tilde{n}-1}$.

We can therefore represent symbolically each solution by the corresponding sequence of signs.

It will be convenient to write also a letter R for each basic arc. For instance, if $\tilde{n} = 5$, the solution corresponding to $y_1 > 0$, $y_2 > 0$, $y_3 < 0$, $y_4 > 0$ will be represented by the sequence

$$\mathtt{R+R+R-R+R}\,. \tag{18.10}$$

(17.169) shows that the signs of the y_i are the relative sides of passage σ'_i. We recall that a *R*-arc originates at $\nu = 1/3$ from the fusion of consecutive *S*-arcs. From Table 17.1 and (17.3), we find that $\sigma'_i = +1$ for a node between two *S*-arcs, while $\sigma'_i = -1$ for an antinode inside a *S*-arc. Thus, the $2^{\tilde{n}-1}$ solutions correspond to the $2^{\tilde{n}-1}$ possible decompositions of the *R*-arc into *S*-arcs (see below Sect. 19.3.3.1).

18.1.3 Stability and Jacobian

From Proposition 18.1.1 we know that $\mathrm{d}y_{\tilde{n}}/\mathrm{d}x_1$ never vanishes, i.e. *the R-arc is never critical.* Reasoning as in Sect. 13.1.4, we find that *the Jacobian never vanishes.*

18.1.4 Small Values of $\tilde{n}$

The system (18.4) can be solved for $\tilde{n} = 1$ to 4. For $\tilde{n} = 1$, we have the single solution $x_1 = 0$. For $\tilde{n} = 2$, the two solutions are

$$y_1 = \pm\frac{1}{\sqrt{2}} \,. \tag{18.11}$$

For $\tilde{n} = 3$, we have the two solutions

$$y_1 = y_2 = \pm 1 \tag{18.12}$$

and the two solutions

$$y_1 = -y_2 = \pm\frac{1}{\sqrt{3}} \,. \tag{18.13}$$

For $\tilde{n} = 4$, there are the two solutions

$$y_1 = y_3 = \sqrt{\frac{7 \pm \sqrt{17}}{8}} \,, \qquad y_2 = \pm\sqrt{\frac{5 \pm \sqrt{17}}{4}} \,, \tag{18.14}$$

where either the upper sign or the lower sign should be taken in both equations ; the two solutions

$$y_1 = \sqrt{\frac{7 \pm \sqrt{13}}{12}} \,, \qquad y_2 = \pm\sqrt{\frac{2}{3}} \,, \qquad y_3 = -\sqrt{\frac{7 \mp \sqrt{13}}{12}} \,; \tag{18.15}$$

and four other solutions obtained by changing all signs.

For $\tilde{n} > 4$, it seems that (18.4) can only be solved numerically. Table 18.2 gives the value of y_1 for $\tilde{n} = 2$ to 6 and for all sign sequences beginning with a +. Thus, only solutions with $y_1 > 0$ are listed; the others are obtained by changing the signs.

Values of y_1 are listed with 8 digits because the R-arcs are strongly unstable: errors are amplified during the computation of the successive y_i.

18.2 *R*-Orbit

In that case, we have a total bifurcation, and the R-region coincides with the whole bifurcation orbit. The number of basic arcs is $\tilde{n} = n$.

The equations are

$$\begin{aligned} &y_i - y_{i-1} + x_i = 0 \,, \\ &y_i(x_{i+1} - x_i) - 1 = 0 \,, \end{aligned} \tag{18.16}$$

where i is to be taken modulo n and takes all values from 1 to n. This is a system of $2n$ equations for $2n$ variables. So here again we expect isolated solutions. As explained in Sect. 18.1, each such solution generates a one-parameter family of orbits.

Table 18.2. *R*-arcs for $\tilde{n} = 2$ to 6.

$\tilde{n}$	sequence	y_1
2	R+R	0.70710678
3	R+R+R	1.00000000
	R+R-R	0.57735027
4	R+R+R+R	1.17914724
	R+R+R-R	0.94010422
	R+R-R+R	0.59967641
	R+R-R-R	0.53185593
5	R+R+R+R+R	1.30656296
	R+R+R+R-R	1.14267956
	R+R+R-R+R	0.95043145
	R+R+R-R-R	0.91921106
	R+R-R+R+R	0.60746124
	R+R-R+R-R	0.59586158
	R+R-R-R+R	0.54119610
	R+R-R-R-R	0.50526000

n	sequence	y_1
6	R+R+R+R+R+R	1.40472828
	R+R+R+R+R-R	1.28123317
	R+R+R+R-R+R	1.14898685
	R+R+R+R-R-R	1.12992127
	R+R+R-R+R+R	0.95398640
	R+R+R-R+R-R	0.94868391
	R+R+R-R-R+R	0.92357558
	R+R+R-R-R-R	0.90676764
	R+R-R+R+R+R	0.61196435
	R+R-R+R+R-R	0.60587822
	R+R-R+R-R+R	0.59651693
	R+R-R+R-R-R	0.59452140
	R+R-R-R+R+R	0.54444346
	R+R-R-R+R-R	0.53958978
	R+R-R-R-R+R	0.51070145
	R+R-R-R-R-R	0.48686874

18.2.1 Properties

The properties are essentially the same as for a *R*-arc (Sect. 18.1.1).

In regard to properties 4 and 5 , we have now

$$y_{i-1} - 2y_i + y_{i+1} + \frac{1}{y_i} = 0 \,, \qquad i = 1, \ldots, n \,. \tag{18.17}$$

This is a system of n equations for n variables $y_1, \ldots, y_n$. It can be rewritten as a system of n quadratic equations:

$$y_i(y_{i-1} - 2y_i + y_{i+1}) + 1 = 0 \,, \qquad i = 1, \ldots, n \,. \tag{18.18}$$

If we introduce homogeneous coordinates $Y_1, \ldots, Y_n, Z$, defined by

$$y_i = Y_i/Z \,, \tag{18.19}$$

we obtain

$$Y_i(Y_{i-1} - 2Y_i + Y_{i+1}) + Z^2 = 0 \,, \qquad i = 1, \ldots, n \,. \tag{18.20}$$

This is a set of n quadratic homogeneous equations. There is a double solution

$$Y_i = 1, \qquad i = 1, \ldots, n \,; \qquad Z = 0 \,. \tag{18.21}$$

This double solution is not valid since it corresponds to infinite values of the inhomogeneous coordinates y_i. It follows that the system (18.18) has at most $2^n - 2$ solutions.

18.2.2 Stability and Jacobian

We proceed as in Sect. 14.1.2. To compute the Jacobian, we order the variables as follows:

$$y_0, x_1, y_1, x_2, y_2, \ldots, y_{n-1}, x_n\ . \tag{18.22}$$

The fundamental equations (18.1) can be written

$$\begin{aligned}
f_1 &= y_1 - y_0 + x_1 = 0\ ,\\
f_2 &= y_1(x_2 - x_1) - 1 = 0\ ,\\
&\ldots\\
f_{2i-1} &= y_i - y_{i-1} + x_i = 0\ ,\\
f_{2i} &= y_i(x_{i+1} - x_i) - 1 = 0\ ,\\
&\ldots\\
f_{2n-1} &= y_0 - y_{n-1} + x_n = 0\ ,\\
f_{2n} &= y_0(x_1 - x_n) - 1 = 0\ .
\end{aligned} \tag{18.23}$$

For a given value of $|W|$, this is a system of $2n$ equations for $2n$ variables. The Jacobian is

$$|J| = \frac{\partial(f_1, \ldots, f_{2n})}{\partial(y_0, x_1, y_1, \ldots, x_n)}\ . \tag{18.24}$$

We can also "unroll" the system of equations (18.23) by adding two variables y_n and x_{n+1}, considered as distinct from y_0 and x_1, modifying the last two equations of (18.23) and adding two equations, so that the end of the new system reads:

$$\begin{aligned}
&\ldots\\
f_{2n-1} &= y_n - y_{n-1} + x_n = 0\ ,\\
f_{2n} &= y_n(x_{n+1} - x_n) - 1 = 0\ ,\\
f_{2n+1} &= y_0 - y_n = 0\ ,\\
f_{2n+2} &= x_1 - x_{n+1} = 0\ .
\end{aligned} \tag{18.25}$$

We consider the system of equations (18.25), with the last two equations deleted. We have then a system of $2n$ equations for $2n+2$ variables. Starting from given values of y_0 and x_1 and applying the equations one by one, we obtain successively y_1, x_2, $\ldots$, x_n, y_n, x_{n+1}. In the same way, starting from variations $\mathrm{d}y_0$ and $\mathrm{d}x_1$, we can compute successively $\mathrm{d}y_1$, $\mathrm{d}x_2, \ldots$, $\mathrm{d}y_n$, $\mathrm{d}x_{n+1}$. For the whole orbit, we have then

$$\begin{pmatrix} \mathrm{d}y_n \\ \mathrm{d}x_{n+1} \end{pmatrix} = \begin{pmatrix} \dfrac{\partial y_n}{\partial y_0} & \dfrac{\partial y_n}{\partial x_1} \\ \dfrac{\partial x_{n+1}}{\partial y_0} & \dfrac{\partial x_{n+1}}{\partial x_1} \end{pmatrix} \begin{pmatrix} \mathrm{d}y_0 \\ \mathrm{d}x_1 \end{pmatrix}\ . \tag{18.26}$$

The stability index is the trace of the matrix divided by 2:

$$z = \frac{1}{2}\left(\frac{\partial y_n}{\partial y_0} + \frac{\partial x_{n+1}}{\partial x_1}\right) . \tag{18.27}$$

Proceeding as in Sect. 14.1.2, we can show that

Proposition 18.2.1. *The Jacobian of a R-orbit vanishes if and only if it is a critical orbit of the first kind (the stability index is $z = 1$).*

Actually any pair of variables can be used to compute the stability index. In practice it will be convenient to use y_0 and y_1 rather than y_0 and x_1. We can then use (18.4) to obtain

$$\begin{pmatrix} \mathrm{d}y_i \\ \mathrm{d}y_{i+1} \end{pmatrix} = \begin{pmatrix} 0 & 1 \\ -1 & u_i \end{pmatrix} \begin{pmatrix} \mathrm{d}y_{i-1} \\ \mathrm{d}y_i \end{pmatrix} \tag{18.28}$$

with

$$u_i = 2 + \frac{1}{y_i^2} . \tag{18.29}$$

We can then compute the matrix

$$\begin{pmatrix} \dfrac{\partial y_n}{\partial y_0} & \dfrac{\partial y_n}{\partial y_1} \\ \dfrac{\partial y_{n+1}}{\partial y_0} & \dfrac{\partial y_{n+1}}{\partial y_1} \end{pmatrix} = \begin{pmatrix} 0 & 1 \\ -1 & u_n \end{pmatrix} \begin{pmatrix} 0 & 1 \\ -1 & u_{n-1} \end{pmatrix} \cdots \begin{pmatrix} 0 & 1 \\ -1 & u_1 \end{pmatrix} \tag{18.30}$$

and the stability index is

$$z = \frac{1}{2}\left(\frac{\partial y_n}{\partial y_0} + \frac{\partial y_{n+1}}{\partial y_1}\right) . \tag{18.31}$$

It can be shown that

Proposition 18.2.2. *R-orbits are always unstable.*

Proof: we will use the fact that, from (18.29):

$$u_i > 2 . \tag{18.32}$$

We write

$$\begin{pmatrix} \mathrm{d}y_{i-1} \\ \mathrm{d}y_i \end{pmatrix} = \begin{pmatrix} a_i & b_i \\ c_i & d_i \end{pmatrix} \begin{pmatrix} \mathrm{d}y_{-1} \\ \mathrm{d}y_0 \end{pmatrix} . \tag{18.33}$$

There is

$$\begin{pmatrix} a_i & b_i \\ c_i & d_i \end{pmatrix} = \begin{pmatrix} 0 & 1 \\ -1 & u_{i-1} \end{pmatrix} \begin{pmatrix} a_{i-1} & b_{i-1} \\ c_{i-1} & d_{i-1} \end{pmatrix} \tag{18.34}$$

or

$$\begin{aligned} &a_i = c_{i-1} , \qquad b_i = d_{i-1} , \\ &c_i = -a_{i-1} + c_{i-1}u_{i-1} , \qquad d_i = -b_{i-1} + d_{i-1}u_{i-1} . \end{aligned} \tag{18.35}$$

We show first that

$$c_i + d_i > 1\,, \qquad c_i + d_i - (a_i + b_i) > 0\,. \tag{18.36}$$

This is true for $i = 1$, because $a_1 = 0$, $b_1 = 1$, $c_1 = -1$, $d_1 = u_0 > 2$. Assume that it is true for $i - 1$. There is $u_{i-1} > 2$, therefore

$$\begin{aligned} c_i + d_i &= u_{i-1}(c_{i-1} + d_{i-1}) - (a_{i-1} + b_{i-1}) \\ &> 2(c_{i-1} + d_{i-1}) - (a_{i-1} + b_{i-1}) > 1\,; \end{aligned} \tag{18.37}$$

and

$$\begin{aligned} c_i + d_i - (a_i + b_i) &= (u_{i-1} - 1)(c_{i-1} + d_{i-1}) - (a_{i-1} + b_{i-1}) \\ &> (c_{i-1} + d_{i-1}) - (a_{i-1} + b_{i-1}) > 0\,. \end{aligned} \tag{18.38}$$

Next we show that

$$c_i \leqq -i\,, \qquad a_i - c_i \geqq 1\,. \tag{18.39}$$

This is true for $i = 1$. Assume that it is true for $i - 1$. There is

$$\begin{aligned} c_i &= -a_{i-1} + c_{i-1}u_{i-1} < -a_{i-1} + 2c_{i-1} = (c_{i-1} - a_{i-1}) + c_{i-1} \\ &\leqq -(i-1) - 1 = -i\,; \end{aligned} \tag{18.40}$$

and

$$a_i - c_i = a_{i-1}) - c_{i-1}(u_{i-1} - 1) > a_{i-1}) - c_{i-1} \geqq 1\,. \tag{18.41}$$

Finally, adding (18.36a) and (18.39b), we obtain

$$a_i + d_i > 2\,. \tag{18.42}$$

Taking $i = n$, we have

$$z = \frac{a_n + d_n}{2} > 1\,. \tag{18.43}$$

It follows that *the Jacobian never vanishes.*

18.2.3 Small Values of n

We solve the system of n equations (18.17).

For $n = 1$, there are no solutions.

18.2.3.1 $n = 2$. There are two solutions. The first is

$$y_0 = \frac{1}{2}\,, \qquad y_1 = -\frac{1}{2}\,, \tag{18.44}$$

and the second is obtained by a shift of the origin of one basic arc.

18.2.3.2 $n = 3$. After a little algebra we obtain the 2 solutions

$$y_0 = \pm\frac{1}{\sqrt{6}}\,, \qquad y_1 = y_2 = \mp\frac{2}{\sqrt{6}}\,, \tag{18.45}$$

and 4 other solutions obtained by shifts of the origin. This gives a total of 6 solutions.

We remark that for $n = 1$ to 3, the number of solutions equals the maximal number $2^n - 2$.

18.2.3.3 $n = 4$. After some more computation, we find 1 solution

$$y_0 = y_1 = \frac{1}{\sqrt{2}}\,, \qquad y_2 = y_3 = -\frac{1}{\sqrt{2}}\,; \tag{18.46}$$

the 2 solutions

$$y_0 = \pm\frac{1+\sqrt{3}}{2}\,, \qquad y_1 = y_3 = \pm 1\,, \qquad y_2 = \pm\frac{1-\sqrt{3}}{2}\,; \tag{18.47}$$

and 9 other solutions obtained by shifts of the origin. This gives a total of 12 solutions. Adding the two spurious solutions with sub-period 2, we have a total of 14, which equals the maximal number $2^4 - 2$.

Numerical values for the solutions (18.44), (18.45), (18.46), (18.47) are listed in Table 18.3.

Table 18.3. R-orbits for $n = 2$ to 6.

n	y_0	y_1	y_2	y_3	y_4	y_5
2	0.500000	−0.500000				
3	0.408248	−0.816497	−0.816497			
	−0.408248	0.816497	0.816497			
4	0.707107	0.707107	−0.707107	−0.707107		
	1.366025	1.000000	−0.366025	1.000000		
	−1.366025	−1.000000	0.366025	−1.000000		
5	1.712938	1.712938	1.129146	−0.340271	1.129146	
	−1.712938	−1.712938	−1.129146	0.340271	−1.129146	
	0.799673	0.799673	−0.450837	0.516750	−0.450837	
	−0.799673	−0.799673	0.450837	−0.516750	0.450837	
	0.652966	0.652966	−0.878507	−1.271686	−0.878507	
	−0.652966	−0.652966	0.878507	1.271686	0.878507	
6	1.224745	0.816497	−0.816497	−1.224745	−0.816497	0.816497
	1.345162	0.973460	−0.425507	0.525667	−0.425507	0.973460
	−1.345162	−0.973460	0.425507	−0.525667	0.425507	−0.973460
	2.193233	1.965259	1.228446	−0.322404	1.228446	1.965259
	−2.193233	−1.965259	−1.228446	0.322404	−1.228446	−1.965259
	1.618034	1.618034	1.000000	−0.618034	−0.618034	1.000000
	−1.618034	−1.618034	−1.000000	0.618034	0.618034	−1.000000
	−0.720707	0.720707	0.774597	−0.462509	0.462509	−0.774597
	0.720707	−0.720707	−0.774597	0.462509	−0.462509	0.774597

18.2.3.4 $n = 5$. A frontal attack on the system (18.17) seems difficult. We remember, however, that a R-orbit originates at $\nu = 1/3$ from the fusion of consecutive S-arcs. Using the notation of Sect. 6.2.1.3, in which each S-arc is represented by its length m, we find that there are 30 possible combinations:

$$5, 41, 32, 311, 221, 2111\,, \tag{18.48}$$

and 24 other combinations deduced by shifts of the origin. We remark then that all of these combinations correspond to symmetric orbits. This suggests that we look for symmetric solutions of the system (18.17).

One of the crossings of the symmetry axis lies in the midpoint of a basic arc. We choose the origin in such a way that this is the first basic arc (i.e. we take $h = 1$, in the notation introduced in Sect. 7.3.2). There is then $y_0 = y_1$, $y_2 = y_4$. After elimination, we obtain the following equation for y_1:

$$5y_1^6 - 20y_1^4 + 17y_1^2 - 4 = 0 , \tag{18.49}$$

which has 6 real solutions, listed in Table 18.3. 24 other solutions are obtained by shifts of the origin. We have thus a total of 30 symmetric solutions. This equals the maximal number of solutions $2^5 - 2$; therefore we have in fact found all solutions.

18.2.3.5 $n = 6$. We follow the same strategy as for $n = 5$. The possible decompositions into S-arcs fall into three groups with respect to the symmetry Σ: (i) symmetric orbits with the two crossings of the symmetry axis in collisions:

$$6, 42, 411, 2211, 21111 ; \tag{18.50}$$

(ii) symmetric orbits with the crossings in midpoints:

$$51, 3111 ; \tag{18.51}$$

(iii) asymmetric orbits:

$$321, 312 . \tag{18.52}$$

We consider each case in turn.

(i) Symmetry axis through collisions. We take the origin in one crossing ($h = 0$). Then $y_1 = y_5$, $y_2 = y_4$. After some algebra, we obtain the solution

$$y_0 = \sqrt{\frac{3}{2}} , \qquad y_1 = y_5 = \sqrt{\frac{2}{3}} , \qquad y_2 = y_4 = -\sqrt{\frac{2}{3}} , \qquad y_3 = -\sqrt{\frac{3}{2}} , \tag{18.53}$$

and the 4 solutions

$$\begin{aligned} y_0 &= \frac{\omega_1}{2\sqrt{2}}\left(\sqrt{7} + \sqrt{7 - 4\sqrt{2}\omega_2}\right) , \\ y_1 = y_5 &= \frac{\omega_1}{4}\left[(\sqrt{2} - \omega_2)\sqrt{7} + (\sqrt{2} + \omega_2)\sqrt{7 - 4\sqrt{2}\omega_2}\right] , \\ y_2 = y_4 &= \frac{\omega_1}{4}\left[(\sqrt{2} - \omega_2)\sqrt{7} - (\sqrt{2} + \omega_2)\sqrt{7 - 4\sqrt{2}\omega_2}\right] , \\ y_3 &= \frac{\omega_1}{2\sqrt{2}}\left(\sqrt{7} - \sqrt{7 - 4\sqrt{2}\omega_2}\right) , \end{aligned} \tag{18.54}$$

where $\omega_1 = \pm 1$, $\omega_2 = \pm 1$, independently of each other.

(ii) Symmetry axis through midpoints. We take one crossing in the midpoint of the first basic arc ($h = 1$). We find the 2 solutions

$$y_0 = y_1 = \pm\frac{\sqrt{5}+1}{2}\,, \qquad y_2 = y_5 = \pm 1\,, \qquad y_3 = y_4 = \pm\frac{1-\sqrt{5}}{2}\,. \tag{18.55}$$

(iii) Asymmetric orbits. It turns out that explicit solutions can also be obtained in this case. The sequence of signs for the asymmetric orbit **321** is: +--+-+. This sequence is nos symmetric under Σ. However, it is it is invariant under a change of the signs of the y_i (as allowed by Property 3, Sect. 18.1.1), followed by a Σ symmetry. This suggests that we look for solutions of the form

$$y_0 = -y_1\,, \qquad y_2 = -y_5\,, \qquad y_3 = -y_4\,. \tag{18.56}$$

We find indeed 2 such solutions:

$$y_0 = -y_1 = \mp\frac{1}{2}\left(\sqrt{\frac{7}{5}} + \frac{1}{\sqrt{15}}\right), y_2 = -y_5 = \pm\sqrt{\frac{3}{5}}\,,$$
$$y_3 = -y_4 = \mp\frac{1}{2}\left(\sqrt{\frac{7}{5}} - \frac{1}{\sqrt{15}}\right)\,. \tag{18.57}$$

The 9 solutions (18.53), (18.54), (18.55), (18.57) are listed in Table 18.3. With the shifted solutions, we have a total of 54 solutions. If we add the two spurious solutions with sub-period 2 and the 6 spurious solutions with sub-period 3, we have a total of 62, which equals the maximal number $2^6 - 2$. Therefore all solutions have been found.

18.2.3.6 $n > 6$. For $n = 7$, there appears for the first time sequences of signs devoid of any symmetry, even when Property 3 is used. One example is the sequence +++-+--. It seems that solutions for $n > 6$ can only be obtained numerically. For instance the solution corresponding to the above sequence was computed as: $y_0 = 0.880142094$, $y_1 = 1.302150709$, $y_2 = 0.956199$, $y_3 = -0.435560$, $y_4 = 0.468576$, $y_5 = -0.761411$, $y_6 = -0.678047$.

18.2.4 Sign Sequences

The example of the R-arcs (see Proposition 18.1.3) suggests that we look at the sequence of the signs of the y_i. A detailed examination of Table 18.3, completed by the addition of the shifted solutions, reveals indeed a remarkable property: *there is exactly one solution for each choice of the signs of* $y_0, \ldots, y_{n-1}$, *with the exception of the two sign sequences +++... and ---... for which there are no solutions.* This property holds true up to $n = 6$, and it is tempting to conjecture that it is true for all n. This conjecture will indeed be proven in Sect. 18.3.

We can therefore represent symbolically each solution by the corresponding sequence of signs. As in the case of R-arcs (Sect. 18.1.2), we write also a letter R for each basic arc. As was done earlier (Sect. 8.5), for the sake of symmetry we repeat the first sign at the end of the sequence. For instance, the solution quoted in the previous Sect. 18.2.3.6 for $n = 7$ is represented by the sequence

$$\texttt{-R+R+R+R-R+R-R-} . \tag{18.58}$$

18.3 Study of the Mapping

We will show now that the equations (18.1) or (18.16) of a R region have a remarkable property: they are equivalent in a sense to the classical *baker transformation*. This equivalence accounts for the properties observed in the previous sections.

Eliminating the x_i from (18.1) or (18.16), we obtain a relation between three consecutive values of the y_i:

$$y_{i-1} - 2y_i + y_{i+1} + \frac{1}{y_i} = 0 . \tag{18.59}$$

Consider a plane with coordinates (y_i, y_{i+1}). (18.59) defines a mapping F in that plane. This becomes obvious if we rename the coordinates

$$\alpha_i = y_i , \qquad \beta_i = y_{i+1} ; \tag{18.60}$$

we have then

$$F : \qquad \alpha_{i+1} = \beta_i , \qquad \beta_{i+1} = -\alpha_i + 2\beta_i - \frac{1}{\beta_i} . \tag{18.61}$$

This is a one-to-one, area-preserving mapping; its inverse can be written explicitly:

$$F^{-1} : \qquad \alpha_i = 2\alpha_{i+1} - \beta_{i+1} - \frac{1}{\alpha_{i+1}} , \qquad \beta_i = \alpha_{i+1} . \tag{18.62}$$

Consider a point $P_0 = (y_0, y_1)$ of the plane. By applying the mapping F repeatedly, we obtain a sequence of points $P_1 = (y_1, y_2), \ldots, P_i = (y_i, y_{i+1})$, ... Equivalently, we may consider that we generate a sequence of values

$$y_1, y_2, \ldots, y_i, \ldots . \tag{18.63}$$

Two cases are possible:

1. All y_i in the sequence (18.63) are different from zero. We obtain an infinite sequence (18.63).
   The results of the previous sections suggest that the sequence of the signs of the y_i plays a fundamental role, and acts as a "signature" of the orbit or arc. We associate therefore to the sequence (18.63) an infinite sequence of binary digits

$$\omega_1, \omega_2, \ldots, \omega_i, \ldots \,, \tag{18.64}$$

where ω_i is defined by

$$\omega_i = \begin{cases} 0 & \text{if } y_i < 0, \\ 1 & \text{if } y_i > 0. \end{cases} \tag{18.65}$$

Incidentally, (17.169) shows that ω_i is related to the relative side of passage σ'_i by

$$\omega_i = \frac{\sigma'_i + 1}{2} \,. \tag{18.66}$$

2. We reach a value $y_i = 0$ for some $i > 0$. Then (18.59) becomes undefined and the sequence (18.63) cannot be continued beyond y_i. (As an extreme case, this may happen for $i = 1$, i.e. the initially given value y_1 equals zero. The sequence (18.63) is then limited to its first term.) The rule (18.65) fails to define ω_j for $j \geqq i$. In that case, we still define an infinite sequence of binary digits (18.64), by writing

$$\omega_i = 1 \,; \qquad \omega_j = 0 \quad \text{for} \quad j > i \,. \tag{18.67}$$

We could also write

$$\omega_i = 0 \,; \qquad \omega_j = 1 \quad \text{for} \quad j > i \,; \tag{18.68}$$

as will be shown below, these two definitions are equivalent.
These definitions can be intuitively justified by considering what happens in the limit of a very small but non-zero y_i. Then (18.59) gives

$$y_{i+1} \simeq -y_i^{-1} \,, \quad y_{i+2} \simeq -2y_i^{-1} \,, \quad \ldots \,, \quad y_{i+j} \simeq -jy_i^{-1} \,, \quad \ldots \tag{18.69}$$

If $y_i > 0$, applying the rule (18.65), we obtain the sequence (18.67). Similarly, if $y_i < 0$, we obtain the sequence (18.68).

Thus in all cases we associate an infinite sequence (18.64) to the point P_0. We consider now the ω_i as the successive binary digits (after the binary point) of a number v. In other words, v is written in binary as

$$v = 0.\omega_1\omega_2 \ldots \omega_i \ldots \tag{18.70}$$

and is formally defined as

$$v = \sum_{i=1}^{\infty} 2^{-i}\omega_i \,. \tag{18.71}$$

We remark that in case 2 above, both sequences (18.67) and (18.68) give the same value for v:

$$v = 0.\omega_1 \ldots \omega_{i-1}0111 \ldots = 0.\omega_1 \ldots \omega_{i-1}1000 \ldots = \sum_{j=1}^{i-1} 2^{-j}\omega_j + 2^{-i} \,. \tag{18.72}$$

Thus, v is uniquely defined for any point P_0, i.e. it is a definite function $v(y_0, y_1)$. From the definition (18.70) or (18.71) we have

$$0 \leqq v \leqq 1 \,. \tag{18.73}$$

We study this function by keeping y_0 constant and varying y_1. The table of variations (Table 18.4) is easily established by going back to the variables x_i, y_i and using (18.6). (This is a generalization of Table 18.1.) When y_1 increases from $-\infty$ to $+\infty$, x_1 decreases from $+\infty$ to $-\infty$; x_2 becomes infinite for $y_1 = 0$, and decreases from $+\infty$ to $-\infty$ in each of the intervals $-\infty < y_1 < 0$, $0 < y_1 < +\infty$; y_2 increases from $-\infty$ to $+\infty$ in each of these two intervals; x_3 decreases 4 times from $+\infty$ to $-\infty$; and so on. The last line but one shows the beginning of the sequence $\omega_1\omega_2\omega_3 \dots$ in the successive intervals; the last line shows values of v when one y_i vanishes.

Table 18.4. Variations of the successive variables when y_0 is constant and y_1 increases from $-\infty$ to $+\infty$.

<table>
<tr><td>y_1</td><td colspan="8">$-\infty$ ↗ 0 ↗ $+\infty$</td></tr>
<tr><td>x_1</td><td colspan="8">$+\infty$ ↘ $-\infty$</td></tr>
<tr><td>x_2</td><td colspan="4">$+\infty$ ↘ $-\infty$</td><td colspan="4">$+\infty$ ↘ $-\infty$</td></tr>
<tr><td>y_2</td><td colspan="4">$-\infty$ ↗ 0 ↗ $+\infty$</td><td colspan="4">$-\infty$ ↗ 0 ↗ $+\infty$</td></tr>
<tr><td>x_3</td><td colspan="2">$+\infty$ ↘ $-\infty$</td><td colspan="2">$+\infty$ ↘ $-\infty$</td><td colspan="2">$+\infty$ ↘ $-\infty$</td><td colspan="2">$+\infty$ ↘ $-\infty$</td></tr>
<tr><td>y_3</td><td colspan="2">$-\infty$ ↗ 0 ↗ $+\infty$</td><td colspan="2">$-\infty$ ↗ 0 ↗ $+\infty$</td><td colspan="2">$-\infty$ ↗ 0 ↗ $+\infty$</td><td colspan="2">$-\infty$ ↗ 0 ↗ $+\infty$</td></tr>
<tr><td>x_4</td><td>↘</td><td>↘</td><td>↘</td><td>↘</td><td>↘</td><td>↘</td><td>↘</td><td>↘</td></tr>
<tr><td>...</td><td colspan="8">...</td></tr>
<tr><td></td><td>000...</td><td>001...</td><td>010...</td><td>011...</td><td>100...</td><td>101...</td><td>110...</td><td>111...</td></tr>
<tr><td>v</td><td colspan="8">$\frac{1}{8}$ $\frac{1}{4}$ $\frac{3}{8}$ $\frac{1}{2}$ $\frac{5}{8}$ $\frac{3}{4}$ $\frac{7}{8}$</td></tr>
</table>

Fig. 18.1 shows v as a function of y_1, for the arbitrarily chosen value $y_0 = 1$; other values of y_0 give similar curves. This figure, and Table 18.4, strongly suggest that v is a continuous and increasing function of y_1, taking all values in the interval

$$0 < v < 1 \,. \tag{18.74}$$

$v(y_1)$ is a highly irregular function, with a "devil's staircase" appearance. (The steps which seem horizontal on the figure are actually slightly inclined.)

Similarly, starting from the point $P_0 = (y_0, y_1)$ and applying F^{-1} repeatedly, we obtain a sequence of points $P_{-1} = (y_{-1}, y_0)$, ..., $P_{-i} = (y_{-i}, y_{-i+1})$, ... Equivalently, we may consider that we generate a sequence of values

$$y_0, y_{-1}, y_{-2}, \dots, y_{-i}, \dots \,. \tag{18.75}$$

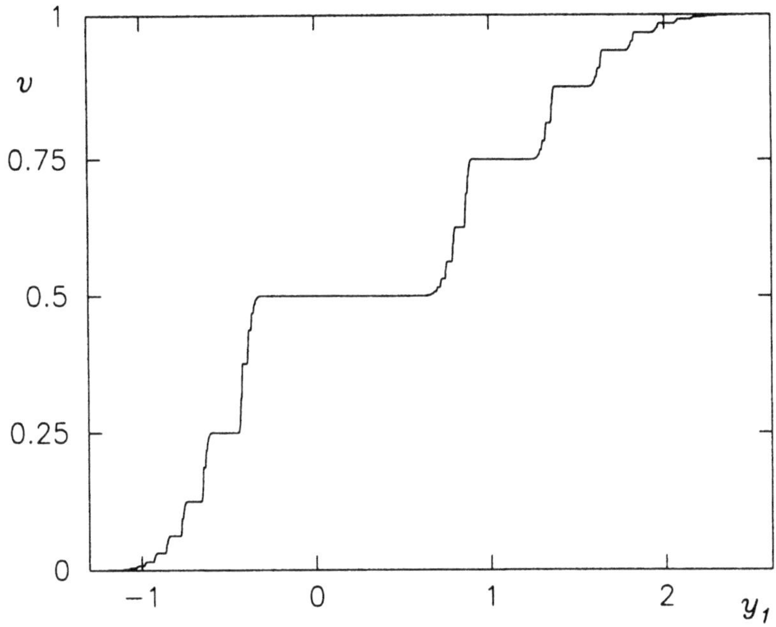

Fig. 18.1. v as a function of y_1, for $y_0 = 0$.

Again two cases are possible: the sequence (18.75) may be infinite, or finite and ending in some $y_{-i} = 0$. (As an extreme case, this may happen for $i = 0$, i.e. the initially given value y_0 equals zero.) Proceeding as above, we define in all cases an infinite sequence

$$\omega_0, \omega_{-1}, \omega_{-2}, \ldots, \omega_{-i}, \ldots \,, \tag{18.76}$$

We consider now the successive terms of (18.76) as the binary digits of a number u:

$$u = 0.\omega_0\omega_{-1}\ldots\omega_{-i}\ldots = \sum_{i=0}^{\infty} 2^{-i-1}\omega_{-i} \,. \tag{18.77}$$

u is uniquely defined for any point P_0. We can study the function $u(y_0, y_1)$ by keeping y_1 constant and varying y_0. We obtain a table similar to Table 18.4, and figures similar to Fig. 18.1, which strongly suggest that u is a continuous and increasing function of y_0, taking all values in the interval

$$0 < u < 1 \,. \tag{18.78}$$

We have thus defined a map φ from the plane (y_0, y_1) to the square $0 < u < 1$, $0 < v < 1$:

$$\varphi : (y_0, y_1) \mapsto (p, q) \,. \tag{18.79}$$

Fig. 18.2 illustrates this map by showing some lines with constant u or v in the (y_0, y_1) plane. These lines are easily computed. $v = 1/2$ corresponds to the horizontal axis $y_1 = 0$. $v = 1/4$ and $v = 3/4$ correspond to $y_2 = 0$ (and $y_1 < 0$, $y_1 > 0$ respectively); therefore the corresponding lines are obtained by applying the mapping F^{-1} to the horizontal axis. Similarly, $v = 1/8$, $3/8$, $5/8$, $7/8$ correspond to $y_3 = 0$ and the corresponding lines are obtained by applying F^{-2} to the horizontal axis. In the same way, $u = 1/2$ corresponds to the vertical axis $y_0 = 0$; and the other lines $u =$ Const. are obtained by applying F or F^2 to the horizontal axis.

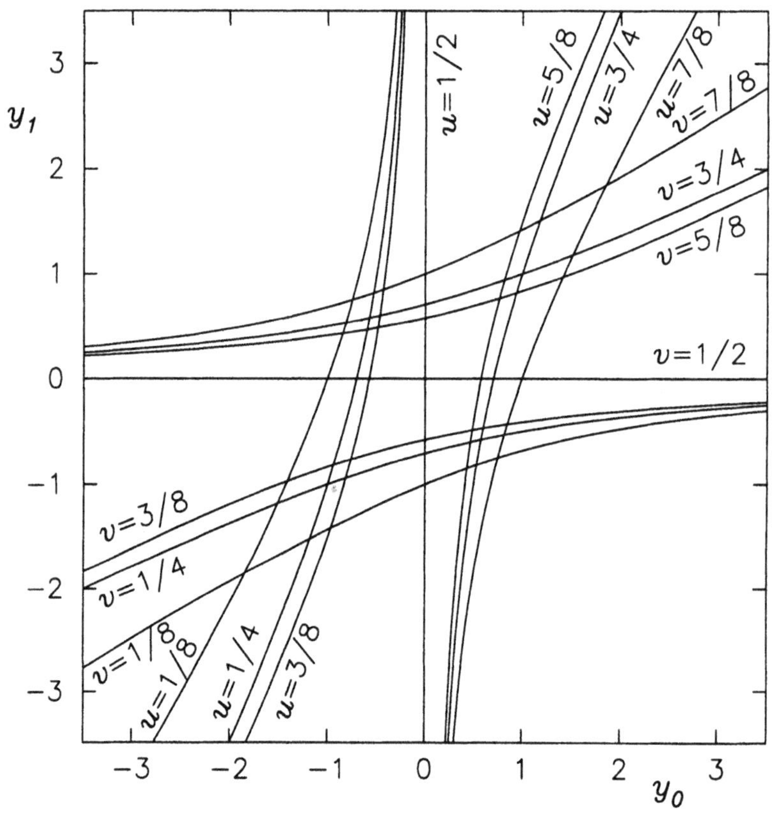

Fig. 18.2. The map φ: curves $u =$ Const. and $v =$ Const. in the (y_0, y_1) plane.

This figure, together with the previous results (Table 18.4 and Fig. 18.1) suggests the following conjecture: *the map φ is continuous and one-to-one* (it is a *homeomorphism*). This conjecture was indeed proved by Devaney (1981). (Devaney uses a slightly different form of the mapping, with x_i and y_i rather than y_i and y_{i+1} as coordinates.)

As a result, we can consider the passage from (y_0, y_1) to (u, v) through the map φ as a simple change of coordinates, and instead of studying the original mapping F in the (y_0, y_1) coordinates, we can consider the corresponding mapping B in the (u, v) coordinates. B is formally defined by

$$B\varphi = \varphi F \,. \tag{18.80}$$

But the mapping B corresponds to a simple Bernoulli shift on the doubly infinite sequence

$$\ldots \omega_{-1}\omega_0 . \omega_1 \omega_2 \ldots \tag{18.81}$$

so that B is simply the well-known baker transformation

$$\begin{aligned} u_{i+1} &= \frac{u_i}{2}\,, \qquad v_{i+1} = 2v_i\,, \qquad \text{if} \quad v_i < \frac{1}{2}\,, \\ u_{i+1} &= \frac{u_i+1}{2}\,, \qquad v_{i+1} = 2v_i - 1\,, \qquad \text{if} \quad v_i > \frac{1}{2}\,. \end{aligned} \tag{18.82}$$

We have thus (Devaney, 1981)

Proposition 18.3.1. *The mapping F is topologically conjugate to the baker transformation.*

All observed properties of the R-arcs and R-orbits can now be simply derived. A R-arc corresponds to $y_0 = 0$, $y_i \neq 0$ for $i = 1, \ldots, \tilde{n} - 1$, $y_{\tilde{n}} = 0$. Therefore the binary representations of u and v are

$$u = 0.1000\ldots\,, \qquad v = 0.\omega_1 \ldots \omega_{\tilde{n}-1}1000\ldots \tag{18.83}$$

Conversely, values of u and v of that form, with arbitrary values ω_1, ..., $\omega_{\tilde{n}-1}$, correspond to a R-arc of order $\tilde{n}$. We obtain thus $2^{\tilde{n}-1}$ distinct R-arcs. Each R-arc corresponds to one sequence of signs. (This was already proved in Sect. 18.1.2 by a different method.) In the (u, v) plane, the representative points lie on $u = 1/2$ and have ordinates $v = 1/2^{\tilde{n}}, 3/2^{\tilde{n}}, 5/2^{\tilde{n}}, \ldots (2\tilde{n}-1)/2^{\tilde{n}}$.

For a R-orbit, the sequence of the ω_i must be periodic. Conversely, any periodic sequence with a period n corresponds to a R-orbit, with two exceptions: the sequences ...000... and ...111.... This is because these sequences correspond to $u = v = 0$ and $u = v = 1$ respectively, and these end values are excluded. As shown by Table 18.4, they correspond to infinite values of all y_i. Another way to show this is to sum (18.59) over all i, obtaining

$$\sum_{i=0}^{n-1} \frac{1}{y_i} = 0\,, \tag{18.84}$$

which shows that the y_i cannot all have the same sign. We obtain thus $2^n - 2$ distinct R-orbits. Each R-orbit corresponds to one sequence of signs.

Finally, all orbits in the baker transformation are highly unstable: any deviation is multiplied by 2 at each iteration. This explains why R-arcs and R-orbits are always unstable.

19. Partial Transition 2.1

We study now in detail the equations obtained for transition 2.1 in Sect. 17.5. A bifurcation orbit or a bifurcating arc is then formed of T-arcs and R-regions. The values of the variables for a T-arc are given explicitly by (17.157) and (17.158). For a R-region, on the other hand, we have only the implicit system of equations (17.154), which we consider now.

We consider first, in the present chapter, the case of a partial transition, and a particular R-arc. To simplify the notations, we renumber the encounters and the basic arcs starting from the origin of the R-arc. The equations are then

$$
\begin{aligned}
&x_i'' - x_i' + w = 0\,, \qquad i = 1, \ldots, \tilde{n}\,,\\
&y_i - y_{i-1} + w(x_i' + x_i'') = 0\,, \qquad i = 1, \ldots, \tilde{n}\,,\\
&y_i(x_{i+1}' - x_i'') - 1 = 0\,, \qquad i = 1, \ldots, \tilde{n} - 1\,;\\
&y_0 = 0\,, \qquad y_{\tilde{n}} = 0\,.
\end{aligned}
\tag{19.1}
$$

These equations form a system of $3\tilde{n}+1$ equations for the $3\tilde{n}+2$ variables w, y_0 to $y_{\tilde{n}}$, x_1' to $x_{\tilde{n}}'$, x_1'' to $x_{\tilde{n}}''$. Therefore we expect one-parameter families of solutions. For a given value of μ, there is a one-to-one correspondence, given by (17.138) to (17.141), between the present variables w, y_i, x_i', x_i'' and the physical variables ΔC, h_i, u_i', u_i''. Thus, the one-parameter families of solutions of (19.1) correspond simply to the ordinary one-parameter families of orbits (see Sect. 2.3).

19.1 Properties

1. The equations (19.1) do not contain any parameter: they are the same for all transitions 2.1, independently of the values I, J, ϵ' which characterize the bifurcation. This differs from type 1 where the equations contained a parameter K (see (13.1)).
2. The quantities y_i for $i = 1, \ldots, \tilde{n}-1$ can never vanish, because of (19.1c). Therefore each of them keeps a constant sign along a family. This is an illustration of the principle of the invariance of the side of passage (Chap. 8).

3. Assume that $w = 0$. (19.1c) shows then that all y_i are equal. From (19.1d) it follows that all y_i are zero. If $\tilde{n} \geqq 2$, this contradicts the previous point. Thus: for $\tilde{n} \geqq 2$, w never vanishes, and has a constant sign along a family.
4. For any solution, by applying the fundamental symmetry Σ of the restricted problem (Sect. 2.7), we obtain another solution:

$$\Sigma: \qquad (y_i, x_i', x_i'') \mapsto (y_{\tilde{n}-i}, -x''_{\tilde{n}+1-i}, -x'_{\tilde{n}+1-i}) \,. \tag{19.2}$$

5. w, defined by (17.140) and (17.141h), has the sign of $-\epsilon' \Delta C$. Thus: the symbolic sign of the branches, as defined in Sect. 6.2.1.3, is the opposite of the sign of w.
6. For $\tilde{n} \geqq 2$, we can eliminate the x_i' and x_i'' by extracting them from (19.1a) and (19.1b) and substituting into (19.1c). We obtain relations involving three successive values of y_i:

$$y_{i-1} - 2y_i + y_{i+1} = -\frac{2w}{y_i} + 2w^2 \,, \qquad i = 1, \ldots, \tilde{n} - 1 \,. \tag{19.3}$$

Taking into account the boundary conditions $y_0 = y_{\tilde{n}} = 0$, and for a given w, we have a system of $\tilde{n} - 1$ equations for $\tilde{n} - 1$ variables $y_1, \ldots, y_{\tilde{n}-1}$. After this system has been solved, the x_i' and x_i'' can be computed by

$$x_i' = \frac{y_{i-1} - y_i}{2w} + \frac{w}{2} \,, \qquad x_i'' = \frac{y_{i-1} - y_i}{2w} - \frac{w}{2} \,. \tag{19.4}$$

7. (19.3) can be rewritten as a system of $\tilde{n} - 1$ quadratic equations:

$$y_i(y_{i-1} - 2y_i + y_{i+1}) - 2w^2 y_i + 2w = 0 \,, \qquad i = 1, \ldots, \tilde{n} - 1 \,. \tag{19.5}$$

It follows that for a given w, there are at most $2^{\tilde{n}-1}$ solutions.
8. We define

$$p_i = y_i \,, \qquad q_i = y_{i+1} \,. \tag{19.6}$$

From (19.3) we have then

$$p_i = q_{i-1} \,, \qquad q_i = -p_{i-1} + 2q_{i-1} - \frac{2w}{q_{i-1}} + 2w^2 \,. \tag{19.7}$$

Thus, the problem is formally equivalent to the study of a plane mapping, which is easily shown to be area-preserving. A R-arc corresponds to an initial point on the q axis ($p_0 = 0$), such that after $\tilde{n} - 1$ iterations of the mapping we obtain a point on the p axis ($q_{\tilde{n}-1} = 0$). Numerical explorations (Hénon, unpublished) of the mapping (19.3) exhibit the characteristic mixture of regular and chaotic orbits found in non-integrable systems. This supports the conjecture that the system (19.1) is not explicitly solvable in general.

19.1.1 Asymptotic Branches for $w \to \pm\infty$

In the limit $w \to \pm\infty$, we enter the region $|\Delta C| \gg \mu^{1/3}$, or $\nu < 1/3$. The solutions in that region are known (Sect. 17.4). For a sequence of consecutive S-arcs, they are given by the equations (17.123), (17.124), (17.127) of Sect. 17.4. If we substitute these expressions into the change of variables (17.138) to (17.141) for a R-region in the case $\nu = 1/3$, we obtain

$$\begin{aligned} y_{i+j} &= -j(m-j)w^2[1+O(w^{-3})] \qquad \text{inside a } S\text{-arc}, \\ y_i &= \frac{2}{m_a+m_b}w^{-1}[1+O(w^{-3})] \qquad \text{at a } SS \text{ node}, \\ x'_{i+j} &= \frac{m-2j+2}{2}w[1+O(w^{-3})]\ , \\ x''_{i+j} &= \frac{m-2j}{2}w[1+O(w^{-3})]\ . \end{aligned} \tag{19.8}$$

It can be verified that these asymptotic values satisfy the equations (19.1).

Conversely, we show now that for $w \to \pm\infty$, the one-parameter families of solutions of (19.1) have no other asymptotic branches than those which correspond to a decomposition of the R-arc into S-arcs, and which are described by the equations (19.8). Consider one particular branch. We assume that the y_i behave asymptotically as powers of $|w|$:

$$y_i \sim b_i|w|^{p_i} \qquad \text{for} \qquad |w| \to \infty\ . \tag{19.9}$$

The b_i are non-zero constants. The p_i are arbitrary constants. The two ends of the R-arc, $y_0 = y_{\tilde{n}} = 0$, can be included in this formulation by taking $p_0 = p_{\tilde{n}} = -\infty$.

Assume that there exists a $p_i > 2$. In (19.3), the five terms behave asymptotically as

$$|w|^{p_{i-1}}, |w|^{p_i}, |w|^{p_{i+1}}, |w|^{1-p_i}, |w|^2\ . \tag{19.10}$$

Terms 4 and 5 are negligible in comparison with term 2. Therefore the latter must be balanced by terms 1 and/or 3, and at least one of the following inequalities is true:

$$p_{i-1} \geqq p_i\ , \qquad p_{i+1} \geqq p_i\ . \tag{19.11}$$

Assume that the second inequality is true. It follows in particular that the encounter $i+1$ cannot be the end of the R-arc. We distinguish several cases.

(i) $p_{i+1} > p_i$. Consider (19.3) with i increased by 1. The terms 1, 4, 5 are negligible in comparison with term 2, which must therefore be balanced by term 3. It follows that

$$p_{i+2} = p_{i+1}\ , \qquad b_{i+2} = 2b_{i+1}\ . \tag{19.12}$$

So encounter $i+2$ is not the end of the R-arc. Increasing i again in (19.3), we obtain in the same way

$$p_{i+3} = p_{i+1}\ , \qquad b_{i+3} = 3b_{i+1}\ , \tag{19.13}$$

and so on. The p are equal and the b constantly increase. This is impossible because at some point we must reach the end of the R-arc.

(ii) $p_{i+1} = p_i$, $b_{i+1} \neq b_i$, and $b_{i+1} - b_i$ has the same sign as b_{i+1}. Increasing i, we obtain

$$\begin{aligned} p_{i+2} &= p_{i+1} , \qquad & b_{i+2} &= b_{i+1} + (b_{i+1} - b_i) , \\ p_{i+3} &= p_{i+1} , \qquad & b_{i+3} &= b_{i+1} + 2(b_{i+1} - b_i) , \end{aligned} \tag{19.14}$$

and so on; this again leads to an impossibility.

(iii) $p_{i+1} = p_i$, $b_{i+1} \neq b_i$, and $b_{i+1} - b_i$ has a sign opposite to that of b_{i+1}. Then $b_i - b_{i+1}$ has the same sign as b_i, and proceeding in the opposite direction (i decreasing), we obtain the same result as in the previous case.

(iv) $p_{i+1} = p_i$, $b_{i+1} = b_i$. We have then

$$\begin{aligned} p_{i+2} &= p_{i+1} , \qquad & b_{i+2} &= b_{i+1} , \\ p_{i+3} &= p_{i+1} , \qquad & b_{i+3} &= b_{i+1} , \end{aligned} \tag{19.15}$$

and so on; this is again impossible.

Therefore $p_i \leqq 2$ for all i.

Assume now that there exists a $p_i < -1$, with $0 < i < \tilde{n}$ (i.e. excluding the two ends). In (19.3), the terms 2 and 5 are negligible in comparison with term 4. Therefore the term 4 must be balanced by terms 1 and/or 3, and at least one of the following inequalities is true:

$$p_{i-1} \geqq 1 - p_i , \qquad p_{i+1} \geqq 1 - p_i . \tag{19.16}$$

But this is impossible since p_i never exceeds 2.

In conclusion, there is

$$-1 \leqq p_i \leqq 2 \qquad \text{for} \qquad i = 1, \ldots, \tilde{n} - 1 . \tag{19.17}$$

We call an encounter a *node* if $p_i = -1$, an *antinode* if $p_i > -1$. We consider a part of the R-arc lying between two consecutive nodes, or between one end of the R-arc and the next node. It consists of m basic arcs and contains $m - 1$ antinodes. The two ends correspond to encounters i and $i + m$. From (19.1b) and (19.1c) we have, for $j = 1, \ldots, m$

$$x'_{i+j} = \frac{w}{2} - \frac{y_{i+j} - y_{i+j-1}}{2w} , \qquad x''_{i+j} = -\frac{w}{2} - \frac{y_{i+j} - y_{i+j-1}}{2w} . \tag{19.18}$$

Using (19.17), we obtain

$$x'_{i+j}, x''_{i+j} = O(w) . \tag{19.19}$$

We assume that the x_i also behave asymptotically as powers of w. We call β'_{i+j} and β''_{i+j} the limits of x'_{i+j}/w and x''_{i+j}/w for $|w| \to \infty$ (note that these numbers can be zero). (19.1b) gives

$$\beta''_{i+j} - \beta'_{i+j} = -1 . \tag{19.20}$$

On the other hand, dividing (19.1c) by wy_{i+j} and using $p_i > -1$, we obtain for $j = 1, \ldots, m - 1$

$$\beta'_{i+j+1} - \beta''_{i+j} = 0. \tag{19.21}$$

Summing over j, we find

$$\beta'_{i+j} = \beta'_{i+1} - j + 1\,, \qquad \beta''_{i+j} = \beta'_{i+1} - j\,. \tag{19.22}$$

We call γ_{i+j} the limit of y_{i+j}/w^2 for $|w| \to \infty$ (note that these numbers can be zero). Dividing (19.1c) by w^2, we obtain

$$\gamma_{i+j} - \gamma_{i+j-1} = -(\beta'_{i+j} + \beta''_{i+j})\,. \tag{19.23}$$

There is also $\gamma_i = \gamma_{i+m} = 0$. Substituting the values (19.22) and summing, we obtain

$$\gamma_{i+j} = -2j\beta'_{i+1} + j^2\,. \tag{19.24}$$

From the condition $\gamma_{i+m} = 0$ we have then

$$\beta'_{i+1} = \frac{m}{2} \tag{19.25}$$

and

$$\beta'_{i+j} = \frac{m}{2} - j + 1\,, \qquad \beta''_{i+j} = \frac{m}{2} - j\,, \qquad \gamma_{i+j} = -j(m-j)\,. \tag{19.26}$$

We have recovered the asymptotic expressions (19.8a), (19.8c), (19.8d). Finally, (19.8b) results from (19.1). This completes the proof.

19.1.2 Variational Equations for $w \to \pm\infty$

We derive here variational equations similar to those which were obtained for type 1 in Sect. 13.1.2. We call N the number of S-arcs which make up the R-arc. We name $U_1, U_2, \ldots, U_\alpha, \ldots, U_N$ the successive S-arcs. We call $i_0, i_1, \ldots, i_\alpha, \ldots, i_N$ the positions of the nodes, with $i_0 = 0$, $i_N = \tilde{n}$, so that the arc U_α extends from $i_{\alpha-1}$ to i_α. The number of basic arcs in U_α is $m_\alpha = i_\alpha - i_{\alpha-1}$.

From the two initial values y_0 and x'_1, using (19.1a), (19.1b) and (19.1c) in turn, we can compute successively the values of x''_1, y_1, x'_2, x''_2, ... We compute now the corresponding variational equations: we assume arbitrary infinitesimal variations $\mathrm{d}y_0$ and $\mathrm{d}x'_1$ and we compute the corresponding variations of x''_1, y_1, ... We have

$$\mathrm{d}x''_i = \mathrm{d}x'_i\,, \qquad \mathrm{d}y_i = \mathrm{d}y_{i-1} - w(\mathrm{d}x'_i + \mathrm{d}x''_i)\,, \qquad \mathrm{d}x'_{i+1} = \mathrm{d}x''_i - \frac{\mathrm{d}y_i}{y_i^2}\,. \tag{19.27}$$

Consider the arc U_α. From (19.8a) we have for the antinodes

$$y_i = \Theta(w^2)\,, \qquad i = i_{\alpha-1}+1, \ldots, i_\alpha - 1\,. \tag{19.28}$$

Starting from the initial variations $\mathrm{d}y_{i_{\alpha-1}}$ and $\mathrm{d}x'_{i_{\alpha-1}+1}$ and applying the relations (19.27), we find that generally

$$\begin{aligned} dx_i'' = dx_i' &= O(w^{-4})dy_{i_{\alpha-1}} + dx'_{i_{\alpha-1}+1}[\,]\,, \qquad i = i_{\alpha-1}+1,\dots,i_\alpha\,, \\ dy_i &= dy_{i_{\alpha-1}}[\,] - 2(i - i_{\alpha-1})w\,dx'_{i_{\alpha-1}+1}[\,]\,, \qquad i = i_{\alpha-1}+1,\dots,i_\alpha\,, \end{aligned} \tag{19.29}$$

where we have used the abbreviation $[\,] = [1 + O(w^{-3})]$. In particular we have the final variations

$$\begin{aligned} dx''_{i_\alpha} &= O(w^{-4})dy_{i_{\alpha-1}} + dx'_{i_{\alpha-1}+1}[\,]\,, \\ dy_{i_\alpha} &= dy_{i_{\alpha-1}}[\,] - 2m_\alpha w\,dx'_{i_{\alpha-1}+1}[\,]\,. \end{aligned} \tag{19.30}$$

We compute now the initial variations of the next arc $U_{\alpha+1}$. We already know dy_{i_α}. Since i_α is a node, we have from (19.8b): $y_{i_\alpha} = \Theta(w^{-1})$. From (19.27c) we obtain

$$dx'_{i_\alpha+1} = -\frac{dy_{i_{\alpha-1}}}{y_{i_\alpha}^2}[\,] + 2m_\alpha w\frac{dx'_{i_{\alpha-1}+1}}{y_{i_\alpha}^2}[\,]\,. \tag{19.31}$$

The equations (19.30b) and (19.31) give the initial variations of $U_{\alpha+1}$ as functions of the initial variations of U_α. We can iterate these equations and we obtain

$$\begin{aligned} dy_{i_\alpha} &= (2w)^{\alpha-1}m_2m_3\dots m_\alpha\frac{dy_0}{y_{i_1}^2\cdots y_{i_{\alpha-1}}^2}[\,] \\ &\qquad -(2w)^\alpha m_1m_2\dots m_\alpha\frac{dx_1'}{y_{i_1}^2\cdots y_{i_{\alpha-1}}^2}[\,]\,, \\ dx'_{i_\alpha+1} &= -(2w)^{\alpha-1}m_2m_3\dots m_\alpha\frac{dy_0}{y_{i_1}^2\cdots y_{i_\alpha}^2}[\,] \\ &\qquad +(2w)^\alpha m_1m_2\dots m_\alpha\frac{dx_1'}{y_{i_1}^2\cdots y_{i_\alpha}^2}[\,]\,. \end{aligned} \tag{19.32}$$

The orders of magnitude are

$$\begin{aligned} \frac{\partial y_{i_\alpha}}{\partial y_0} &= \Theta(w^{3\alpha-3})\,, \qquad \frac{\partial y_{i_\alpha}}{\partial x_1'} = \Theta(w^{3\alpha-2})\,, \\ \frac{\partial x'_{i_\alpha+1}}{\partial y_0} &= \Theta(w^{3\alpha-1})\,, \qquad \frac{\partial x'_{i_\alpha+1}}{\partial x_1'} = \Theta(w^{3\alpha})\,. \end{aligned} \tag{19.33}$$

The variations are strongly amplified after each node.

19.1.3 Asymptotic Branches for $w \to 0$

In the limit $w \to 0$, we enter the region $|\Delta C| \ll \mu^{1/3}$, or $\nu > 1/3$. The solutions in that region are known (Sect. 17.6 and Chap. 18). For a R-arc, they are given by (17.191). In these equations, the x_{i+j} and y_{i+j} are numerical constants, given for small values of n in Sect. 18.1.4, and more generally obtained by solving the system (18.1). We call these constants x^R_{i+j}, y^R_{i+j} to avoid a confusion with the notations of the present chapter.

If we substitute these expressions into the change of variables (17.138) to (17.141) for a R-region in the case $\nu = 1/3$, we obtain

$$x'_{i+j}, x''_{i+j} = \frac{x^R_{i+j}}{\sqrt{2w}}[1 + O(|w|^{3/2}], \quad y_{i+j} = y^R_{i+j}\sqrt{2w}[1 + O(|w|^{3/2}] . \tag{19.34}$$

It can be verified that, if the numerical values x^R_{i+j}, y^R_{i+j} satisfy the equations (18.1), then the asymptotic values (19.34) satisfy the equations (19.1).

Conversely, we show now that for $w \to 0$, the one-parameter families of solutions of (19.1) have no other asymptotic branches than those which correspond to a R-arc, and which are described by the equations (19.34). Consider one particular branch. We assume that the y_i behave asymptotically as powers of $|w|$:

$$y_i \sim b_i |w|^{p_i} \qquad \text{for} \qquad |w| \to 0 . \tag{19.35}$$

The b_i are non-zero constants. The p_i are arbitrary constants. The two ends of the arc, $y_0 = y_{\bar{n}} = 0$, can be included in this formulation by taking $p_0 = p_{\bar{n}} = +\infty$.

Assume that there exists a $p_i < 1/2$. In (19.3), the five terms behave asymptotically as

$$|w|^{p_{i-1}}, |w|^{p_i}, |w|^{p_{i+1}}, |w|^{1-p_i}, |w|^2 . \tag{19.36}$$

Terms 4 and 5 are negligible in comparison with term 2. Therefore the latter must be balanced by terms 1 and/or 3, and at least one of the following inequalities is true:

$$p_{i-1} \leqq p_i , \qquad p_{i+1} \leqq p_i . \tag{19.37}$$

Assume that the second inequality is true. It follows in particular that the encounter $i+1$ cannot be the end of the R-arc. We distinguish several cases.

(i) $p_{i+1} < p_i$. Consider (19.3) with i increased by 1. The terms 1, 4, 5 are negligible in comparison with term 2, which must therefore be balanced by term 3. It follows that

$$p_{i+2} = p_{i+1} , \qquad b_{i+2} = 2b_{i+1} . \tag{19.38}$$

So encounter $i+2$ is not the end of the R-arc. Increasing i again in (19.3), we obtain in the same way

$$p_{i+3} = p_{i+1} , \qquad b_{i+3} = 3b_{i+1} , \tag{19.39}$$

and so on. The p are equal and the b constantly increase. This is impossible because at some point we must reach the end of the R-arc.

(ii) $p_{i+1} = p_i$, $b_{i+1} \neq b_i$, and $b_{i+1} - b_i$ has the same sign as b_{i+1}. Increasing i, we obtain

$$\begin{aligned} p_{i+2} &= p_{i+1} , \qquad b_{i+2} = b_{i+1} + (b_{i+1} - b_i) , \\ p_{i+3} &= p_{i+1} , \qquad b_{i+3} = b_{i+1} + 2(b_{i+1} - b_i) , \end{aligned} \tag{19.40}$$

and so on; this again leads to an impossibility.

(iii) $p_{i+1} = p_i$, $b_{i+1} \neq b_i$, and $b_{i+1} - b_i$ has a sign opposite to that of b_{i+1}. Then $b_i - b_{i+1}$ has the same sign as b_i, and proceeding in the opposite direction (i decreasing), we obtain the same result as in the previous case.

(iv) $p_{i+1} = p_i$, $b_{i+1} = b_i$. We have then

$$\begin{aligned} p_{i+2} &= p_{i+1} \,, & b_{i+2} &= b_{i+1} \,, \\ p_{i+3} &= p_{i+1} \,, & b_{i+3} &= b_{i+1} \,, \end{aligned} \tag{19.41}$$

and so on; this is again impossible.

Therefore $p_i \geqq 1/2$ for all i. Assume that there exists a $p_i > 1/2$ with $0 < i < \tilde{n}$. Then (19.3), the terms 1, 2, 3, 5 are negligible in comparison with term 4, which is impossible. We have thus shown that

$$p_i = 1/2 \qquad \text{for} \qquad 0 < i < \tilde{n} \,. \tag{19.42}$$

This can also be written

$$y_i = \Theta(|w|^{1/2}) \qquad \text{for} \qquad 0 < i < \tilde{n} \,. \tag{19.43}$$

The equations (19.1a) and (19.1b) give

$$x_i' = \frac{w}{2} - \frac{1}{2w}(y_i - y_{i-1}) \,, \qquad x_i'' = -\frac{w}{2} - \frac{1}{2w}(y_i - y_{i-1}) \,, \tag{19.44}$$

from which we have

$$x_i', x_i'' = O(|w|^{-1/2}) \,. \tag{19.45}$$

Here we have only a $O()$ relation; it can happen that the order of x_i', x_i'' is smaller than $|w|^{-1/2}$. This happens for instance in the case `2.1P1` (see Sect. 19.2.1), and more generally when there exists a symmetry entailing $y_{i-1} = y_i$.

We have proved the relations (17.161), which had been guessed from the examination of small values of $\tilde{n}$.

We show now that *for $\tilde{n} > 1$, w can tend toward 0 only through positive values*. Assume that a branch exists with $w \to 0^-$. We call γ_i the limit of $y_i/|w|^{1/2}$ for $w \to 0^-$. There is $\gamma_0 = 0$, $\gamma_{\tilde{n}} = 0$ from (19.1d). From (19.43) above, we have $\gamma_i \neq 0$ for $0 < i < \tilde{n}$. Dividing (19.3) by $|w|^{1/2}$ and letting $w \to 0^-$, we have

$$\gamma_{i-1} - 2\gamma_i + \gamma_{i+1} = \frac{2}{\gamma_i} \,, \qquad i = 1, \ldots, \tilde{n} - 1 \tag{19.46}$$

which can be written

$$\gamma_{i+1} - \gamma_i = \gamma_i - \gamma_{i-1} + \frac{2}{\gamma_i} \,. \tag{19.47}$$

Assume $\gamma_1 > 0$. Then $\gamma_1 - \gamma_0 > 0$. From (19.47) we have $\gamma_2 - \gamma_1 > 0$; hence $\gamma_2 > 0$. Using (19.47) again, we obtain $\gamma_3 - \gamma_2 > 0$, $\gamma_3 > 0$; and so on. The γ_i form a positive increasing sequence. But this is in contradiction with $\gamma_{\tilde{n}} = 0$.

If $\gamma_1 < 0$, an identical reasoning holds. We have thus shown that for $\tilde{n} > 1$, there exists no branch with $w \to 0^-$.

19.1.4 *R*-Jacobian

As shown in Sect. 17.5, Step 7, the Jacobian for the whole bifurcation orbit or bifurcating arc is equal, within a non-zero multiplicative constant, to the product of the R-Jacobians of the individual R-arcs. We must therefore compute the R-Jacobian for the R-arc defined by (19.1).

We proceed as in Sect. 13.1.4. We order the variables as follows:

$$x_1', x_1'', y_1, x_2', x_2'', y_2, \ldots, y_{\tilde{n}-1}, x_{\tilde{n}}', x_{\tilde{n}}'', y_{\tilde{n}} \,. \tag{19.48}$$

The fundamental equations (19.1) can be written

$$\begin{aligned}
f_1 &= x_1'' - x_1' + w = 0 \,, \\
f_2 &= y_1 + w(x_1' + x_1'') = 0 \,, \\
f_3 &= y_1(x_2' - x_1'') - 1 = 0 \,, \\
&\ldots \\
f_{3i-2} &= x_i'' - x_i' + w = 0 \,, \\
f_{3i-1} &= y_i - y_{i-1} + w(x_i' + x_i'') = 0 \,, \\
f_{3i} &= y_i(x_{i+1}' - x_i'') - 1 = 0 \,, \\
&\ldots \\
f_{3\tilde{n}-2} &= x_{\tilde{n}}'' - x_{\tilde{n}}' + w = 0 \,, \\
f_{3\tilde{n}-1} &= y_{\tilde{n}} - y_{\tilde{n}-1} + w(x_{\tilde{n}}' + x_{\tilde{n}}'') = 0 \,, \\
f_{3\tilde{n}} &= y_{\tilde{n}} = 0 \,.
\end{aligned} \tag{19.49}$$

We have eliminated y_0, substituting $y_0 = 0$ in the first equation, but we have kept $y_{\tilde{n}}$ and the last equation $y_{\tilde{n}} = 0$.

For a given value of w, this is a system of $3\tilde{n}$ equations for $3\tilde{n}$ variables. The R-Jacobian is

$$\frac{\partial(f_1, \ldots, f_{3\tilde{n}})}{\partial(x_1', x_1'', y_1, \ldots, x_{\tilde{n}}', x_{\tilde{n}}'', y_{\tilde{n}})} \,. \tag{19.50}$$

19.1.5 Stability

We consider the system of equations (19.49) for a given value of w, , with the last equation deleted. We have then a system of $3\tilde{n} - 1$ equations for $3\tilde{n}$ variables. Starting from a given value x_1' and applying the equations one by one, we obtain successively x_1'', y_1, ..., $y_{\tilde{n}}$. In the same way, starting from a variation dx_1', we can compute successively dx_1'', dy_1, ..., $dy_{\tilde{n}}$. We define a *critical R-arc* as a R-arc for which there is

$$\frac{dy_{\tilde{n}}}{dx_1'} = 0 \,. \tag{19.51}$$

Proceeding as in Sect. 13.1.4, we can show that

Proposition 19.1.1. *A R-Jacobian vanishes if and only if the corresponding R-arc is critical.*

For the whole bifurcation orbit or bifurcating arc, we have then

Proposition 19.1.2. *In a partial transition 2.1, the Jacobian vanishes if and only if the bifurcation orbit or bifurcating arc contains at least one critical R-arc.*

We have from (19.49)

$$\frac{\mathrm{d}x_i''}{\mathrm{d}x_1'} = \frac{\mathrm{d}x_i'}{\mathrm{d}x_1'} ,$$
$$\frac{\mathrm{d}y_i}{\mathrm{d}x_1'} = \frac{\mathrm{d}y_{i-1}}{\mathrm{d}x_1'} - 2w\frac{\mathrm{d}x_i'}{\mathrm{d}x_1'} ,$$
$$\frac{\mathrm{d}x_{i+1}'}{\mathrm{d}x_1'} = \frac{\mathrm{d}x_i'}{\mathrm{d}x_1'} - \frac{1}{y_i^2}\frac{\mathrm{d}y_i}{\mathrm{d}x_1'} , \tag{19.52}$$

from which we can compute the successive derivatives

$$\frac{\mathrm{d}y_1}{\mathrm{d}x_1'} = -2w , \tag{19.53}$$
$$\frac{\mathrm{d}y_2}{\mathrm{d}x_1'} = -4w - \frac{4w^2}{y_1^2} , \tag{19.54}$$
$$\frac{\mathrm{d}y_3}{\mathrm{d}x_1'} = -6w - \frac{8w^2}{y_1^2} - \frac{8w^2}{y_2^2} - \frac{8w^3}{y_1^2 y_2^2} , \tag{19.55}$$
$$\ldots$$

These equations will be used below.

19.2 Small Values of $\tilde{n}$

The characteristics of the families will be represented in the (w, y_1) plane.

19.2.1 $\tilde{n} = 1$

The equations are immediately solved into

$$x_1' = \frac{w}{2} , \qquad x_1'' = \frac{-w}{2} . \tag{19.56}$$

For $w \to 0$, the condition $\Delta C = \Theta(\mu^{1/3})$ ceases to be verified: we move away from transition 2.1 and into the region $1/3 < \nu < 1/2$ studied in Chap. 18.

The characteristic is the line $y_1 = 0$.

(19.53) shows that there is no critical R-arc for $w \neq 0$.

19.2.2 $\tilde{n} = 2$

For $\tilde{n} \geqq 2$, the simplest method of solution of the fundamental equations consists in solving the equations (19.3) for the y_i, and then computing the x'_i and x''_i from (19.4).

In the case $\tilde{n} = 2$, we obtain a single equation for y_1:

$$y_1^2 + w^2 y_1 - w = 0 \tag{19.57}$$

and we obtain the two explicit solutions

$$y_1 = \frac{-w^2 \pm \sqrt{w^4 + 4w}}{2}, \qquad -x''_1 = x'_2 = \frac{w^2 \pm \sqrt{w^4 + 4w}}{4w},$$
$$-x'_1 = x''_2 = \frac{-3w^2 \pm \sqrt{w^4 + 4w}}{4w}. \tag{19.58}$$

The asymptotic branches for $w \to \pm\infty$ are

$$x'_1 \sim \frac{w}{2}, \quad x''_1 \sim -\frac{w}{2}, \quad y_1 \sim \frac{1}{w}, \quad x'_2 \sim \frac{w}{2}, \quad x''_2 \sim -\frac{w}{2}, \tag{19.59}$$

and

$$x'_1 \sim w, \quad x''_1 \sim \frac{1}{2w^2}, \quad y_1 \sim -w^2, \quad x'_2 \sim -\frac{1}{2w^2}, \quad x''_2 \sim -w. \tag{19.60}$$

The asymptotic branches for $w \to 0^+$ are

$$x'_1 \sim \pm\frac{1}{2\sqrt{w}}, \qquad x''_1 \sim \pm\frac{1}{2\sqrt{w}}, \qquad y_1 \sim \mp\sqrt{w},$$
$$x'_2 \sim \mp\frac{1}{2\sqrt{w}}, \qquad x''_2 \sim \mp\frac{1}{2\sqrt{w}}. \tag{19.61}$$

The characteristics are represented on Fig. 19.1. The asymptotic branches for $w \to \pm\infty$ are easily identified with the help of (19.8). We recall that the symbolic sign of the branches is the opposite of the sign of w (Sect. 19.1).

The limit point $y_1 = 0$, $w = 0$ does not belong to the characteristic: there is $w \neq 0$ from Property 3, Sect. 19.1. (We remark also that (19.61) would give infinite values of x'_1, etc. for $w = 0$.)

The notation used for the branches is the same as the notation introduced in Sect. 6.2.1.3: each S-arc is represented by a number equal to its m value. The sign of a branch is the sign of $\epsilon'\Delta C$, i.e. the sign of $-w$ (Sect. 19.1).

The stability can be determined from (19.54):

$$\frac{\mathrm{d}y_2}{\mathrm{d}x'_1} = -2w(w^3 + 4) \mp 2w^2\sqrt{w(w^3 + 4)}. \tag{19.62}$$

This vanishes only for $w^3 = -4$, i.e. $w = -2^{2/3}$ (extremum). (The value $w = 0$ is forbidden; see Sect. 19.1).

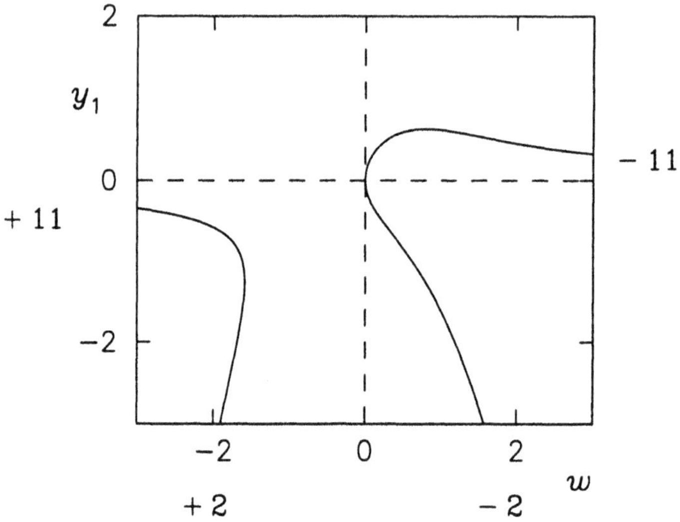

Fig. 19.1. Characteristics for 2.1P2 transitions.

19.2.3 $\tilde{n} = 3$

From (19.1d) and (19.3) we obtain two equations for y_1 and y_2:

$$-2y_1 + y_2 = -\frac{2w}{y_1} + 2w^2 \,, \qquad y_1 - 2y_2 = -\frac{2w}{y_2} + 2w^2 \,. \tag{19.63}$$

Subtracting, we have

$$(y_2 - y_1)(3y_1y_2 + 2w) = 0 \tag{19.64}$$

and the two solutions

$$\begin{aligned} &y_1 = y_2 = -w^2 \pm \sqrt{w^4 + 2w} \,, && x_2' = -x_2'' = \frac{w}{2} \,, \\ &x_1'' = -x_3' = \mp \frac{\sqrt{w^4 + 2w}}{2w} \,, && x_1' = -x_3'' = w \mp \frac{\sqrt{w^4 + 2w}}{2w} \end{aligned} \tag{19.65}$$

as well as the two other solutions

$$\begin{aligned} &y_1 = \frac{-3w^2 \pm \sqrt{9w^4 + 24w}}{6} \,, && y_2 = \frac{-3w^2 \mp \sqrt{9w^4 + 24w}}{6} \,, \\ &x_1' = \frac{9w^2 \mp \sqrt{9w^4 + 24w}}{12w} \,, && x_1'' = \frac{-3w^2 \mp \sqrt{9w^4 + 24w}}{12w} \,, \\ &x_2' = \frac{3w^2 \pm \sqrt{9w^4 + 24w}}{6w} \,, && x_2'' = \frac{-3w^2 \pm \sqrt{9w^4 + 24w}}{6w} \,, \\ &x_3' = \frac{3w^2 \mp \sqrt{9w^4 + 24w}}{12w} \,, && x_3'' = \frac{-9w^2 \mp \sqrt{9w^4 + 24w}}{12w} \,. \end{aligned} \tag{19.66}$$

The characteristics are represented on Fig. 19.2. Here again the origin is excluded.

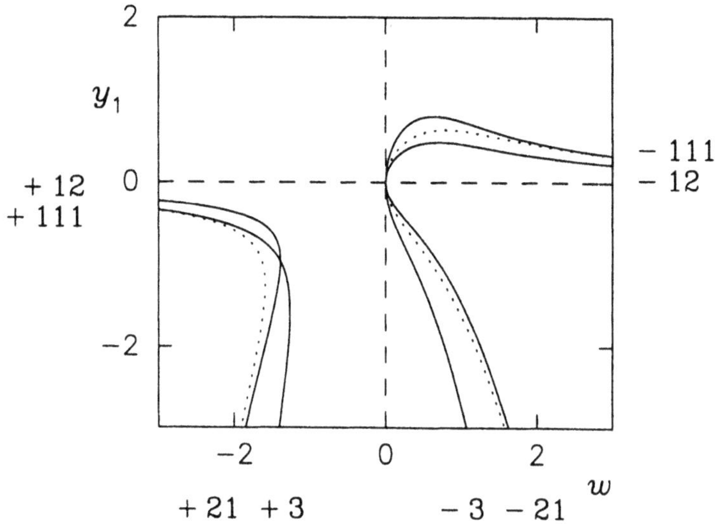

Fig. 19.2. Characteristics for 2.1P3 transitions (solid lines). Also shown: 2.1P2 transitions (dotted lines).

The stability can be determined from (19.55). For the first solution (19.65) we obtain

$$\frac{\mathrm{d}y_3}{\mathrm{d}x_1'} = 4w[-(w^3+2)^2 \mp w(w^3+3)\sqrt{w^4+2w}] . \tag{19.67}$$

This expression vanishes for $w^3 = -2$ (extremum) and for $w^3 = -8/3$ (extremum et intersection).

For the second solution (19.66) we obtain

$$\frac{\mathrm{d}y_3}{\mathrm{d}x_1'} = -6w(3w^3+8) . \tag{19.68}$$

This vanishes for $w^3 = -8/3$ (intersection).

19.2.4 $\tilde{n} > 3$

The solutions found for $\tilde{n} = 2$ and $\tilde{n} = 3$ (equations (19.58), (19.65), (19.66)) suggest that for larger values of $\tilde{n}$ there might exist also solutions of the form

$$y_i = b_i(-w^2 \pm \sqrt{w^4 + cw}) \tag{19.69}$$

where the b_i and c are constants. It can be shown, however, that no such solutions exist for $\tilde{n} = 4$. It seems that the system (19.1) for $\tilde{n} > 3$ can only be solved numerically.

19.3 Positional Method

We introduce now a "positional method" similar to that of Sect. 13.3.

Here again, it is easily shown that *two characteristics in the* (w, y_1) *plane which correspond to different values of* n *never intersect.* This can be verified on Fig. 19.2 for $\tilde{n} = 2$ and 3. The characteristics up to $\tilde{n} - 1$ divide the (w, y_1) plane into regions, and we have the equivalent of Propositions 13.3.1 and 13.3.2:

Proposition 19.3.1. *Two branches can be joined only if (i) they lie in the same region and (ii) the two symmetrical branches lie in the same region.*

Here again, we will determine in which region a branch lies by studying the relative position of the asymptotic branches. We need to know all asymptotic branches in the (w, y_1) plane. For $w \to \pm\infty$, we have shown in Sect. 19.1.1 that there are no other asymptotic branches than those which correspond to a decomposition of the R-arc into S-arcs. For $w \to 0$, similarly, we have shown in Sect. 19.1.3 that there are no other asymptotic branches than those which correspond to a R-arc. A last possibility would consist in a branch $y_1 \to \pm\infty$, $w \to w_0$, i.e. a branch with a vertical asymptote in the (w, y_1) plane. But in that case, (19.3) would give successively $y_2 \simeq 2y_1$, $y_3 \simeq 3y_1$, $\dots$, $y_i \simeq iy_1$, and we would never reach $y_{\tilde{n}} = 0$: this is impossible.

19.3.1 Branch Order for $w \to \pm\infty$

This limit corresponds to a passage to the case $\nu < 1/3$. It was shown in Sect. 19.1.1 that there are no other asymptotic branches than those which correspond to a decomposition of the R-arc into S-arcs.

19.3.1.1 Variations. For a given large value of $|w|$, using the condition $y_0 = 0$ and starting from any given value of x'_1, we can compute the successive values x''_1, y_1, x'_2, x''_2, y_2, $\dots$ from (19.1). We consider the value of x'_1 which corresponds to some particular branch of order n (so that the computation ends in $y_n = 0$). We apply now a small variation $\mathrm{d}x'_1$, and we compute the corresponding variations of x''_1, y_1, $\dots$. We can use the computations made in Sect. 19.1.2, setting $\mathrm{d}y_0 = 0$. We obtain in particular

$$\frac{\mathrm{d}y_{i_\alpha}}{\mathrm{d}x'_1} = -\frac{(2w)^\alpha m_1 m_2 \dots m_\alpha}{y_{i_1}^2 y_{i_2}^2 \cdots y_{i_{\alpha-1}}^2}[1 + O(w^{-3})] \,. \tag{19.70}$$

We will need the order of magnitude and the sign

$$\frac{\mathrm{d}y_{i_\alpha}}{\mathrm{d}x'_1} = \Theta(w^{3\alpha-2}) \,, \qquad \mathrm{sign}\left(\frac{\mathrm{d}y_{i_\alpha}}{\mathrm{d}x'_1}\right) = -(\mathrm{sign} w)^\alpha \,. \tag{19.71}$$

We have referred the variations to $\mathrm{d}x'_1$. On the figures showing the characteristics, however, y_1 is used. From (19.27) we find that these two variables are simply related by

$$\mathrm{d}y_1 = -2w\mathrm{d}x_1' \,. \tag{19.72}$$

Therefore we have

$$\frac{\mathrm{d}y_{i_\alpha}}{\mathrm{d}y_1} = \Theta(w^{3\alpha-3}) \,, \qquad \mathrm{sign}\left(\frac{\mathrm{d}y_{i_\alpha}}{\mathrm{d}y_1}\right) = (\mathrm{sign} w)^{\alpha+1} \,. \tag{19.73}$$

19.3.1.2 Relative Positions of Two Branches with Initial Common Arcs. We consider now two branches for which the arcs U_1 to U_α are the same, but the continuation is different: either the arcs $U_{\alpha+1}$ are different in the two branches, or the arc $U_{\alpha+1}$ does not exist in one of the two branches (the R-arc ends in i_α).

The asymptotic expressions of x_i', x_i'', y_i up to $i = i_\alpha$ given by (19.8) are the same for the two branches; therefore these quantities differ by $O(w^{-1})$ between the two branches. For $|w|$ large enough, this can be made as small as desired. Therefore these differences will be called $\mathrm{d}x_1'$, etc., and the above results on small variations can be applied.

If $U_{\alpha+1}$ exists, we have from (19.8b)

$$y_{i_\alpha} = \frac{2}{m_\alpha + m_{\alpha+1}} w^{-1}[1 + O(w^{-3})] \,. \tag{19.74}$$

If If $U_{\alpha+1}$ does not exist, i.e. the R-arc ends in i_α, we have $y_{i_\alpha} = 0$.

Thus, if $w > 0$, the order of the values of y_{i_α} for the two branches is given by the order of the two values of $m_{\alpha+1}$ in the sequence

$$0; \; \ldots, 3, 2, 1, \tag{19.75}$$

where the case of the R-arc ending in i_α is represented symbolically by $m_{\alpha+1} = 0$. If $w < 0$, the order is reversed:

$$1, 2, 3, \ldots; \; 0 \,. \tag{19.76}$$

Finally, we use (19.73b) to determine the order of the values of y_1 for the two branches: if $w > 0$, or $w < 0$ and α is even, that order is defined by (19.75); if $w < 0$ and α is odd, it is defined by (19.76).

19.3.1.3 Relative Positions of Two Branches with Different First Arcs. The method of the previous section does not work when the two branches differ already in their first arc (this can be taken to correspond to $\alpha = 0$). In that case we compare directly the values of y_1. We have from (19.8a)

$$y_1 = -(m_1 - 1)w^2 + O(w^{-1}) \,. \tag{19.77}$$

(This is true also if $m_1 = 1$, i.e. y_1 corresponds to a node.) Therefore the order of the values of y_1 is given by the order of the two values of m_1 in the sequence

$$\ldots, 3, 2, 1 \,. \tag{19.78}$$

19.3.1.4 Packets. From (19.74) we have, for $\alpha > 0$,

$$\mathrm{d}y_{i_\alpha} = \Theta(w^{-1}) . \tag{19.79}$$

From (19.73a) we have then

$$\mathrm{d}y_1 = \Theta(w^{2-3\alpha}) . \tag{19.80}$$

For $\alpha = 0$ (previous section), the difference between the values of y_1 for the two branches is $\Theta(w^2)$. Therefore (19.80) covers this case also, with the proviso that $\mathrm{d}y_1$ is not a small quantity anymore.

As in Sect. 13.3.2.5, we find thus that the branches are organized hierarchically. The characteristics of all branches having a given first arc in common are at a distance $O(w^{-1})$ of each other; they form a *first-order packet.* Inside such a packet, the characteristics of all branches having in common their first two arcs are at a distance $O(w^{-4})$ of each other, and form a *second-order packet*; and so on.

19.3.1.5 Results for $\tilde{n} \leqq 5$. The rules established in the preceding sections allow us to determine completely how the characteristics of all branches are ordered in y_1. Figs. 19.3 and 19.4 show this ordering for $w \to +\infty$ and $w \to -\infty$, respectively, up to $\tilde{n} = 5$. In these figures, y_1 increases from bottom to top, as in Figs. 19.1 and 19.2, and packets of first order, second order, etc. have been separated into different columns, with their filiation indicated by line segments.

19.3.2 Branch Order for $w \to 0$

This limit corresponds to a passage to the case $\nu > 1/3$. It was shown in Sect. 19.1.3 that for $w \to 0$, there are no other branches than those which correspond to R-arcs. We can therefore use the results of Section 18.1. The variable y_i used in that Section has the same sign as y_i here (see Sects. 17.5 and 17.6, steps 2 and 3).

In the case $w < 0$ ($\epsilon' \Delta C > 0$), there exists only one R-arc, for $\tilde{n} = 1$; for $\tilde{n} > 1$ there is no solution, therefore no branch $w \to 0^-$ (Sect. 19.1.3).

In the case $w > 0$ ($\epsilon' \Delta C < 0$), there are $2^{\tilde{n}-1}$ branches. Exactly one branch corresponds to each choice of the signs of the y_i. As shown by Table 18.1, the branches are simply ordered. Consider two branches which have the same signs for $y_1, \ldots, y_i$, but a different continuation: either the signs are different for y_{i+1}, or $y_{i+1} = 0$ for one of the branches (the R-arc ends in $i+1$). Then the order of the values of y_1 for the two branches is given by the order of the two signs of y_{i+1} in the sequence

$$-, 0, + . \tag{19.81}$$

The case where the R-arc ends in $i+1$ has been represented symbolically by 0. The rule applies even if $i = 0$, i.e. if the two branches differ already in y_1.

Fig. 19.5 shows the ordering of the branches, up to $\tilde{n} = 4$. Note that this is essentially a lexicographic order, with the order of the three symbols given by (19.81).

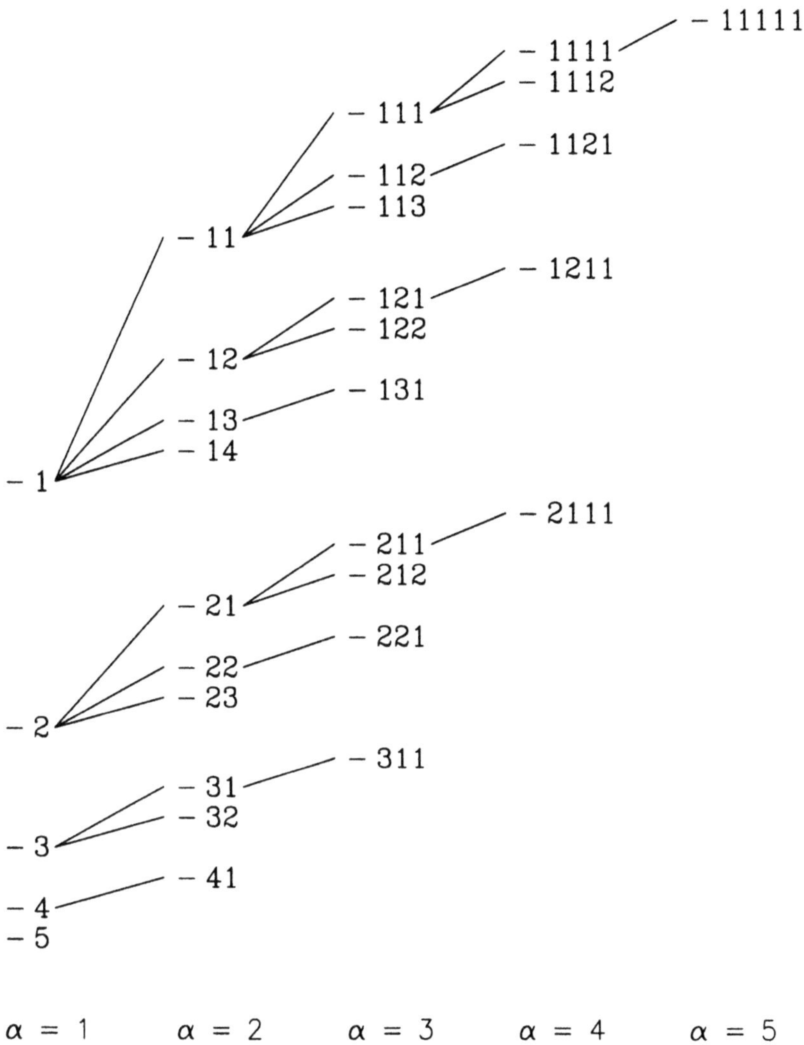

Fig. 19.3. Order of the branches for $w \to +\infty$, $\tilde{n} \leqq 5$. (y_1 increases upwards.)

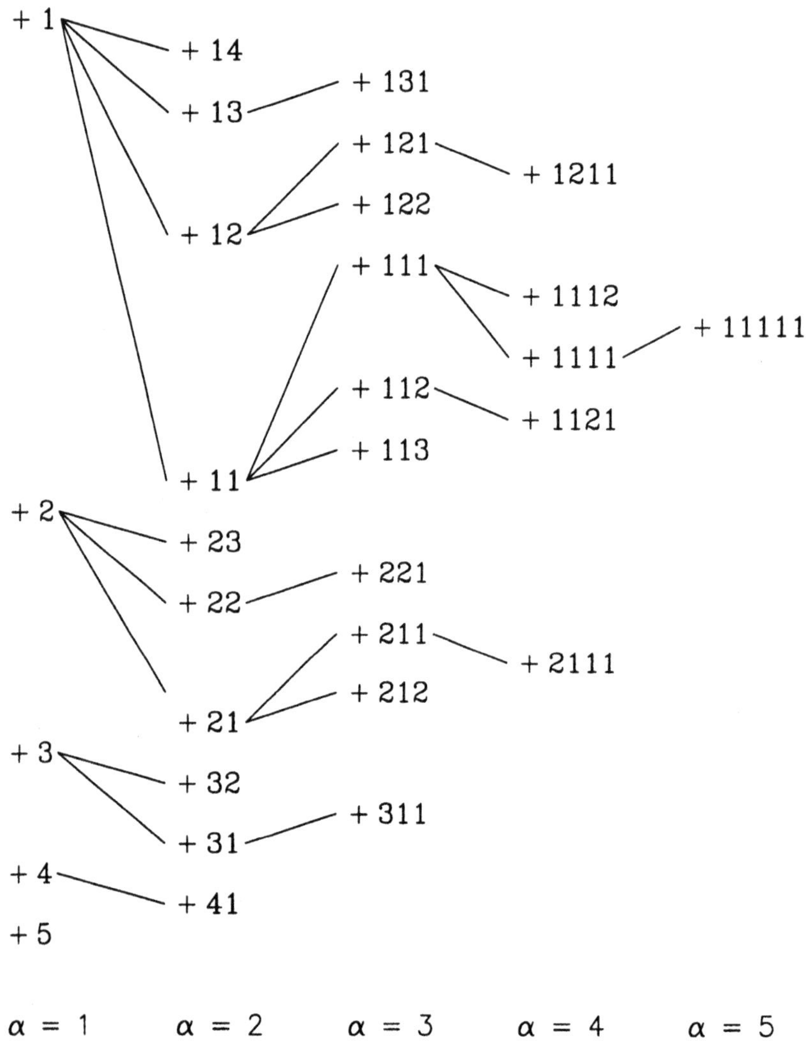

Fig. 19.4. Order of the branches for $w \to -\infty$, $\tilde{n} \leqq 5$. (y_1 increases upwards.)

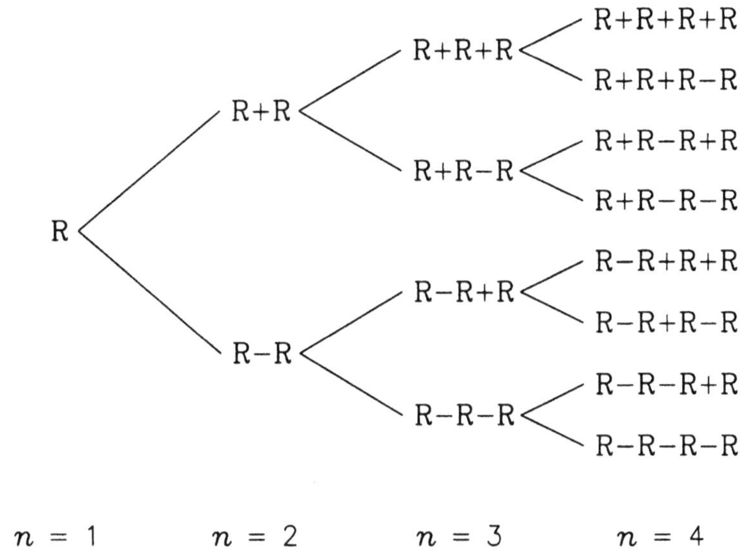

Fig. 19.5. Order of the branches for $w \to 0^+$, $\tilde{n} \leqq 4$. (y_1 increases upwards.)

19.3.3 Results

We use now the positional method systematically to determine the junctions.

19.3.3.1 The Case $w > 0$. We represent schematically the characteristics in the $(w > 0, y_1)$ half-plane, with on the left side the asymptotic branches for $w \to 0^+$, ordered as in Fig. 19.5, and on the right side the asymptotic branches for $w \to +\infty$, ordered as in Fig. 19.3. We apply now Proposition 19.3.1 to determine the junctions.

For $\tilde{n} = 1$, there are only the two branches 'R' on the left and '- 1' on the right; these two branches must be joined.

For $\tilde{n} = 2$, we represent the already known junction as a solid line (Fig. 19.6). It divides the half-plane into two regions. In each of these regions, there is one branch on the left and one branch on the right; these two branches must be joined (dashed lines in Fig. 19.6).

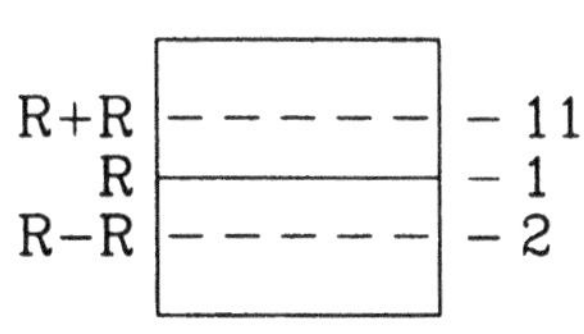

Fig. 19.6. Junctions for $w > 0$, $\tilde{n} \leqq 2$.

For $\tilde{n} = 3$, we again represent the already known junctions as solid lines (Fig. 19.7). They divide the half-plane into four regions. Again in each region there are two branches, which must therefore be joined (dashed lines in Fig. 19.7).

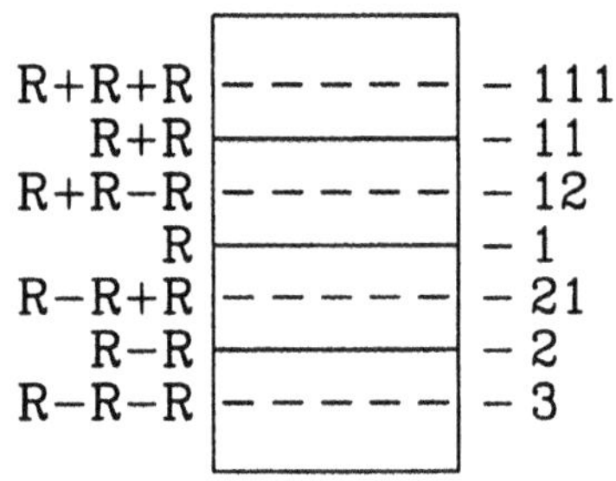

Fig. 19.7. Junctions for $w > 0$, $\tilde{n} \leqq 3$.

There is a strong suggestion that this regular construction will continue indefinitely; and indeed it is not difficult to prove it. On the left side, it is clear that the ordering described in Sect. 19.3.2 results in exactly one new branch in each region. On the right side, the same property can be proved by considering the relative sides of passage, i.e. the signs of the y_i (see (17.143)), and the construction of the ordering described in Sect. 19.3.1 .

Thus the junctions of R-arcs across the transition 2.1 are trivially established for all $\tilde{n}$ in the case $w > 0$. In fact the positional method is not necessary in that case; it is sufficient to consider the relative sides of passage, i.e. essentially to apply Broucke's principle (together with the fact that the sign of ΔC does not change).

For $\tilde{n} = 1, 2, 3$, we recover the results of Sect. 19.2.

19.3.3.2 The Case $w < 0$. This case is less simple. All relative sides of passage are negative, and therefore Broucke's principle gives no information on the junctions.

We represent schematically the characteristics in the $(w < 0, y_1)$ half-plane, with on the left side the asymptotic branches for $w \to -\infty$, ordered as in Fig. 19.4, and on the right side the single asymptotic branch for $w \to 0^-$. We apply now Proposition 19.3.1 to determine the junctions.

For $\tilde{n} = 1$, there are only the two branches '+ 1' on the left and 'R' on the right; these two branches must be joined.

For $\tilde{n} > 1$, there are no right branches; the left branches, corresponding to $w \to -\infty$, must be joined between themselves.

For $\tilde{n} = 2$, we represent the already known junction as a solid line (Fig. 19.8). There are two other branches, which must be joined.

For $\tilde{n} = 3$, again we represent the already known junction as solid lines (Fig. 19.9). There are 4 new branches, which lie in the same region. They form

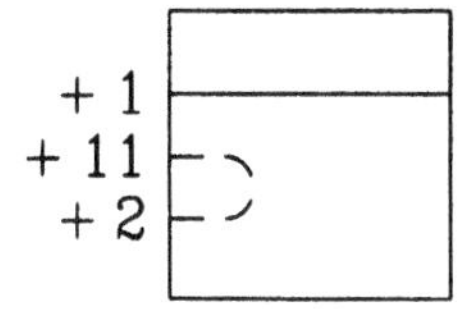

Fig. 19.8. Junctions for $w < 0$, $\tilde{n} \leqq 2$.

a *trident* (Sect. 13.3.3.2). In the present case, the R-arc is identical with the whole bifurcating arc (there are no T-arcs), and we can use Restriction 7.3.2. Then the two branches + 111 and + 3 are made of symmetric orbits. The two other branches, + 12 and + 21, are made of asymmetric orbits, but are changed into each other by the symmetry. We can reason as in Chap. 7 and conclude that the two symmetric branches must be joined, and the two asymmetric branches must be joined.

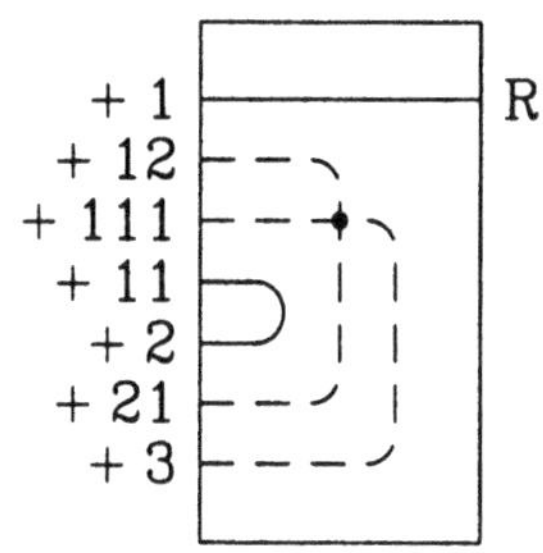

Fig. 19.9. Junctions for $w < 0$, $\tilde{n} \leqq 3$.

Note that Fig. 19.9 agrees with Fig. 19.2.

For $\tilde{n} = 4$, using the same method, we find that the 8 new branches form two tridents (Fig. 19.10). All junctions are established.

Since Figs. 19.8 to 19.10 are topological in nature, they can be distorted. We notice then that they can be made simpler by "unfolding" them: we consider all branches for $\tilde{n} \geqq 2$ (i.e. all branches except '+ 1' and 'R') and we let the lower half of these branches slide around the bottom of the frame in a counter-clockwise direction, until they lie on the right side. Fig. 19.10 is thus transformed into Fig. 19.11, which has a simpler appearance. (Note, however, that this figure is not anymore a sketch of the (w, y_1) plane.)

This regularity can be explained. By studying the process of construction, one can show that when passing from $\tilde{n} - 1$ to $\tilde{n}$, the new branches (for $\tilde{n}$) and the old branches (for 2 to $\tilde{n} - 1$) alternate regularly as follows: 2 new branches, 2 old branches, 2 new branches, ..., 2 new branches.

We use this new representation for the next case, $\tilde{n} = 5$ (Fig. 19.12, left). The new branches are separated into 4 groups of 4. Two of these groups

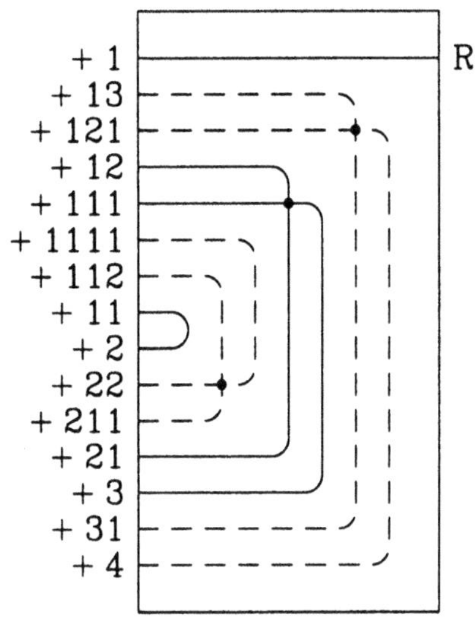

Fig. 19.10. Junctions for $w < 0$, $\tilde{n} \leqq 4$.

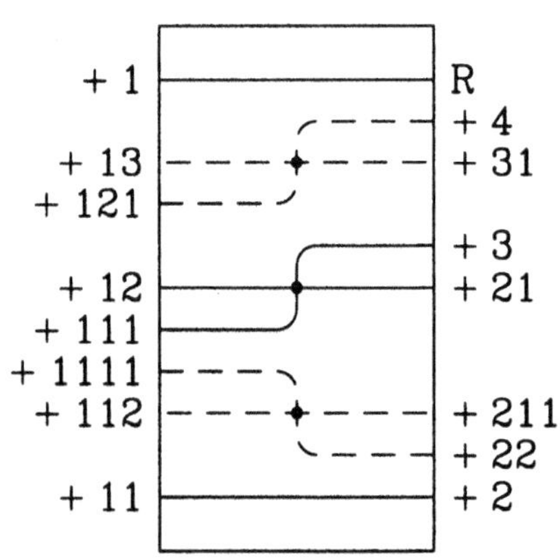

Fig. 19.11. Junctions for $w < 0$, $\tilde{n} \leqq 4$: unfolded figure.

correspond to tridents; their junctions are established (dashed lines). But the two other groups contain only asymmetric orbits; one is made of + 1211, + 122, + 32, + 311, and the other consists of + 11121, + 113, + 23, + 221. Moreover, the symmetry Σ exchanges these two groups, and therefore no new information is obtained from the second half of Proposition 19.3.1. Thus the positional method is unable to establish all junctions. We must resort to a numerical computation, which shows that the junctions are as indicated by the dotted lines on Fig. 19.12 (left).

For $\tilde{n} = 6$, the new branches form 6 groups of 4 and 1 group of 8 (Fig. 19.12, right). Using the second half of Proposition 19.3.1, we can split this last group into 2 groups of 4, so that we have 8 groups of 4 branches to solve. 4 of these groups form a trident and are immediately solved (dashed lines). For the 4 other groups, numerical computation must again be used, and produces the junctions shown as dotted lines.

Figs. 19.11 and 19.12 exhibit some regularity. Going down, we find that the junctions involve successive groups of 4 branches, except at the top and at the bottom where only two branches are involved. This can be stated in a different way. A given branch is completely defined by specifying whether each collision ($i = 1$ to $\tilde{n} - 1$) is a node or an antinode. Then the following rule is found to hold up to $\tilde{n} = 6$: *in two branches which are joined, only the collisions 1 and* $\tilde{n} - 1$ *can differ.* For instance, in the junction between + 1311 and + 132, only the collision $\tilde{n} - 1$ changes from node to antinode. In the junction between + 15 and + 51, the collision 1 goes from node to antinode and the collision $\tilde{n} - 1$ goes from antinode to node.

It would be of interest to prove this rule for all $\tilde{n}$. On the other hand, even if proved, this rule would not be sufficient to establish all junctions.

19.4 Results for Bifurcations of Type 2

So far in this chapter we have considered only a single R-arc. We go back now to the partial transition 2.1, which may contain more than one R-arc.

19.4.1 The Case $w > 0$

A consequence of the regular structure found in Sect. 19.3.3.1 is that for $w > 0$, every characteristic of a R-arc runs from $w = 0$ to $w = +\infty$. In other words, a R-arc simply traverses the transition 2.1. On the other hand, the T-arcs are not affected either by transition 2.1 (Sect. 17.6). It follows that the whole bifurcation orbit or bifurcating arc passes through the partial transition 2.1 without being affected. Recalling the definition of a partial transition 2.1 (Sect. 17.5), and the fact that $w > 0$ corresponds to a symbolic sign - (Sect. 19.1, Property 5), we obtain, for partial and total bifurcations respectively:

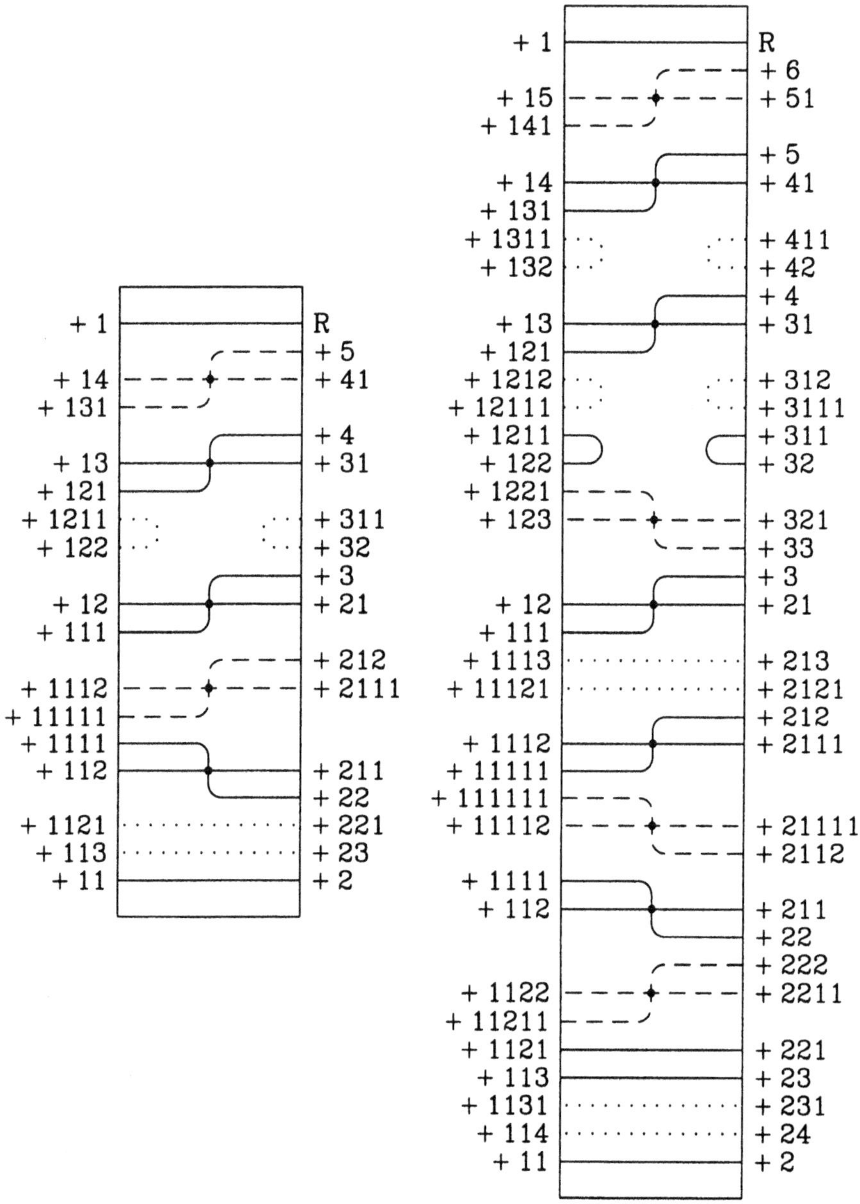

Fig. 19.12. Junctions for $w < 0$, $\tilde{n} \leqq 5$ (left) and $\tilde{n} \leqq 6$ (right).

Proposition 19.4.1. *All branches with a - sign in Table 8.12 pass through the transition 2.1 at $\nu = 1/3$ and go on to higher values of ν.*

Proposition 19.4.2. *All branches with a - sign and including T-arcs (symbols* i *and* e*) in Table 8.18 pass through the transition 2.1 at $\nu = 1/3$ and go on to higher values of ν.*

Nothing happens to these branches at transition 2.1; their junctions will only be determined at transition 2.2, for $\nu = 1/2$ (Chaps. 21 and 22).

19.4.2 The Case $w < 0$

This case corresponds to a symbolic sign + for the branches, as defined in Sect. 6.2.1.3. As a consequence, there are no T-arcs. The partial transition 2.1 corresponds to a partial bifurcation. We have a particular case of Fig. 17.1c where only S-arcs are present. It follows that the whole bifurcating arc forms a single R-arc, with $\tilde{n} = n$.

Thus, the junctions found in Sect. 19.3.3.2 are immediately applicable to the branches with a + sign in the partial bifurcation of type 2, without any change.

In the particular case $\tilde{n} = n = 1$, the characteristic runs from $w = -\infty$ to $w = 0$ (Sect. 19.2.1). The situation is the same as for the general case $w > 0$ (Sect. 19.4.1). So we have

Proposition 19.4.3. *The branch + 1 in Table 8.12 passes through the transition 2.1 at $\nu = 1/3$ and goes on to higher values of ν.*

For $\tilde{n} = n > 1$, on the other hand, the characteristics are joined between themselves (Sect. 19.3.3.2). We have thus:

Proposition 19.4.4. *All branches with a + sign and $n \geqq 2$ in Table 8.12 are joined between themselves at transition 2.1.*

For these branches, therefore, ν increases from 0 to a maximal value 1/3, and then decreases again to 0. It never goes on to higher values.

Table 19.1 lists the junctions up to $n = 6$, in the same format as Table 8.12, which was obtained using only qualitative methods. It can be verified that there is agreement for all branches which appear in both tables (sign + and $2 \leqq n \leqq 4$).

Table 19.1. Partial bifurcation, type 2, n = 2 to 6: junctions between the + branches, determined by the quantitative study.

2P2-S
+ 2
+ 11

2P3--S
+ 3
+ 111

2P3--A
+ 21
+ 12

2P4---S
+ 22
+ 1111
+ 4
+ 121

2P4---A
+ 211
+ 112
+ 31
+ 13

2P5----S
+ 131
+ 5
+ 11111
+ 212

2P5----A
+ 14
+ 41
+ 122
+ 1211
+ 32
+ 311
+ 1112
+ 2111
+ 1121
+ 221
+ 113
+ 23

2P6-----S
+ 141
+ 6
+ 1221
+ 33
+ 111111
+ 2112
+ 11211
+ 222

2P6-----A
+ 15
+ 51
+ 132
+ 1311
+ 42
+ 411
+ 12111
+ 1212
+ 3111
+ 312
+ 123
+ 321
+ 1113
+ 213
+ 11121
+ 2121
+ 11112
+ 21111
+ 1122
+ 2211
+ 1131
+ 231
+ 114
+ 24

20. Total Transition 2.1

The case of a total transition 2.1 corresponds to a total bifurcation, where the R-region is a R-orbit which corresponds to the whole bifurcation orbit, made of n basic arcs. The equations are then

$$x_i'' - x_i' + w = 0 ,$$
$$y_i - y_{i-1} + w(x_i' + x_i'') = 0 ,$$
$$y_i(x_{i+1}' - x_i'') - 1 = 0 , \qquad (20.1)$$

where i is to be taken modulo n and takes all values from 1 to n.

These equations form a system of $3n$ equations for the $3n + 1$ variables w, y_0 to y_{n-1}, x_1' to x_n', x_1'' to x_n''. As in the case of partial bifurcations (Chap. 19), we expect one-parameter families of solutions, which correspond to ordinary one-parameter families of orbits.

20.1 Properties

The properties are mostly the same as in the case of a partial transition 2.1 (Sect. 19.1). In the case of property 3, we find again from (20.1c) that if $w = 0$, all y_i are equal. From (20.1a) and (20.1c) we obtain then

$$x_{i+1}' - x_i' = \frac{1}{y_i} , \qquad (20.2)$$

and summing over the whole orbit,

$$0 = \frac{n}{y_i} , \qquad (20.3)$$

which is impossible. Therefore w never vanishes, and has a constant sign along a family.

In regard to properties 6 and 7 , we have now

$$y_{i-1} - 2y_i + y_{i+1} = -\frac{2w}{y_i} + 2w^2 , \qquad i = 1, \dots, n . \qquad (20.4)$$

This is a system of n equations for n variables $y_1, \dots, y_n$. It can be rewritten as a system of n quadratic equations:

$$y_i(y_{i-1} - 2y_i + y_{i+1}) - 2w^2 y_i + 2w = 0 , \qquad i = 1, \dots, n . \qquad (20.5)$$

It follows that for a given w, there are at most 2^n solutions. But if we introduce homogeneous coordinates $Y_1, \ldots, Y_n, Z$, defined by

$$y_i = Y_i/Z \,, \tag{20.6}$$

we obtain

$$Y_i(Y_{i-1} - 2Y_i + Y_{i+1}) - 2w^2 Y_i Z + 2wZ^2 = 0 \,, \qquad i = 1, \ldots, n \,. \tag{20.7}$$

One particular solution is

$$Y_i = 1, \qquad i = 1, \ldots, n \,; \qquad Z = 0 \,. \tag{20.8}$$

This solution is not valid since it corresponds to infinite values of the y_i. It follows that the system (20.5) has at most $2^n - 1$ solutions.

In regard to property 8, the bifurcation orbit corresponds to a cycle of period n of the mapping (19.3).

20.1.1 Asymptotic Branches

For $w \to \pm\infty$, the equations (19.8) are still applicable. It can be shown that the one-parameter families of solutions of (20.1) have no other asymptotic branches than those which correspond to a decomposition of the R-orbit into S-arcs; the proof is the same as in Sect. 19.1.1, with slight modifications.

For $w \to 0$, the equations (19.34) are still applicable. It can be shown that the one-parameter families of solutions of (20.1) have no other asymptotic branches than those which correspond to a R-arc; the proof is the same as in Sect. 19.1.3, with some modifications. It can be shown also that for $n > 1$, there exists no branch with $w \to 0^-$.

20.1.2 Stability and Jacobian

We proceed as in Sect. 18.2.2. To compute the Jacobian, we order the variables as follows:

$$y_0, x_1', x_1'' . y_1, x_2', x_2'', y_2, \ldots, y_{n-1}, x_n', x_n'' \,. \tag{20.9}$$

The fundamental equations (20.1) can be written

$$\begin{aligned}
f_1 &= x_1'' - x_1' + w = 0 \,, \\
f_2 &= y_1 - y_0 + w(x_1' + x_1'') = 0 \,, \\
f_3 &= y_1(x_2' - x_1'') - 1 = 0 \,, \\
&\ldots \\
f_{3i-2} &= x_i'' - x_i' + w = 0 \,, \\
f_{3i-1} &= y_i - y_{i-1} + w(x_i' + x_i'') = 0 \,, \\
f_{3i} &= y_i(x_{i+1}' - x_i'') - 1 = 0 \,, \\
&\ldots
\end{aligned}$$

$$
\begin{aligned}
f_{3n-2} &= x''_n - x'_n + w = 0 \,, \\
f_{3n-1} &= y_0 - y_{n-1} + w(x'_n + x''_n) = 0 \,, \\
f_{3n} &= y_0(x'_1 - x''_n) - 1 = 0 \,.
\end{aligned}
\tag{20.10}
$$

For a given value of w, this is a system of $3n$ equations for $3n$ variables. The Jacobian is

$$
|J| = \frac{\partial(f_1, \ldots, f_{3n})}{\partial(y_0, x'_1, x''_1, y_1, \ldots, x' - n, x''_n)} \,. \tag{20.11}
$$

We can also "unroll" the system of equations (20.10) by adding two variables y_n and x'_{n+1}, considered as distinct from y_0 and x'_1, modifying the last two equations of (20.10) and adding two equations, so that the end of the new system reads:

$$
\begin{aligned}
&\ldots \\
f_{3n-1} &= y_n - y_{n-1} + w(x'_n + x''_n) = 0 \,, \\
f_{3n} &= y_n(x'_{n+1} - x''_n) - 1 = 0 \,, \\
f_{3n+1} &= y_0 - y_n = 0 \,, \\
f_{3n+2} &= x'_1 - x'_{n+1} = 0 \,.
\end{aligned}
\tag{20.12}
$$

We consider the system of equations (20.12), with the last two equations deleted. We have then a system of $3n$ equations for $3n+2$ variables. Starting from given values of y_0 and x'_1 and applying the equations one by one, we obtain successively x''_1, y_1, $\ldots$, x'_n, x''_n, y_n, x'_{n+1}. In the same way, starting from variations $\mathrm{d}y_0$ and $\mathrm{d}x'_1$, we can compute successively $\mathrm{d}x''_1$, $\mathrm{d}y_1, \ldots, \mathrm{d}y_n$, $\mathrm{d}x'_{n+1}$. For the whole orbit, we have then

$$
\begin{pmatrix} \mathrm{d}y_n \\ \mathrm{d}x'_{n+1} \end{pmatrix} = \begin{pmatrix} \dfrac{\partial y_n}{\partial y_0} & \dfrac{\partial y_n}{\partial x'_1} \\ \dfrac{\partial x'_{n+1}}{\partial y_0} & \dfrac{\partial x'_{n+1}}{\partial x'_1} \end{pmatrix} \begin{pmatrix} \mathrm{d}y_0 \\ \mathrm{d}x'_1 \end{pmatrix} \,. \tag{20.13}
$$

The stability index is the trace of the matrix divided by 2:

$$
z = \frac{1}{2}\left(\frac{\partial y_n}{\partial y_0} + \frac{\partial x'_{n+1}}{\partial x'_1} \right) \,. \tag{20.14}
$$

Proceeding as in Sect. 14.1.2, we can show that

Proposition 20.1.1. *In a total transition 2.1, the Jacobian vanishes if and only if the R-orbit is a critical orbit of the first kind (the stability index is* $z = 1$*).*

Actually any pair of variables can be used to compute the stability index. In practice it will be convenient to use y_0 and y_1 rather than y_0 and x'_1. We can then use (19.3) to obtain

$$
\begin{pmatrix} \mathrm{d}y_i \\ \mathrm{d}y_{i+1} \end{pmatrix} = \begin{pmatrix} 0 & 1 \\ -1 & u_i \end{pmatrix} \begin{pmatrix} \mathrm{d}y_{i-1} \\ \mathrm{d}y_i \end{pmatrix} \tag{20.15}
$$

with

$$u_i = 2 + \frac{2w}{y_i^2} \,. \tag{20.16}$$

We can then easily compute the matrix

$$\begin{pmatrix} \dfrac{\partial y_n}{\partial y_0} & \dfrac{\partial y_n}{\partial y_1} \\ \dfrac{\partial y_{n+1}}{\partial y_0} & \dfrac{\partial y_{n+1}}{\partial y_1} \end{pmatrix} = \begin{pmatrix} 0 & 1 \\ -1 & u_n \end{pmatrix} \begin{pmatrix} 0 & 1 \\ -1 & u_{n-1} \end{pmatrix} \cdots \begin{pmatrix} 0 & 1 \\ -1 & u_1 \end{pmatrix} \tag{20.17}$$

and the stability index is

$$z = \frac{1}{2}\left(\frac{\partial y_n}{\partial y_0} + \frac{\partial y_{n+1}}{\partial y_1}\right) \,. \tag{20.18}$$

It can be shown that

Proposition 20.1.2. *R-orbits are always unstable for* $w > 0$.

Proof: we will use the fact that, from (20.16),

$$u_i > 2 \,. \tag{20.19}$$

We can then use the same proof as in Sect. 18.2.2 to show that $z > 1$. It follows that *the Jacobian never vanishes for* $w > o$.

20.2 Small Values of n

We will represent the characteristics in the (w, x_1') plane. A point of this plane is not sufficient to define an orbit: the three values w, y_0, x_1' are needed. As a consequence, the intersection of two characteristics on the figures does not necessarily correspond to a true intersection of the two families; it can be merely a projection effect.

Here again, the simplest method of solution consists in solving the equations (20.4) for the y_i, and then computing the x_i' and x_i'' from (19.4).

20.2.1 $n = 1$

This case corresponds to a bifurcation 2T1 and will be treated separately in Sect. 23.1. However, it is of interest to consider it also here, in the context of transition 2.1.

The equations are immediately solved into

$$y_0 = \frac{1}{w} \,, \qquad x_1' = -x_1'' = \frac{w}{2} \,. \tag{20.20}$$

The stability index is

$$z = \frac{u_0}{2} = 1 + w^3 \,. \tag{20.21}$$

The Jacobian vanishes only for $w = 0$.

20.2.2 $n = 2$

This case was already studied by Guillaume (1971, pp. 139–145).

From (19.3) we obtain the two equations

$$y_0 - y_1 = -\frac{w}{y_1} + w^2\,, \qquad y_1 - y_0 = -\frac{w}{y_0} + w^2\,. \tag{20.22}$$

Adding and subtracting, we obtain

$$y_0 + y_1 = -w^2\,, \qquad (y_0 - y_1)(2y_0y_1 + w) = 0\,. \tag{20.23}$$

The solution $y_0 = y_1$ is spurious: it corresponds to a 2T1 orbit described twice. Therefore we have $y_0y_1 = -w/2$. Solving for y_0, y_1 and computing the x_i' and x_i'' from (19.4), we obtain the two solutions

$$\begin{aligned} y_0 &= \frac{-w^2 \pm \sqrt{w^4 + 2w}}{2}\,, \qquad & y_1 &= \frac{-w^2 \mp \sqrt{w^4 + 2w}}{2}\,, \\ x_1' = -x_2'' &= \frac{w^2 \pm \sqrt{w^4 + 2w}}{2w}\,, \qquad & x_1'' = -x_2' &= \frac{-w^2 \pm \sqrt{w^4 + 2w}}{2w}\,. \end{aligned} \tag{20.24}$$

In the $\pm$ and $\mp$ symbols, either the upper signs or the lower ones should be selected everywhere.

The asymptotic branches for $w \to \pm\infty$ are for the upper signs

$$\begin{aligned} & y_0 \sim \frac{1}{2w}\,, \quad y_1 \sim -w^2\,, \\ & x_1' \sim w\,, \quad x_1'' \sim \frac{1}{2w^2}\,, \quad x_2' \sim -\frac{1}{2w^2}\,, \quad x_2'' \sim -w\,, \end{aligned} \tag{20.25}$$

and for the lower signs

$$\begin{aligned} & y_0 \sim -w^2\,, \quad y_1 \sim \frac{1}{2w}\,, \\ & x_1' \sim -\frac{1}{2w^2}\,, \quad x_1'' \sim -w\,, \quad x_2' \sim w\,, \quad x_2'' \sim \frac{1}{2w^2}\,. \end{aligned} \tag{20.26}$$

The asymptotic branches for $w \to 0^+$ are

$$\begin{aligned} & y_0 \sim \pm\sqrt{\frac{w}{2}}\,, \quad y_1 \sim \mp\sqrt{\frac{w}{2}}\,, \qquad x_1' \sim \pm\frac{1}{\sqrt{2w}}\,, \\ & x_1'' \sim \pm\frac{1}{\sqrt{2w}}\,, \qquad x_2' \sim \mp\frac{1}{\sqrt{2w}}\,, \qquad x_2'' \sim \mp\frac{1}{\sqrt{2w}}\,. \end{aligned} \tag{20.27}$$

The characteristics are represented on Fig. 20.1. The asymptotic branches for $w \to \pm\infty$ are easily identified with the help of (19.8).

This is essentially Fig. V–2 in Guillaume (1971, p. 140+). (N in Guillaume is proportional to $-w$ here.) (See also Guillaume 1973b.)

The stability index is obtained from (20.17) and (20.18):

$$z = \frac{u_0u_1}{2} - 1 = 8w^3 + 17\,. \tag{20.28}$$

There is $z = 1$ for $w^3 = -2$. This corresponds to the extremum in w.

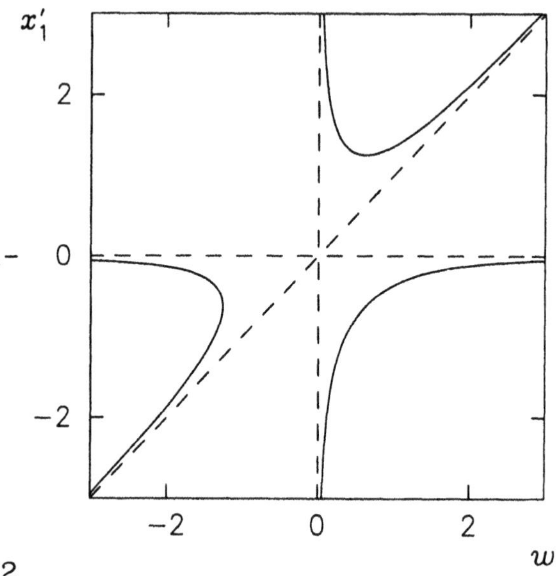

Fig. 20.1. Characteristics for 2.1T2 transitions.

20.2.3 $n = 3$

This case was also studied by Guillaume (1971, pp. 158–164).
(19.3) gives the three relations

$$
\begin{aligned}
y_2 - 2y_0 + y_1 &= -\frac{2w}{y_0} + 2w^2 \,, \\
y_0 - 2y_1 + y_2 &= -\frac{2w}{y_1} + 2w^2 \,, \\
y_1 - 2y_2 + y_0 &= -\frac{2w}{y_2} + 2w^2 \,.
\end{aligned} \tag{20.29}
$$

Subtracting one relation from another, we obtain

$$
\begin{aligned}
&(y_2 - y_1)(3y_1y_2 + 2w) = 0 \,, \\
&(y_0 - y_2)(3y_2y_0 + 2w) = 0 \,, \\
&(y_1 - y_0)(3y_0y_1 + 2w) = 0 \,.
\end{aligned} \tag{20.30}
$$

One solution is $y_0 = y_1 = y_2 = 1/w$. However, this is a spurious solution corresponding to a 2T1 orbit described three times. Thus not all y_i can be equal. On the other hand, assume that no two y_i are equal. From (20.30b) and (20.30c) we have then

$$
3y_2y_0 + 2w = 0 \,, \qquad 3y_0y_1 + 2w \,, \tag{20.31}
$$

from which $y_1 = y_2$ follows: contradiction. Therefore two y_i must be equal, while the third is different. The equations are then easily solved. In the case $y_0 \neq y_1 = y_2$, we find the two solutions

$$\begin{aligned} y_0 &= \frac{-3w^2 \pm \sqrt{9w^4+12w}}{6}\,, \qquad y_1 = y_2 = \frac{-3w^2 \mp \sqrt{9w^4+12w}}{3}\,, \\ x_1' &= -x_3'' = \frac{3w^2 \pm \sqrt{9w^4+12w}}{4w}\,, \\ x_1'' &= -x_3' = \frac{-w^2 \pm \sqrt{9w^4+12w}}{4w}\,, \qquad x_2' = -x_2'' = \frac{w}{2}\,. \end{aligned} \tag{20.32}$$

The four other solutions are obtained by rotating the indices.

The asymptotic branches of the two solutions (20.32) for $w \to \pm\infty$ are for the upper signs

$$\begin{aligned} y_0 &\sim \frac{1}{3w}\,, \qquad y_1 = y_2 \sim -2w^2\,, \\ x_1' &= -x_3'' \sim \frac{3w}{2}\,, \qquad x_1'' = -x_3' \sim \frac{w}{2}\,, \qquad x_2' = -x_2'' = \frac{w}{2}\,. \end{aligned} \tag{20.33}$$

and for the lower signs

$$\begin{aligned} y_0 &\sim -w^2\,, \qquad y_1 = y_2 \sim \frac{2}{3w}\,, \\ x_1' &= -x_3'' \sim -\frac{1}{2w^2}\,, \qquad x_1'' = -x_3' \sim -w\,, \qquad x_2' = -x_2'' = \frac{w}{2}\,. \end{aligned} \tag{20.34}$$

The asymptotic branches for $w \to 0^+$ are

$$\begin{aligned} y_0 &\sim \pm\sqrt{\frac{w}{3}}\,, \qquad y_1 = y_2 \sim \mp\sqrt{\frac{4w}{3}}\,, \\ x_1' &= -x_3'' \sim \pm\sqrt{\frac{3}{4w}}\,, \qquad x_1'' = -x_3' \sim \pm\sqrt{\frac{3}{4w}}\,, \qquad x_2' = -x_2'' = \frac{w}{2}\,. \end{aligned} \tag{20.35}$$

The characteristics are represented on Fig. 20.2. The asymptotic branches for $w \to \pm\infty$ are identified with the help of (19.8).

The stability index is obtained from (20.17) and (20.18), after a little computation:

$$\begin{aligned} z &= \frac{u_0 u_1 u_2 - u_0 - u_1 - u_2}{2} \\ &= \frac{81w^6 + 351w^3 + 332 \mp (27w^4 + 27w)\sqrt{9w^4+12w}}{8}\,. \end{aligned} \tag{20.36}$$

The equation $z = 1$ becomes

$$(3w^3+4)(2w^3+3)^2 = 0\,. \tag{20.37}$$

The case $w^3 = -4/3$ corresponds to an extremum. For the case $w^3 = -3/2$, (20.36) gives

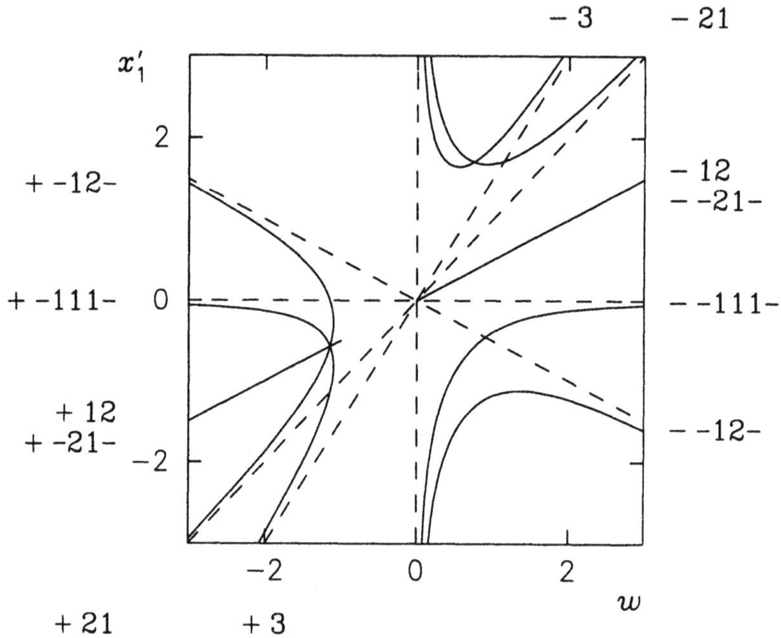

Fig. 20.2. Characteristics for **2.1T3** transitions.

$$z = \frac{-49 \mp 81}{32} \tag{20.38}$$

which shows that we must take the lower sign in (20.32) in order to have $z = 1$. This case corresponds to an intersection with an orbit **2T1** described three times.

20.2.4 $n = 4$

A frontal attack on the equations (20.1) for $n = 4$ seems difficult. We will simplify the task by noting that the asymptotic branches for $w \to \pm\infty$ must correspond to a decomposition into S-arcs. There are 24 possible branches of this kind (Table 6.5): **± 4**, **± 31**, **± 211**, and the other branches obtained by a shift of the origin. All of these branches correspond to symmetric orbits (see Table 7.5). Since we are interested in families of solutions which come from large values of w, we can limit our search for solutions of (20.1) to symmetric orbits.

The two crossings of the symmetry axis can either lie both in collisions, at the junctions of two basic arcs, or both in the midpoints of basic arcs. We consider these two cases in turn.

20.2.4.1 First Case: Symmetry Axis Through Collisions. This case was studied by Guillaume (1971, pp. 146–157). We assume first that the origin is at one crossing of the symmetry axis ($h = 0$). We have then from the symmetry: $y_3 = y_1$, $x'_3 = -x''_2$, $x''_3 = -x'_2$, $x'_4 = -x''_1$, $x''_4 = -x'_1$. The equations (20.1) can be solved with a little algebra. After elimination of the spurious solutions corresponding to a 2T1 orbit described four times and to a 2T2 orbit described twice, one finds the solutions

$$y_0 = \frac{q + \omega_1\sqrt{q^2+4w}}{2}\,, \quad y_1 = y_3 = q + w^2\,, \quad y_2 = \frac{q - \omega_1\sqrt{q^2+4w}}{2}\,,$$
$$x'_1 = -x''_4 = \frac{-q + \omega_1\sqrt{q^2+4w}}{4w}\,,$$
$$x''_1 = -x'_4 = \frac{-q - 4w^2 + \omega_1\sqrt{q^2+4w}}{4w}\,,$$
$$x'_2 = -x''_3 = \frac{q + 4w^2 + \omega_1\sqrt{q^2+4w}}{4w}\,, \quad x''_2 = -x'_3 = \frac{q + \omega_1\sqrt{q^2+4w}}{4w}\,, \tag{20.39}$$

with

$$q = \frac{-5w^2 + \omega_2\sqrt{9w^4+8w}}{2} \tag{20.40}$$

and $\omega_1 = \pm 1$, $\omega_2 = \pm 1$, independently of each other. Thus (20.39) and (20.40) provide 4 solutions.

The characteristics of the 4 solutions (20.39) are represented on Fig. 20.3. The asymptotic branches for $w \to \pm\infty$ are identified with the help of (19.8). (Here there is a disagreement between our results and those of Guillaume (1971, p. 150++). Guillaume's results indicate that the among the 4 branches of the `2T4----0` subset (Table 8.18), `+ 4` is joined to `+ 121` and `+ -1111-` is joined to `+ -22-`. On the other hand, our results indicate that `+ 4` is joined to `+ -22-` and `+ -1111-` is joined to `+ 121`. It seems that Guillaume did not realize that the quantity $p^2 + 4w$ becomes negative in the case $\omega_2 = +1$, in the interval $-9/4 < w^3 < -1$.)

4 other solutions are obtained by shifting the origin by one basic arc, i.e. rotating the indices. This corresponds to $h = 2$. The characteristics are represented on Fig. 20.4.

The stability index is obtained from (20.17) and (20.18):

$$z = \frac{2 - (u_0 + u_2)(u_1 + u_3) + u_0 u_1 u_2 u_3}{2}\,, \tag{20.41}$$

For the solutions $h = 0$ and $h = 2$, this gives

$$z = 1 + \frac{[p^2 + 4w][(p + 4w^2)^2 + 4w][(p + 4w^2)^2 + 2w]}{2w^3}\,. \tag{20.42}$$

$z - 1$ is the product of three factors. The first factor, $p^2 + 4w$, vanishes for $\omega_2 = +1$, $w^3 = -1$ and for $\omega_2 = +1$, $w^3 = -9/4$. The second factor vanishes for $\omega_2 = +1$, $w^3 = -1$. The third factor vanishes for $w^3 = -8/9$ (for both signs of ω_2). All these cases correspond to extremums, which are visible on Figs. 20.3 and 20.4.

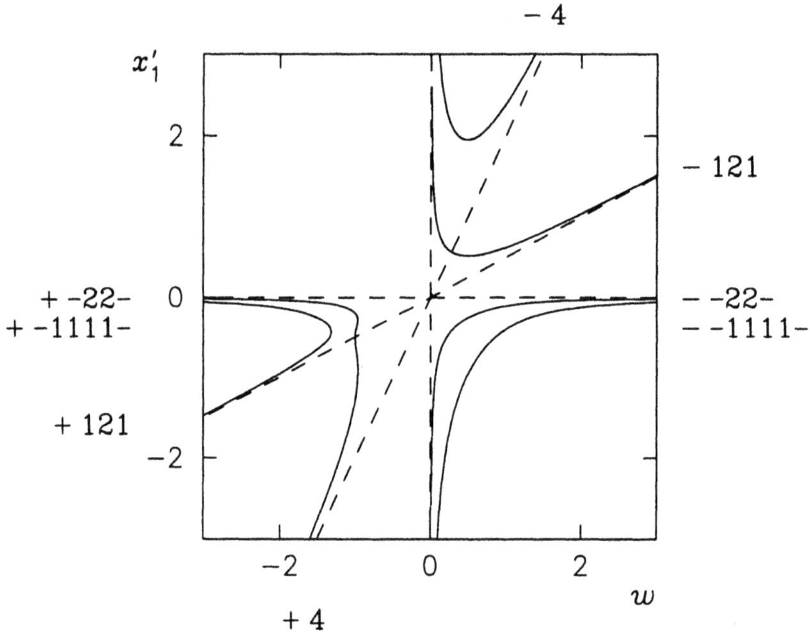

Fig. 20.3. Characteristics for $h = 0$ in the **2.1T4** transitions.

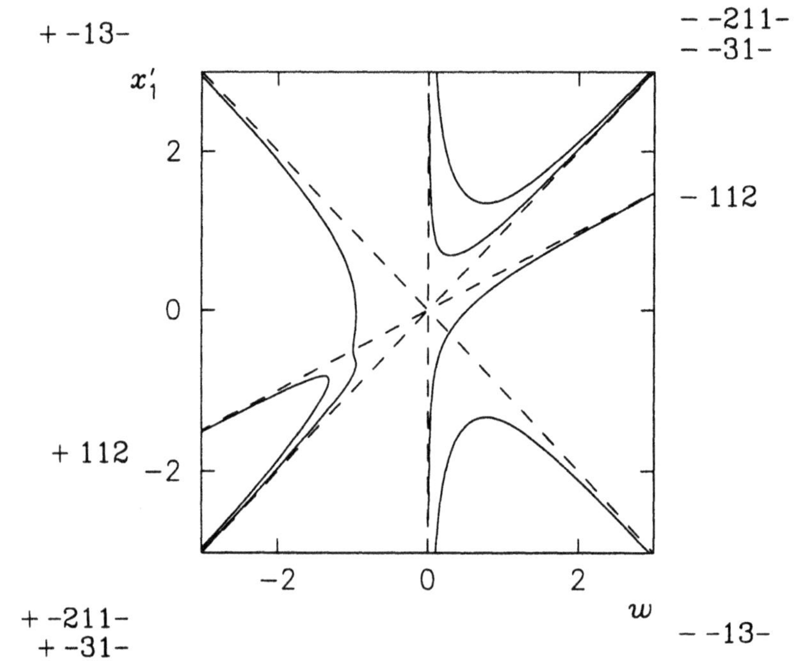

Fig. 20.4. Characteristics for $h = 2$ in the **2.1T4** transitions.

20.2.4.2 Second Case: Symmetry Axis Through Midpoints. We assume first that there is a crossing of the symmetry axis in the midpoint of the first basic arc ($h = 1$). We obtain then the two solutions

$$y_0 = y_1 = -w^2 \pm \sqrt{w^4 + w}\,, \qquad y_2 = y_3 = -w^2 \mp \sqrt{w^4 + w}\,,$$
$$x_1' = -x_1'' = \frac{w}{2}\,, \qquad x_2' = -x_4'' = \frac{w^2 \pm 2\sqrt{w^4 + w}}{2w}\,,$$
$$x_2'' = -x_4' = \frac{-w^2 \pm 2\sqrt{w^4 + w}}{2w}\,, \qquad x_3' = -x_3'' = \frac{w}{2}\,. \tag{20.43}$$

2 other solutions are obtained by rotating the indices; this corresponds to $h = 3$.

The characteristics of these 4 solutions ($h = 1$ and $h = 3$) are represented on Fig. 20.5.

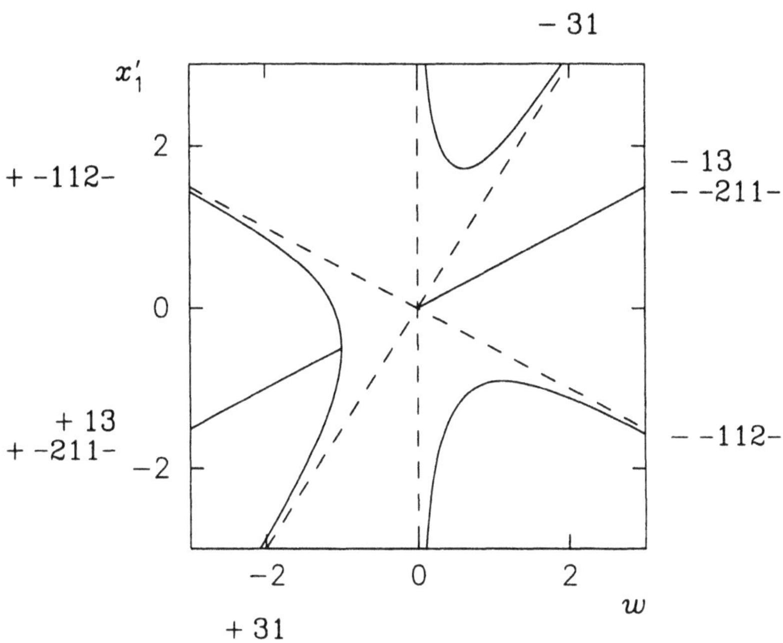

Fig. 20.5. Characteristics for $h = 1$ and $h = 3$ in the 2.1T4 transitions.

The stability index is computed from (20.41); we obtain

$$z = 1 + 96(w^3 + 1)^2\,. \tag{20.44}$$

There is $z = 1$ only for $w = -1$, which corresponds to an extremum (Fig. 20.5).

We remark that in all cases $n = 1, 2, 3, 4$, when the spurious solutions are included, the maximal number of solutions $2^n - 1$ (Sect. 20.1) is reached.

(In the case $n = 4$, this proves that there are no other solutions than the symmetric solutions which we have computed.)

20.2.5 $n > 4$

Here again the solutions found for $n = 2$ to 4 suggest that for larger values of n there might exist also solutions of the form (19.69). It is found, however, that no such solutions exist for $n = 5$. It seems that the system (20.1) for $n > 4$ can only be solved numerically.

20.2.5.1 $w > 0$: Relative Sides of Passage. The positional method used in Sect. 19.3 works only for a partial transition. However, it was noted in Sect. 19.3.3 that in the case $w > 0$, the positional method is not necessary: the consideration of the relative sides of passage, i.e. Broucke's principle, is sufficient to establish all junctions. The same approach can be used for the total transition.

We consider first the asymptotic branches for $w \to +\infty$. The orbit consists in a sequence of S-arcs. y_i is positive in a node, negative in an antinode (Sect. 18.1.2). Therefore the sequence of signs defines completely the orbit. But if the sequence of signs has a sub-period (for instance +R-R-R+R-R-R+), the corresponding orbit also has a subperiod, and therefore must be excluded. Thus, *for a given n, there is one and only one branch for each sequence of signs having no subperiod.*

We consider now the branches for $w \to 0^+$. The case $n = 1$ is excluded (see Sect. 23.1). For $n > 1$, the study of the R-orbits (Sects. 18.2, 18.3) shows that there exists one and only one orbit for each sequence of signs having no subperiod.

Therefore, all junctions can indeed be trivially established for total transitions 2.1 in the case $w > 0$. Each orbit made of a sequence of S-arcs is joined to a R-orbit, which is immediately determined by considering the signs.

As in the case of a partial transition (Sect. 19.3.3.1), we have the result that *all branches with a - sign in Table 8.18 pass through the transition 2.1 at $\nu = 1/3$ and go on to higher values of ν.* Their junctions will only be determined at transition 2.2, for $\nu = 1/2$ (Chap. 22).

20.2.5.2 $w < 0$: Numerical Computation. In that case, all relative sides of passage are negative, and therefore Broucke's principle gives no information on the junctions.

The case $n = 1$ is excluded here; it will be treated in Sect. 23.1.

For $n > 1$, there are no R-orbits, and therefore no branches $w \to 0^-$. It follows that the branches $w \to -\infty$ must be joined between themselves: *all branches with a + sign in Table 8.18 are joined in the transition 2.1.*

The cases $n = 2$ to 4 have been solved analytically in the previous sections. For $n > 4$, a numerical computation seems necessary. A program similar to the one described in Sect. 14.2.2 for the total bifurcation of type 1 was used. The computation of a given branch is begun at a large value of $|w|$. An iterative relaxation method, based on the asymptotic decomposition of the orbit into S arcs and on the asymptotic equations (19.8), works well in that case.

The branch is followed towards decreasing $|w|$. Usually at some point the iteration does not converge anymore, because the decomposition into S arcs ceases to be a good approximation. We shift then to a trial-and-error shooting method, which works well for moderate values of $|w|$ where the branches are well separated.

20.3 Results for Bifurcations of Type 2

20.3.1 The Case $w > 0$

It was shown in Sect. 20.2.5.1 that for $w > 0$, every characteristic of a R-orbit runs from $w = 0$ to $w = +\infty$. In other words, a R-orbit simply traverses the total transition 2.1. The R-orbit coincides with the whole bifurcation orbit. Thus we obtain

Proposition 20.3.1. *All branches with a - sign and containing no T-arcs (symbols* i *and* e*) in Table 8.18 pass through the transition 2.1 at $\nu = 1/3$ and go on to higher values of ν.*

This result complements the one found in Sect. 19.4.

20.3.2 The Case $w < 0$

Since the R-orbit coincides with the whole bifurcation orbit, the junctions found in Sect.20.2.5.2 are immediately applicable to the branches with a + sign in the total bifurcation of type 2.

Table 20.1 lists the junctions up to $n = 6$, in the same format as Table 8.18, which was obtained using only qualitative methods. It can be verified that there is agreement for all branches which appear in both tables (sign + and $2 \leqq n \leqq 4$).

Table 20.1. Total bifurcation, type 2, $n = 2$ to 6: junctions between the + branches, determined by the quantitative study.

2T2---0	+ 2	+ -11-
2T3----0	+ 3	+ -111-
2T3----1	+ -21-	+ 12
2T3----2	+ -12-	+ 21
2T4-----0	+ 4	+ -22-
	+ -1111-	+ 121
2T4-----1	+ -211-	+ 13
2T4-----2	+ -31-	+ -13-
	+ 211	+ 112
2T4-----3	+ 31	+ -112-
2T5------0	+ 131	+ -131-
	+ 212	+ -11111-
	+ 5	+ -212-
2T5------1	+ -2111-	+ -221-
	+ 122	+ 1121
	+ -32-	+ 14
2T5------2	+ 113	+ 23
	+ -1211-	+ 2111
	+ -14-	+ -311-
2T5------3	+ 311	+ 32
	+ -1121-	+ 1112
	+ -41-	+ -113-
2T5------4	+ -1112-	+ -122-
	+ 221	+ 1211
	+ -23-	+ 41
2T6-------0	+ 141	+ -141-
	+ -2112-	+ -222-
	+ 11211	+ 2112
	+ -111111-	+ 1221
	+ 6	+ -33-
2T6-------1	+ 1131	+ -21111-
	+ 15	+ -312-
2T6-------2	+ 114	+ 24
	+ -3111-	+ -321-
	+ 11121	+ -12111-
	+ 21111	+ 1122
	+ -15-	+ -42-
2T6-------3	+ 1113	+ 3111
	+ -114-	+ -411-
2T6-------4	+ -1113-	+ -123-
	+ 411	+ 42
	+ 11112	+ 2211
	+ 12111	+ -11121-
	+ -24-	+ -51-
2T6-------5	+ -11112-	+ 1311
	+ -213-	+ 51
2T6-------A	+ 321	+ 312
	+ 132	+ -1311-
	+ -1131-	+ 231
	+ 213	+ 123
	+ -1212-	+ -1122-
	+ -2211-	+ -2121-

21. Partial Transition 2.2

We study now in detail the equations (17.205) obtained in Sect. 17.7 for transition 2.2, corresponding to $\nu = 1/2$.

We consider first in the present chapter the case of a partial transition. The equations are then

$$
\begin{aligned}
&X_i^2 - Z_i - W = 0\,, \qquad i = 1, \ldots, n\,,\\
&Y_i - Y_{i-1} - Z_i X_i = 0\,, \qquad i = 1, \ldots, n\,,\\
&Y_i(X_{i+1} - X_i) - 1 = 0\,, \qquad i = 1, \ldots, n-1\,;\\
&Y_0 = 0\,, \qquad Y_n = 0\,.
\end{aligned} \tag{21.1}
$$

These equations form a system of $3n+1$ equations for the $3n+2$ variables W, Y_0 to Y_n, X_1 to X_n, Z_1 to Z_n. Therefore we expect one-parameter families of solutions. For a given value of μ, there is a one-to-one correspondence, given by (17.198) to (17.199), between the present variables W, Y_i, X_i, Z_i and the physical variables ΔC, h_i, u_i, Δa_i. Thus, the one-parameter families of solutions of (21.1) correspond simply to the ordinary one-parameter families of orbits (see Sect. 2.3).

21.1 Properties

1. The quantities Y_i for $i = 1, \ldots, n-1$ can never vanish, because of (21.1c). Therefore each of them keeps a constant sign along a family.
2. For $n > 1$, there is

$$
Z_i < 0\,, \qquad i = 1, \ldots, n\,. \tag{21.2}
$$

Proof: Assume that there exists a $Z_i \geqq 0$. We consider first the case where $X_i \geqq 0$, $Y_i > 0$. From (21.1c) we have then $X_{i+1} > X_i > 0$; from (21.1a), $Z_{i+1} > Z_i \geqq 0$; from (21.1b), $Y_{i+1} > Y_i > 0$; from (21.1c), $X_{i+2} > X_{i+1}$; and so on. The Y_i form an increasing sequence. This contradicts (21.1e). We consider next the case $X_i \geqq 0$, $Y_i < 0$. Then from (21.1b) we have $Y_{i-1} \leqq Y_i < 0$; from (21.1c), $X_{i-1} > X_i \geqq 0$; from (21.1a), $Z_{i-1} > Z_i \geqq 0$; from (21.1b), $Y_{i-2} < Y_{i-1}$; and so on. The Y_i towards lower i form a decreasing sequence. This contradicts (21.1d).
If $X_i \leqq 0$, the same proof holds with all signs inverted for the Y_i and X_i.

Finally, in the case $i = n$, there is $Y_i = 0$. We reason in a similar way, using Y_{i-1} instead of Y_i.

3. It follows immediately from (21.1a) and the previous Property 2 that, for $n > 1$,

$$W > 0 \,. \tag{21.3}$$

Since W has the sign of $-\epsilon'\Delta C$ from (17.198) and 17.199), it follows that *only branches with a - sign are involved in transition 2.2.* This agrees with what was found in Sects. 19.3.3 and 20.2.5. Branches with a + sign are joined at transition 2.1 and never reach transition 2.2. The only exception is the case $n = 1$, which will be treated separately in Sect. 23.2.
The result (21.3) can be refined. From (21.1a) and Property 2 we obtain

$$|X_i| < W^{1/2} \,, \qquad |Z_i| < W \,. \tag{21.4}$$

From (21.1b) (for $i = 1$) and (21.1d), we derive then

$$|Y_1| < W^{3/2} \,. \tag{21.5}$$

On the other hand, (21.1c) gives

$$|Y_i| > \frac{1}{2} W^{-1/2} \,. \tag{21.6}$$

Combining these results, we obtain

$$W > 2^{-1/2} \,. \tag{21.7}$$

4. For any solution, by applying the fundamental symmetry Σ of the restricted problem (Sect. 2.7), we obtain another solution:

$$\Sigma : \qquad (Y_i, X_i, Z_i) \mapsto (Y_{n-i}, -X_{n+1-i}, Z_{n+1-i}) \,. \tag{21.8}$$

5. For any solution, there exists a symmetrical solution obtained by changing the signs of all variables Y_i and X_i. We call this symmetry Σ':

$$\Sigma' : \qquad (Y_i, X_i, Z_i) \mapsto (-Y_i, -X_i, Z_i) \,. \tag{21.9}$$

By combining Σ and Σ', we obtain

$$\Sigma'\Sigma : \qquad (Y_i, X_i, Z_i) \mapsto (-Y_{n-i}, X_{n+1-i}, Z_{n+1-i}) \,. \tag{21.10}$$

6. Eliminating the Z_i, we can write the equations (21.1) as

$$X_{i+1} = X_i + \frac{1}{Y_i} \,, \qquad Y_{i+1} = Y_i + X_{i+1}(X_{i+1}^2 - W) \,. \tag{21.11}$$

For a given W, this is an area-preserving mapping of the (X, Y) plane. Numerical explorations (Hénon, unpublished) exhibit the mixture of regular and chaotic orbits characteristic of non-integrable systems. This supports the conjecture that the system of equations (21.1) is not explicitly solvable in general.

21.1.1 Asymptotic Branches for $W \to +\infty$

In the limit $W \to +\infty$, we enter the region $|\Delta C| \gg \mu^{1/2}$, or $\nu < 1/2$. The solutions in that region are known (Sect. 17.6); they are given by the equations (17.187). to (17.194) of Sect. 17.6. If we substitute these expressions into the changes of variables (17.198) and (17.199) for the case $\nu = 1/2$, we find:

- T-arc:

  $$Z_1 = -\frac{g}{2W}[1 + O(W^{-1})] \,, \tag{21.12}$$

  where g is defined by (17.195) and (17.196); and

  $$X_1 = \pm\text{sign}(a-1)W^{1/2}[1 + O(W^{-2})] \,, \tag{21.13}$$

  with a + sign for a T^i arc and a − sign for a T^e arc. With the new notations T^f and T^g (Sect. 17.1), this becomes

  $$X_1 = \pm W^{1/2}[1 + O(W^{-1})] \,, \tag{21.14}$$

  with a + sign for a T^f arc and a − sign for a T^g arc.
- R-arc:

  $$\begin{aligned} Z_i &= -W[1 + O(W^{-2})] \,, \\ Y_i &= W^{1/2}y_j[1 + O(W^{-1})] \,, \\ X_i &= W^{-1/2}x_j[1 + O(W^{-1})] \,, \end{aligned} \tag{21.15}$$

  where we have defined

  $$j = i - i_{\alpha-1} \,. \tag{21.16}$$

  j is the relative position inside the R-arc. y_j, x_j are the numbers computed in Chap. 18 (see Tables 18.2 and 18.3).
- TT node:

  $$Y_i = \pm\frac{1}{2}W^{-1/2}[1 + O(W^{-2})] \,, \tag{21.17}$$

  with a + sign for a T^gT^f node and a − sign for a T^fT^g node.
- TR or RT node:

  $$Y_i = \pm W^{-1/2}[1 + O(W^{-1})] \,, \tag{21.18}$$

  with a + sign for a T^gR or RT^f node, and a − sign for a T^fR or RT^g node.

It can be verified that these asymptotic values satisfy the equations (21.1).

We will need a finer expression for a TR node, distinguishing between different R-arcs. This expression is easily obtained by substituting (21.14) for X_i and (21.15c) for X_{i+1} into (21.1c):

$$Y_i = \frac{1}{X_{i+1} - X_i} = \frac{1}{W^{-1/2}x_{i+1} \mp W^{1/2} + O(W^{-3/2})}$$
$$= \mp W^{-1/2} - W^{-3/2}x_{i+1} + O(W^{-5/2}) \qquad (21.19)$$

with a + sign for a $T^g R$ node and a − sign for a $T^f R$ node.

Conversely, we show now that for $W \to +\infty$, the one-parameter families of solutions of (21.1) have no other asymptotic branches than those which correspond to a sequence of T-arcs and R-arcs, and which are described by (21.12) to 21.19). Consider one particular branch. We assume that the Y_i and the X_i behave asymptotically as powers of W:

$$Y_i \sim a_i W^{p_i} , \qquad X_i \sim b_i W^{q_i} \qquad \text{for} \qquad W \to +\infty . \qquad (21.20)$$

The a_i and b_i are non-zero constants. The p_i and q_i are arbitrary constants. The two ends of the bifurcating arc, $Y_0 = Y_n = 0$, can be included in this formulation by taking $p_0 = p_n = -\infty$.

We rewrite the equations (21.1) as

$$X_{i+1} - X_i = \frac{1}{Y_i} , \qquad (21.21)$$

$$Y_i - Y_{i-1} = X_i^3 - X_i W . \qquad (21.22)$$

1. Assume that there exists a $q_i > 1/2$. In (21.22), term 4 is negligible compared to term 3. Therefore the latter must be balanced by terms 1 and/or 2, and at least one of the following inequalities is true:

$$p_{i-1} \geqq 3q_i , \qquad p_i \geqq 3q_i . \qquad (21.23)$$

Assume that the second inequality is true (a similar reasoning holds in the other case). It follows in particular that the encounter i cannot be the end of the bifurcating arc. We distinguish several cases.

(i) $p_i > p_{i-1}$. Term 2 in (21.22) is negligible, therefore terms 1 and 3 must balance:

$$p_i = 3q_i , \qquad a_i = b_i^3 . \qquad (21.24)$$

In (21.21), term 3 is negligible compared to term 2; therefore

$$q_{i+1} = q_i , \qquad b_{i+1} = b_i . \qquad (21.25)$$

Consider (21.22) with i increased by 1. Term 4 is still negligible; terms 2 and 3 are of the same order. We obtain

$$p_{i+1} = 3q_i , \qquad a_{i+1} = 2b_i^3 . \qquad (21.26)$$

Continuing in the same way, we have

$$p_{i+j} = 3q_i , \qquad a_{i+j} = (j+1)b_i^3 . \qquad (21.27)$$

The p are equal and the a constantly increase. This is impossible because at some point we must reach the end of the bifurcating arc.

(ii) $p_i < p_{i-1}$. Term 1 in (21.22) is negligible. Proceeding as in the previous case but in the opposite direction, we obtain again an impossibility.

(iii) $p_i = p_{i-1} = 3q_i$. We have from (21.22)

$$a_i - a_{i-1} = b_i^3 \ . \tag{21.28}$$

Assume first that a_i has the same sign as b_i. Using (21.21) and (21.22), we obtain

$$q_{i+j} = q_i \ , \qquad b_{i+j} = b_i \ , \qquad p_{i+j} = 3q_i \ , \qquad a_{i+j} = a_i + jb_i^3 \ . \tag{21.29}$$

This is again an increasing sequence: we have reached an impossibility. If a_i has a sign opposite to b_i, then a_{i-1} has the same sign as $-b_i$, and the same reasoning in the opposite direction again gives an impossibility.

(iv) $p_i = p_{i-1} > 3q_i$. We have

$$a_i = a_{i-1} \ . \tag{21.30}$$

Continuing, we obtain

$$q_{i+j} = q_i \ , \qquad b_{i+j} = b_i \ , \qquad p_{i+j} = p_{i-1} \ , \qquad a_{i+j} = a_{i-1} \ . \tag{21.31}$$

Here again, we will never reach $Y_n = 0$, and thus we have an impossibility.

Therefore $q_i \leqq 1/2$ for all i.

2. Assume now that there exists a q_i in the interval $-1/2 < q_i < 1/2$. In (21.22), term 3 is negligible compared to term 4. Therefore at least one of the following inequalities is true:

$$p_{i-1} \geqq 1 + q_i \ , \qquad p_i \geqq 1 + q_i \ . \tag{21.32}$$

Assume that the second inequality is true. We distinguish again several cases.

(i) $p_i > p_{i-1}$. We have

$$p_i = 1 + q_i \ , \qquad a_i = -b_i \ . \tag{21.33}$$

In (21.21), term 3 is negligible compared to term 2; therefore

$$q_{i+1} = q_i \ , \qquad b_{i+1} = b_i \ . \tag{21.34}$$

Continuing, we obtain

$$p_{i+j} = 1 + q_i \ , \qquad a_{i+j} = -(j+1)b_i \ , \tag{21.35}$$

which is impossible.

(ii) $p_i = p_{i-1} = 1 + q_i$. This case is treated as above.

(iii) $p_i = p_{i-1} > 1 + q_i$. This case is treated as above.

Therefore no q_i can lie between $-1/2$ and $+1/2$.

3. Assume that there exists a $q_i = 1/2$ with $b_i^2 \neq 1$. In (21.22), the sum of the terms 3 and 4 has as asymptotic value: $b_i(b_i^2 - 1)W^{3/2}$. It follows that at least one of the following inequalities is true:

$$p_{i-1} \geqq \frac{3}{2} \ , \qquad p_i \geqq \frac{3}{2} \ . \tag{21.36}$$

Proceeding as in the previous cases, we reach again an impossibility.

We have thus shown that for any given i, either $q_i \leqq -1/2$, or $q_i = 1/2$ and $b_i = \pm 1$. We recognize in the second case the T-arcs, and we recover Equ. (21.14).

4. We consider a sequence of basic arcs for which $q_i \leqq -1/2$. We assume that this sequence is maximal, i.e. it is not possible to extend it on one side or the other. For the interior encounters of this sequence, (21.21) shows that $p_i \geqq 1/2$. Assume that $p_i > 1/2$ for one of them. In (21.22) with i replaced by $i+1$, terms 3 and 4 are negligible compared to term 2. It follows that

$$p_{i+1} = p_i \,, \qquad a_{i+1} = a_i \,. \tag{21.37}$$

(21.21) gives then: $q_{i+2} \leqq 1/2$. Continuing in the same way, we obtain

$$p_{i+j} = p_i \,, \qquad a_{i+j} = a_i \,, \qquad \text{and} \qquad q_{i+j} \leqq 1/2 \,. \tag{21.38}$$

We never reach $Y_n = 0$; this is impossible.

It follows that $p_i = 1/2$ for all interior encounters of the sequence. We recognize a R-arc, and we recover the asymptotic equations (21.15).

5. At the junction of a T-arc and a R-arc, there is $q_i = 1/2$ with $b_i = \pm 1$ for the T-arc, and $q_{i+1} \leqq -1/2$ for the T-arc. (21.21) shows then that

$$p_i = -1/2 \,, \qquad a_i = \mp 1 \,. \tag{21.39}$$

We recover Equ. (21.18).

6. At the junction of two T-arcs with opposite signs, we recover similarly Equ. (21.17).

21.1.2 Variational Equations for $W \to +\infty$

We derive here variational equations similar to those which were obtained for type 1 in Sects. 13.1.2 and 19.1.2.

We name $U_1, U_2, \ldots, U_\alpha, \ldots, U_N$ the successive R-arcs or T-arcs.

We call $i_0, i_1, \ldots, i_\alpha, \ldots, i_N$ the positions of the nodes, with $i_0 = 0$, $i_N = n$, so that the arc U_α extends from $i_{\alpha-1}$ to i_α. The number of basic arcs in U_α is $m_\alpha = i_\alpha - i_{\alpha-1}$. ($m_\alpha = 1$ if U_α is a T arc.)

From the two initial values Y_0 and X_1, using (21.1a), (21.1b) and (21.1c) in turn, we can compute successively the values of Z_1, Y_1, X_2, Z_2, ... We compute now the corresponding variational equations: we assume arbitrary infinitesimal variations $\mathrm{d}Y_0$ and $\mathrm{d}X_1$ and we compute the corresponding variations of Z_1, Y_1, ... We have

$$\begin{aligned} &\mathrm{d}Z_i = 2X_i \mathrm{d}X_i \,, \qquad \mathrm{d}Y_i = \mathrm{d}Y_{i-1} + Z_1 \mathrm{d}X_i + X_i \mathrm{d}Z_i \,, \\ &\mathrm{d}X_{i+1} = \mathrm{d}X_i - \frac{\mathrm{d}Y_i}{Y_i^2} \,. \end{aligned} \tag{21.40}$$

For a given arc U_α, we compute now the final variations $\mathrm{d}X_{i_\alpha}$, $\mathrm{d}Y_{i_\alpha}$ as functions of the initial variations $\mathrm{d}Y_{i_{\alpha-1}}$, $\mathrm{d}X_{i_{\alpha-1}+1}$. We consider first the case where U_α is a T-arc. Then $m_\alpha = i_\alpha - i_{\alpha-1} = 1$, and using (21.40a) and (21.40b) we have

$$\mathrm{d}X_{i_\alpha} = \mathrm{d}X_{i_{\alpha-1}+1}\,, \qquad \mathrm{d}Y_{i_\alpha} = \mathrm{d}Y_{i_{\alpha-1}} + (Z_{i_\alpha} + 2X_{i_\alpha}^2)\mathrm{d}X_{i_{\alpha-1}+1} \tag{21.41}$$

or, using the asymptotic expressions (21.12) and (21.13),

$$\mathrm{d}Y_{i_\alpha} = \mathrm{d}Y_{i_{\alpha-1}} + 2W\mathrm{d}X_{i_{\alpha-1}+1}[1 + O(W^{-2})]\,. \tag{21.42}$$

We consider now the case where U_α is a R-arc. (21.40a) and (21.40b) can be combined into

$$\mathrm{d}Y_i = \mathrm{d}Y_{i-1} + (Z_1 + 2X_i^2)\mathrm{d}X_i\,. \tag{21.43}$$

Using the asymptotic expressions (21.15), we can rewrite this equation and (21.40c) as

$$\begin{aligned} \mathrm{d}Y_i &= \mathrm{d}Y_{i-1} - W\mathrm{d}X_i[1 + O(W^{-2})]\,, \\ \mathrm{d}X_{i+1} &= \mathrm{d}X_i - \frac{\mathrm{d}Y_i}{Wy_j^2}[1 + O(W^{-2})]\,, \end{aligned} \tag{21.44}$$

where $j = i - i_{\alpha-1}$. Starting from the initial variations $\mathrm{d}Y_{i_{\alpha-1}}$ and $\mathrm{d}X_{i_{\alpha-1}+1}$ and applying these relations, we find that generally

$$\begin{aligned} \mathrm{d}X_i &= -W^{-1}a_j\mathrm{d}Y_{i_{\alpha-1}}[\,] + b_j\mathrm{d}X_{i_{\alpha-1}+1}[\,]\,, \qquad i = i_{\alpha-1}+1,\dots,i_\alpha\,, \\ \mathrm{d}Y_i &= c_j\mathrm{d}Y_{i_{\alpha-1}}[\,] - Wd_j\mathrm{d}X_{i_{\alpha-1}+1}[\,]\,, \qquad i = i_{\alpha-1}+1,\dots,i_\alpha\,, \end{aligned} \tag{21.45}$$

where we have used the abbreviation $[\,] = \left[1 + O(W^{-1})\right]$, and a_j, b_j, c_j, d_j are numerical coefficients defined recursively by

$$\begin{aligned} &a_1 = 0\,, \qquad b_1 = 1\,, \qquad c_1 = 1\,, \qquad d_1 = 1\,, \\ &a_{j+1} = a_j + \frac{c_j}{y_j^2}\,, \qquad b_{j+1} = b_j + \frac{d_j}{y_j^2}\,, \\ &c_{j+1} = c_j + a_{j+1}\,, \qquad d_{j+1} = d_j + b_{j+1}\,. \end{aligned} \tag{21.46}$$

Note that all a_j, b_j, c_j, d_j are positive.

In particular we have the final variations

$$\begin{aligned} \mathrm{d}X_{i_\alpha} &= -W^{-1}a_{m_\alpha}\mathrm{d}Y_{i_{\alpha-1}}[\,] + b_{m_\alpha}\mathrm{d}X_{i_{\alpha-1}+1}[\,]\,, \\ \mathrm{d}Y_{i_\alpha} &= c_{m_\alpha}\mathrm{d}Y_{i_{\alpha-1}}[\,] - Wd_{m_\alpha}\mathrm{d}X_{i_{\alpha-1}+1}[\,]\,. \end{aligned} \tag{21.47}$$

These equations cover also the case of a T-arc, described by (21.41a) and (21.42), if we take in that case

$$a_1 = 0\,, \qquad b_1 = 1\,, \qquad c_1 = 0\,, \qquad d_1 = -2\,. \tag{21.48}$$

We compute now the initial variations of the next arc $U_{\alpha+1}$. We already know $\mathrm{d}Y_{i_\alpha}$. Since i_α is a node, we have from (21.17) and (21.18): $Y_{i_\alpha} = \Theta(W^{-1/2})$. From (21.40c) we obtain

$$\mathrm{d}X_{i_\alpha+1} = -\frac{\mathrm{d}Y_{i_\alpha}}{Y_{i_\alpha}^2}[\,] = -\frac{c_{m_\alpha}}{Y_{i_\alpha}^2}\mathrm{d}Y_{i_{\alpha-1}}[\,] + \frac{Wd_{m_\alpha}}{Y_{i_\alpha}^2}\mathrm{d}X_{i_{\alpha-1}+1}[\,]\,. \tag{21.49}$$

The equations (21.47b) and (21.49) give the initial variations of $U_{\alpha+1}$ as functions of the initial variations of U_α. We can iterate these equations and we obtain

$$\mathrm{d}Y_{i_\alpha} = W^{\alpha-1}\frac{c_{m_1} d_{m_2} d_{m_3}\dots d_{m_\alpha}}{Y_{i_1}^2\dots Y_{i_{\alpha-1}}^2}\mathrm{d}Y_0[\,] - W^\alpha \frac{d_{m_1} d_{m_2}\dots d_{m_\alpha}}{Y_{i_1}^2\dots Y_{i_{\alpha-1}}^2}\mathrm{d}X_1[\,]\,,$$

$$\mathrm{d}X_{i_\alpha+1} = -W^{\alpha-1}\frac{c_{m_1} d_{m_2} d_{m_3}\dots d_{m_\alpha}}{Y_{i_1}^2\dots Y_{i_\alpha}^2}\mathrm{d}Y_0[\,] + W^\alpha \frac{d_{m_1} d_{m_2}\dots d_{m_\alpha}}{Y_{i_1}^2\dots Y_{i_\alpha}^2}\mathrm{d}X_1[\,]\,. \tag{21.50}$$

The orders of magnitude are

$$\frac{\partial Y_{i_\alpha}}{\partial Y_0} = \Theta(W^{2\alpha-2})\,, \qquad \frac{\partial Y_{i_\alpha}}{\partial X_1} = \Theta(W^{2\alpha-1})\,,$$
$$\frac{\partial X_{i_\alpha+1}}{\partial Y_0} = \Theta(W^{2\alpha-1})\,, \qquad \frac{\partial X_{i_\alpha+1}}{\partial X_1} = \Theta(W^{2\alpha})\,. \tag{21.51}$$

The variations are strongly amplified after each node.

c_{m_1} is always positive. On the other hand, d_{m_α} is negative if U_α is a T-arc, positive if it is a R-arc.

21.1.3 Jacobian

We proceed as in Sects. 13.1.4 and 19.1.4. To compute the Jacobian, we order the variables as follows:

$$X_1, Z_1, Y_1, X_2, Z_2, Y_2, \dots, Y_{n-1}, X_n, Z_n, Y_n\,. \tag{21.52}$$

The fundamental equations (21.1) can be written

$$f_1 = {X_1}^2 - Z_1 - W = 0\,,$$
$$f_2 = Y_1 - Z_1 X_1 = 0\,,$$
$$f_3 = Y_1(X_2 - X_1) - 1 = 0\,,$$
$$\dots$$
$$f_{3i-2} = {X_i}^2 - Z_i - W = 0\,,$$
$$f_{3i-1} = Y_i - Y_{i-1} - Z_i X_i = 0\,,$$
$$f_{3i} = Y_i(X_{i+1} - X_i) - 1 = 0 = 0\,,$$
$$\dots$$
$$f_{3n-2} = {X_n}^2 - Z_n - W = 0\,,$$
$$f_{3n-1} = Y_n - Y_{n-1} - Z_n X_n = 0\,,$$
$$f_{3n} = Y_n = 0\,. \tag{21.53}$$

We have eliminated Y_0, substituting $Y_0 = 0$ in the first equation, but we have kept Y_n and the last equation $Y_n = 0$.

For a given value of W, this is a system of $3n$ equations for $3n$ variables. The Jacobian is

$$|J| = \frac{\partial(f_1, \ldots, f_{3n})}{\partial(X_1, Z_1, Y_1, \ldots, X_n, Z_n, Y_n)} . \tag{21.54}$$

21.1.4 Stability

We consider the system of equations (21.53), with the last equation deleted. We have then a system of $3n - 1$ equations for $3n$ variables. Starting from a given value X_1 and applying the equations one by one, we obtain successively Z_1, Y_1, ..., Y_n. In the same way, starting from a variation $\mathrm{d}X_1$, we can compute successively $\mathrm{d}Z_1$, $\mathrm{d}Y_1$, ..., $\mathrm{d}Y_n$. We define a *critical bifurcating arc* as a bifurcating arc for which there is

$$\frac{\mathrm{d}Y_n}{\mathrm{d}X_1} = 0 . \tag{21.55}$$

Proceeding as in Sect. 13.1.4, we can show that

Proposition 21.1.1. *In a partial transition 2.2, the Jacobian vanishes if and only if the bifurcating arc is critical.*

We have from (21.53)

$$\begin{aligned} \frac{\mathrm{d}Y_i}{\mathrm{d}X_1} &= \frac{\mathrm{d}Y_{i-1}}{\mathrm{d}X_1} + (3X_i^2 - W)\frac{\mathrm{d}X_i}{\mathrm{d}X_1} , \\ \frac{\mathrm{d}X_{i+1}}{\mathrm{d}X_1} &= \frac{\mathrm{d}X_i}{\mathrm{d}X_1} - \frac{1}{Y_i^2}\frac{\mathrm{d}Y_i}{\mathrm{d}X_1} , \end{aligned} \tag{21.56}$$

from which we can compute the successive derivatives

$$\frac{\mathrm{d}Y_1}{\mathrm{d}X_1} = 3X_1^2 - W , \tag{21.57}$$

$$\frac{\mathrm{d}Y_2}{\mathrm{d}X_1} = (3X_1^2 - W) + (3X_2^2 - W) - \frac{(3X_1^2 - W)(3X_2^2 - W)}{Y_1^2} , \tag{21.58}$$

$$\begin{aligned} \frac{\mathrm{d}Y_3}{\mathrm{d}X_1} &= (3X_1^2 - W) + (3X_2^2 - W) + (3X_3^2 - W) \\ &\quad - \frac{(3X_1^2 - W)(3X_2^2 - W)}{Y_1^2} - \frac{(3X_2^2 - W)(3X_3^2 - W)}{Y_2^2} \\ &\quad - (3X_1^2 - W)(3X_3^2 - W)\left(\frac{1}{Y_1^2} + \frac{1}{Y_2^2}\right) \\ &\quad + \frac{(3X_1^2 - W)(3X_2^2 - W)(3X_3^2 - W)}{Y_1^2 Y_2^2} , \end{aligned} \tag{21.59}$$

...

21.1.5 Branch Notation

In Chap. 18, the R-arcs have been represented by

$$\texttt{R±R±R...±R} \tag{21.60}$$

where there are $\tilde{n}$ letters R. Each of them stands for one basic arc, and each + or - sign is the sign of one y_i. Similarly, a R-orbit was represented by

$$\pm R\pm R\pm R\ldots\pm R\pm\,, \tag{21.61}$$

where there are $\tilde{n}$ letters R.

Now we generalize this notation to an arbitrary bifurcating arc or bifurcation orbit of type 2, made of R-regions and T-arcs. It will be represented by a sequence of alternating letters and signs. Each letter represents one basic arc and is either R, inside a R-region, f for a T^f-arc, or g for a T^g-arc. Each sign is the sign of the corresponding y_i as before. An example of a bifurcating arc is thus: f-g+R+R-R-g.

Not all sequences are permitted. As shown by Table 17.1, a f symbol is always preceded by a + sign (or by the beginning of the bifurcating arc) and followed by a - sign (or by the end of the bifurcating arc), and the converse is true for a g symbol. If theses conditions are satisfied, then the sequence represents one and only one branch.

21.2 Small Values of n

The characteristics of the families will be represented in the (W, X_1) plane.

Because of the symmetry Σ' (Sect. 21.1), it will be sufficient to study the case $X_1 > 0$, i.e. the upper half of the plane. (For $n > 1$, we deduce from (21.1)b for $i = 1$ and from Property 1, Sect. 21.1, that X_1 never vanishes.)

21.2.1 $n = 1$

This case belongs to the bifurcation 2P1 and will be treated separately in Sect. 23.2. But it is of interest to consider it also here, in the context of the transition 2.2.

The equations reduce to

$$X_1^2 - Z_1 - W = 0\,, \qquad Z_1 X_1 = 0\,. \tag{21.62}$$

Therefore we have two solutions:

$$X_1 = 0\,, \qquad Z_1 = -W\,, \tag{21.63}$$

and

$$Z_1 = 0\,, \qquad X_1 = \pm\sqrt{W}\,. \tag{21.64}$$

Using the asymptotic values (Sect. 21.1.1), we identify easily the 4 branches (Fig. 21.1). The left and right branches correspond to a R-arc made of a single basic arc; the upper and lower branches correspond to a T^f and a T^g arc respectively.

For $W \to 0$, the condition $\Delta C = \Theta(\mu^{1/2})$ is not satisfied any more: we move out of transition 2.2. This continuation will be studied in Sect. 23.2.

(21.57) shows that there is no critical arc for $W \neq 0$.

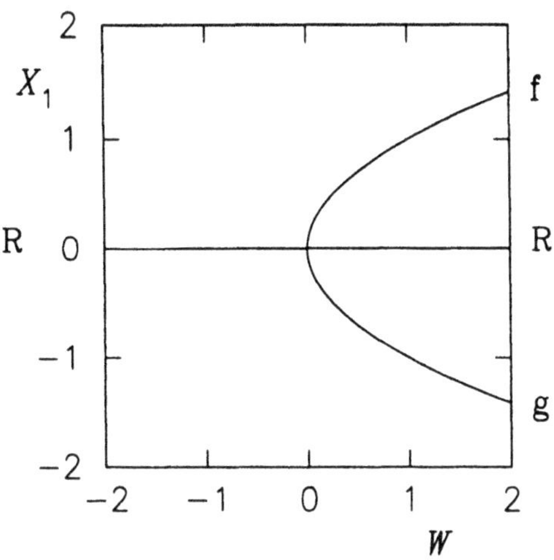

Fig. 21.1. Characteristics for 2.2P1 transitions.

21.2.2 $n = 2$

The equations are

$$X_1^2 - Z_1 - W = 0\,, \qquad Y_1 - Z_1X_1 = 0\,, \qquad Y_1(X_2 - X_1) - 1 = 0\,,$$
$$X_2^2 - Z_2 - W = 0\,, \qquad -Y_1 - Z_2X_2 = 0\,. \tag{21.65}$$

This system can be explicitly solved. Combining the equations, we obtain

$$0 = X_1Z_1 + X_2Z_2 = (X_1 + X_2)(X_1^2 - X_1X_2 + X_2^2 - W)\,. \tag{21.66}$$

Thus a first solution corresponds to $X_1 + X_2 = 0$. Restricting our attention to solutions with $X_1 > 0$, we find

$$X_1 = -X_2 = \sqrt{\frac{W \mp \sqrt{W^2 - 2}}{2}}\,, \qquad Z_1 = Z_2 = \frac{-W \mp \sqrt{W^2 - 2}}{2}\,,$$
$$Y_1 = -\frac{\sqrt{W \pm \sqrt{W^2 - 2}}}{2}\,. \tag{21.67}$$

The asymptotic branches for $W \to +\infty$ are, for the upper signs,

$$X_1 = -X_2 \sim \sqrt{\frac{1}{2W}}\,, \qquad Z_1 = Z_2 \sim -W\,, \qquad Y_1 \sim -\sqrt{\frac{W}{2}}\,, \tag{21.68}$$

and for the lower signs

$$X_1 = -X_2 \sim \sqrt{W}\,, \qquad Z_1 = Z_2 \sim -\frac{1}{2W}\,, \qquad Y_1 \sim -\frac{1}{2\sqrt{W}}\,. \tag{21.69}$$

Using the asymptotic expressions of Sect. 21.1.1, the numerical values of y_1 obtained in Sect. 18.1.4 and the branch notation defined in Sect. 21.1.5, we find that these two branches correspond to R-R and T-T respectively.

The stability can be determined from (21.58):

$$\frac{dY_2}{dX_1} = -8W(W^2-2) \pm (8W^2-12)\sqrt{W^2-2}\,. \qquad (21.70)$$

This vanishes for $W = \sqrt{2}$, which corresponds to an extremum in W, and for $W = 3/2$ with the upper sign, which corresponds to the intersection with the second family (presently to be described).

A second solution is obtained by setting the second factor in (21.66) equal to 0:

$$X_1^2 - X_1X_2 + X_2^2 - W = 0\,. \qquad (21.71)$$

Using (21.65)a to (21.65)c to eliminate X_1, Y_1, X_2, we obtain an equation for Z_1:

$$Z_1^4 + WZ_1^3 + Z_1^2 + WZ_1 + 1 = 0\,. \qquad (21.72)$$

It will be convenient here to use Z_1 as independent variable instead of W. The other variables are expressed as functions of Z_1, in the case $X_1 > 0$, by

$$W = -\frac{Z_1^4 + Z_1^2 + 1}{Z_1(Z_1^2+1)}\,, \qquad X_1 = \frac{1}{\sqrt{-Z_1(Z_1^2+1)}}\,,$$
$$Y_1 = -\sqrt{\frac{-Z_1}{Z_1^2+1}}\,, \qquad X_2 = Z_1\sqrt{\frac{-Z_1}{Z_1^2+1}}\,, \qquad Z_2 = \frac{1}{Z_1}\,. \qquad (21.73)$$

The variations as Z_1 runs from $-\infty$ to 0 are shown by Table 21.1.

Table 21.1. Variations of the variables when Z_1 increases from $-\infty$ to 0.

Z_1	$-\infty$	$\nearrow$	-1	$\nearrow$	0
W	$+\infty$	$\searrow$	$\frac{3}{2}$	$\nearrow$	$+\infty$
X_1	0	$\nearrow$	$\frac{1}{\sqrt{2}}$	$\nearrow$	$+\infty$
Y_1	0	$\searrow$	$-\frac{1}{\sqrt{2}}$	$\nearrow$	0
X_2	$-\infty$	$\nearrow$	$-\frac{1}{\sqrt{2}}$	$\nearrow$	0
Z_2	0	$\searrow$	-1	$\searrow$	$-\infty$

There are two branches $W \to +\infty$, corresponding to $Z_1 \to -\infty$ and $Z_1 \to 0$ respectively. The asymptotic values are, for the first branch:

$$X_1 \sim W^{-3/2}\,, \qquad Z_1 \sim -W\,, \qquad Y_1 \sim -W^{-1/2}\,,$$
$$X_2 \sim -W^{1/2}\,, \qquad Z_2 \sim -W^{-1}\,, \qquad (21.74)$$

and for the second branch:

$$X_1 \sim W^{1/2}\,, \qquad Z_1 \sim -W^{-1}\,, \qquad Y_1 \sim -W^{-1/2}\,,$$
$$X_2 \sim -W^{-3/2}\,, \qquad Z_2 \sim -W\,. \tag{21.75}$$

Using the asymptotic expressions of Sect. 21.1.1, we find that these two branches correspond to R-T and T-R respectively.

The stability is determined from (21.58):

$$\frac{dY_2}{dX_1} = -\frac{2(Z_1^2-1)^2(Z_1^4+3Z_1^2+1)}{Z_1^3(Z_1^2+1)}\,. \tag{21.76}$$

This has a double root for $Z_1^2 = 1$, or $W = 3/2$, corresponding to both an extremum and an intersection with the other family.

The solutions are shown on Fig. 21.2). Here it is sufficient to show the quarter-plane $W > 0$, $X_1 > 0$. We have a *trident* (Sect. 13.3.3.2), formed by the two branches of symmetric orbits R-R and T-T and the two branches of asymmetric orbits R-T and T-R.

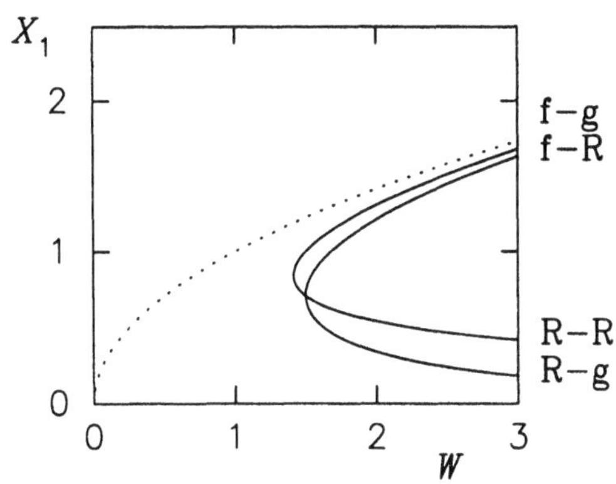

Fig. 21.2. Characteristics for 2.2P2 transitions. Also shown: 2.2P1 transitions (dotted lines).

21.2.3 $n = 3$

The equations are

$$X_1^2 - Z_1 - W = 0\,, \qquad Y_1 - Z_1X_1 = 0\,, \qquad Y_1(X_2 - X_1) - 1 = 0\,,$$
$$X_2^2 - Z_2 - W = 0\,, \qquad Y_2 - Y_1 - Z_2X_2 = 0\,, \qquad Y_2(X_3 - X_2) - 1 = 0\,,$$
$$X_3^2 - Z_3 - W = 0\,, \qquad -Y_2 - Z_3X_3 = 0\,. \tag{21.77}$$

It seems that the general solution can only be obtained numerically. The equations can be solved, however, in the particular case of symmetric orbits. We have then, from (21.8): $X_1 = -X_3$, $Z_1 = Z_3$, $Y_1 = Y_2$, $X_2 = 0$, and the equations reduce to

$$X_1^2 - Z_1 - W = 0\,, \qquad Y_1 - Z_1 X_1 = 0\,, \qquad -Y_1 X_1 - 1 = 0\,,$$
$$-Z_2 - W = 0\,, \tag{21.78}$$

from which we obtain, in the case $X_1 > 0$,

$$X_1 = -X_3 = \sqrt{\frac{W \mp \sqrt{W^2-4}}{2}}\,, \qquad Z_1 = Z_3 = \frac{-W \mp \sqrt{W^2-4}}{2}\,,$$
$$Y_1 = Y_2 = -\sqrt{\frac{W \pm \sqrt{W^2-4}}{2}}\,, \qquad Z_2 = -W\,. \tag{21.79}$$

The asymptotic branches for $W \to \pm\infty$ are, for the upper signs,

$$X_1 = -X_3 \sim W^{-1/2}\,, \qquad Z_1 = Z_3 \sim -W\,,$$
$$Y_1 = Y_2 \sim -W^{1/2}\,, \qquad X_2 = 0\,, \qquad Z_2 = -W\,, \tag{21.80}$$

and for the lower signs

$$X_1 = -X_3 \sim W^{1/2}\,, \qquad Z_1 = Z_3 \sim -W^{-1}\,,$$
$$Y_1 = Y_2 \sim -W^{-1/2}\,, \qquad X_2 = 0\,, \qquad Z_2 = -W\,. \tag{21.81}$$

Using the asymptotic expressions of Sect. 21.1.1, we find that these two branches correspond to R-R-R and T-R-T respectively.

The stability is determined from (21.59):

$$\frac{dY_3}{dX_1} = -2W^3(W^2-4) \pm 2(W^4 - 2W^2 - 6)\sqrt{W^2-4}\,. \tag{21.82}$$

This vanishes for $W = 2$, which corresponds to an extremum in W, and for

$$W = \sqrt{\frac{3(1+\sqrt{3})}{2}} \tag{21.83}$$

with the upper sign; numerical computations show that this corresponds to an intersection with another family.

21.3 Positional Method

We introduce now a "positional method" similar to those of Sects. 13.3 and 19.3.

Here again, it is easily shown that *two characteristics in the* (W, X_1) *plane which correspond to different values of* n *never intersect.* This can be verified on Fig. 21.2 for $n = 1$ and 2. The characteristics up to $n - 1$ divide the (W, X_1) plane into regions, and we have

Proposition 21.3.1. *Two branches can be joined only if (i) they lie in the same region and (ii) the two symmetrical branches lie in the same region.*

21.3.1 Branch Order for $W \to +\infty$

Here again, we will determine in which region a branch lies by studying the relative position of the asymptotic branches, which in the present case correspond to $W \to +\infty$. We need to know all asymptotic branches in the (W, X_1) plane. For $W \to +\infty$, we have shown in Sect. 21.1.1 that there are no other asymptotic branches than those which correspond to a sequence of T-arcs and R-arcs. It was shown in Sect. 21.1 that $W > 2^{-1/2}$, and therefore there are no branches with $W \to 0$. A last possibility would consist in a branch $X_1 \to \pm\infty$, $W \to W_0$, i.e. a branch with a vertical asymptote in the (W, X_1) plane. But in that case, the equations (21.1) would give successively $Z_1 \simeq X_1^2$, $Y_1 \simeq X_1^3$, $X_2 \simeq X_1$, $Z_2 \simeq X_1^2$, $Y_2 \simeq 2X_1^3$, ..., $Y_i \simeq iX_1^3$, and we would never reach $Y_n = 0$: this is impossible.

21.3.1.1 Variations. For a given large value of W, using the condition $Y_0 = 0$ and starting from any given value of X_1, we can compute the successive values Z_1, Y_1, X_2, Z_2, Y_2, ... from (21.1). We consider the value of X_1 which corresponds to some particular branch of order n (so that the computation ends in $Y_n = 0$). We apply now a small variation $\mathrm{d}X_1$, and we compute the corresponding variations of Z_1, Y_1, ... We can use the computations made in Sect. 21.1.2, setting $\mathrm{d}Y_0 = 0$. We obtain in particular

$$\frac{\mathrm{d}Y_{i_\alpha}}{\mathrm{d}X_1} = -W^\alpha \frac{d_{m_1} d_{m_2} \dots d_{m_\alpha}}{Y_{i_1}^2 \dots Y_{i_{\alpha-1}}^2} [\,] \,. \tag{21.84}$$

We will need the order of magnitude and the sign

$$\frac{\mathrm{d}Y_{i_\alpha}}{\mathrm{d}X_1} = \Theta(W^{2\alpha-1}) \,, \qquad \operatorname{sign}\left(\frac{\mathrm{d}Y_{i_\alpha}}{\mathrm{d}X_1}\right) = -(-1)^\zeta \,, \tag{21.85}$$

where ζ is te number of T-arcs in the bifurcating arc.

21.3.1.2 Relative Positions of Two Branches with Initial Common Arcs. We consider now two branches for which the arcs U_1 to U_α are the same, but the continuation is different: either the arcs $U_{\alpha+1}$ are different in the two branches, or the arc $U_{\alpha+1}$ does not exist in one of the two branches (the bifurcating arc ends in i_α).

The asymptotic expressions of X_i, Z_i, Y_i up to $i = i_\alpha$ given by (21.12) to (21.18) are the same for the two branches; therefore these quantities differ by $O(W^{-1/2})$ between the two branches. For W large enough, this can be made as small as desired. Therefore these differences will be called $\mathrm{d}X_1$, etc., and the above results on small variations can be applied.

We consider first the case where U_α is a R-arc. Then if $U_{\alpha+1}$ exists, it must be a T-arc. From (21.17) we know that Y_{i_α} is positive (resp. negative) if $U_{\alpha+1}$ is a T^f arc (resp. T^g arc). If $U_{\alpha+1}$ does not exist, i.e. if the bifurcating arc ends in i_α, we have $Y_{i_\alpha} = 0$. Thus the order of the values of Y_{i_α} for the two branches is given by the order of the corresponding symbols (for $U_{\alpha+1}$) in the sequence

$$\text{g; 0; f ,} \tag{21.86}$$

where the case of the bifurcating arc ending in i_α is represented by a 0.

We consider next the case where U_α is a T^g arc. If $U_{\alpha+1}$ exists, it must be a T^f arc or a R-arc. If it is a T^f arc, we have from (21.17):

$$Y_{i_\alpha} = \frac{1}{2}W^{-1/2} + O(W^{-5/2}) \,. \tag{21.87}$$

If it is a R-arc, we have from (21.19)

$$Y_{i_\alpha} = W^{-1/2} - W^{-3/2}x_1 + O(W^{-5/2}) \,. \tag{21.88}$$

Here x_1 is the value which appears in Table 18.1 and which corresponds to the first basic arc of the R-arc. The order of the values of x_1 for various R-arcs can be read from that table. If we consider for instance all R-arcs up to $n = 3$, the x_1 increase along the following sequence:

$$\text{R+R+R, R+R, R+R−R, R, R−R+R, R−R, R−R−R .} \tag{21.89}$$

This is the same lexicographic order as in Fig. 19.5, read from top to bottom.

Since x_1 is multiplied by a negative factor in (21.88), and considering also (21.87), we find that the order of the values of Y_{i_α} for the two branches is given by the order of the corresponding symbols in the sequence

$$\text{0; f; R−R−R, R−R, R−R+R, R, R+R−R, R+R, R+R+R .} \tag{21.90}$$

This sequence should be appropriately enlarged if R-arcs with $n > 3$ are present.

Finally, if U_α is a T^g arc, a similar argument leads to the sequence

$$\text{R−R−R, R−R, R−R+R, R, R+R−R, R+R, R+R+R; g; 0 .} \tag{21.91}$$

We have thus determined the order of the values of Y_{i_α} in all cases. Using (21.85b), we obtain then the order of the values of X_1, i.e. the relative position of the two branches in the (W, X_1) plane: if ζ, the number of T-arcs in $\{U_1, \ldots, U_\alpha\}$, is odd, the sequence (21.86), (21.86), or (21.86)can be used as it is; if ζ is even, the sequence must be inverted.

21.3.1.3 Relative Positions of Two Branches with Different First Arcs. The method of the previous section does not work when the two branches differ already in their first arc (this can be taken to correspond to $\alpha = 0$). In that case we compare directly the values of X_1, which are given by (21.13) if U_1 is a T-arc, (21.15c) if it is a R-arc. We find that, taking again as an example all R-arcs up to $n = 3$, the order of the values of X_1 for the two branches is given by the order of the symbols corresponding to U_1 in the sequence

$$\text{g; R+R+R, R+R, R+R−R, R, R−R+R, R−R, R−R−R; f .} \tag{21.92}$$

21.3.1.4 Packets. From (21.87) and (21.88) we have, for $\alpha > 0$,

$$\mathrm{d}Y_{i_\alpha} = \Theta(W^{-3/2}) \tag{21.93}$$

if $U_{\alpha+1}$ exists and is a R-arc in the two branches; and

$$\mathrm{d}Y_{i_\alpha} = \Theta(W^{-1/2}) \tag{21.94}$$

in all other cases.

From (21.85a) we have then

$$\mathrm{d}X_1 = \Theta(W^{-1/2-2\alpha}) \tag{21.95}$$

or

$$\mathrm{d}X_1 = \Theta(W^{1/2-2\alpha}) \tag{21.96}$$

respectively.

For $\alpha = 0$ (previous section), the difference between the values of Y_1 for the two branches is $\Theta(W^{-1/2})$ if U_1 is a R-arc in the two branches, $\Theta(W^{1/2})$ in all other cases. Therefore (21.95) and (21.96) cover this case also, with the proviso that $\mathrm{d}y_1$ is not necessarily a small quantity anymore.

Thus here again we find that the branches are organized hierarchically. The characteristics of all branches having a given first arc in common are at a distance $O(W^{-3/2})$ of each other; they form a *first-order packet.* Inside such a packet, the characteristics of all branches having in common their first two arcs are at a distance $O(w^{-7/2})$ of each other, and form a *second-order packet*; and so on.

21.3.1.5 Results for $n \leqq 3$. The rules established in the preceding sections allow us to determine completely how the characteristics of all branches are ordered in X_1. Fig. 21.3 shows this ordering up to $n = 3$. X_1 increases from bottom to top, as in Figs. 21.1 and 21.2, and packets of first order, second order, etc. have been separated into different columns, with their filiation indicated by line segments. The left column reproduces (21.92).

The figure has a symmetry: it is invariant if we exchange top and bottom,the + and - signs, and the symbols **f** and **g**. This is a consequence of the Σ' symmetry (Sect. 21.1, Property 5).

21.3.1.6 Other Method. Instead of building the branches by successive packets, each time adding a T- or R-arc, we can build them by adding each time one basic arc. Fig. 21.3 is then transformed into Fig. 21.4. Successive columns correspond now to successive values of n. A regular structure emerges. (The vertical order of the branches is of course the same as in Fig. 21.3.)

This structure can be explained by reworking the building rules established in Sects. 21.3.1.2 and 21.3.1.3. We consider first the case where two branches have initial common basic arcs up to arc i, but the continuation is different. Using the asymptotic expressions (21.17), (21.18), and (21.15b), we find that the order of the values of Y_i for the two branches is then given by

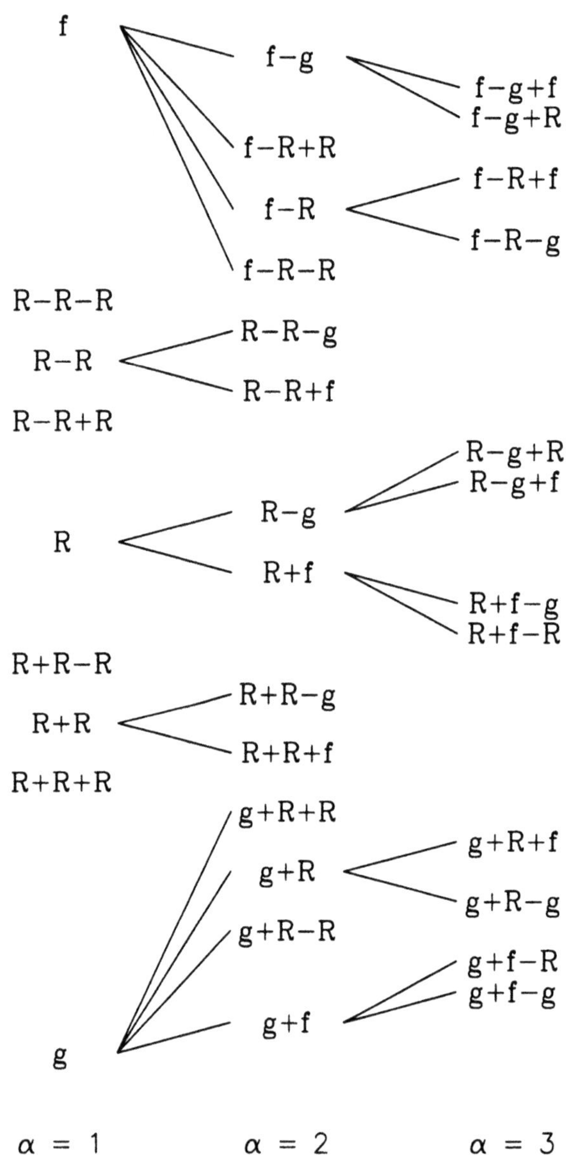

Fig. 21.3. Order of the branches for $W \to +\infty$, $n \leqq 3$, found using packets. (X_1 increases upwards.)

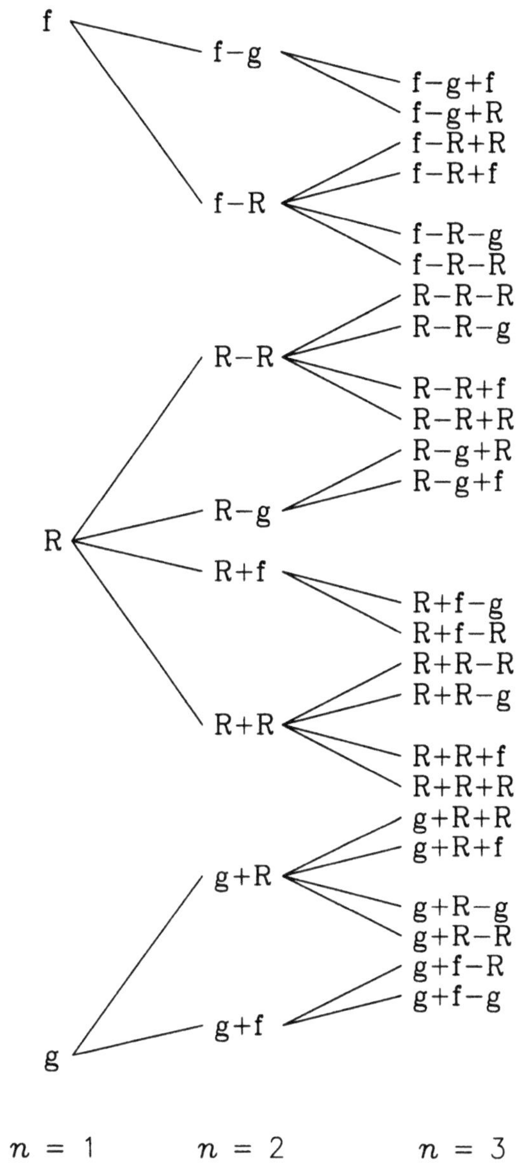

Fig. 21.4. Order of the branches for $W \to +\infty$, $n \leqq 3$, found using basic arcs. (X_1 increases upwards.)

the order of the symbols representing the sign of Y_i and the next basic arc in the sequence

$$\texttt{-R, -g, 0, +f, +R}\ , \tag{21.97}$$

where the case of the bifurcating arc ending in i_α is represented by a 0.

If the basic arc i is a T^g arc, the following sign cannot be a -, and therefore only the last 3 elements of the sequence are present. Similarly, if the basic arc i is a T^f arc, only the first 3 elements of the sequence are present.

We deduce the relative position of the two branches in the (W, X_1) plane as in Sect.21.3.1.2: if ζ, the number of T-arcs among the basic arcs 1 to i, is odd, the sequence (21.97)can be used as it is; if ζ is even, the sequence must be inverted.

If the two branches have different first basic arcs, the order of the values of X_1 is given by the order of the symbols representing the first basic arc in the sequence

$$\texttt{g, R, f}\ . \tag{21.98}$$

This new construction method shows that the number of asymptotic branches for a given $n \geqq 2$ is

$$8 \times 3^{n-2}\ . \tag{21.99}$$

21.3.2 Results

21.3.2.1 Symmetries. We use now the positional method systematically to determine the junctions. We will make use of the symmetry Σ; it inverts the sequence of symbols and exchanges the symbols `f` and `g`. We will also need the symmetry Σ': it exchanges the signs + and - and the symbols `f` and `g`. Finally the symmetry $\Sigma\Sigma'$ inverts the sequence and changes the signs.

An orbit cannot be symmetric under Σ' (except in the singular case $n = 1$) since the signs are changed. But an orbit can be symmetric under $\Sigma\Sigma'$, provided that n is odd. An example is: `R-g+f-g+R`. (For n even, this is not possible since the central sign changes.) Exactly as was the case with Σ (Sect. 7.3), this property is invariant along a characteristic. Therefore a branch of symmetric orbits under $\Sigma\Sigma'$ must be joined to another branch of symmetric orbits under $\Sigma\Sigma'$, and a branch of asymmetric orbits must be joined to another branch of asymmetric orbits. This property will be used below.

21.3.2.2 $n = 1$. In the special case $n = 1$, there are three branches for $W \to +\infty$, whose relative position is given by the first column in Fig. 21.4, and in addition there is one branch for $W \to -\infty$ (Fig. 21.1). They form a trident.

21.3.2.3 $n = 2$. In the generic case $n > 1$, we have $W > 0$ (Sect. 21.1), and there are only branches for $W \to +\infty$. Since the figure is symmetric with respect to the axis $X_1 = 0$ (symmetry Σ'), it will be sufficient to represent schematically the characteristics in the quarter-plane $W > 0$, $X_1 > 0$ (as in Fig. 21.2).

For $n = 2$, we represent the already known junctions as solid lines (Fig. 21.5). There are 4 new branches, all of which lie in the same region. These branches form a trident, and their junctions can be established (dashed lines).

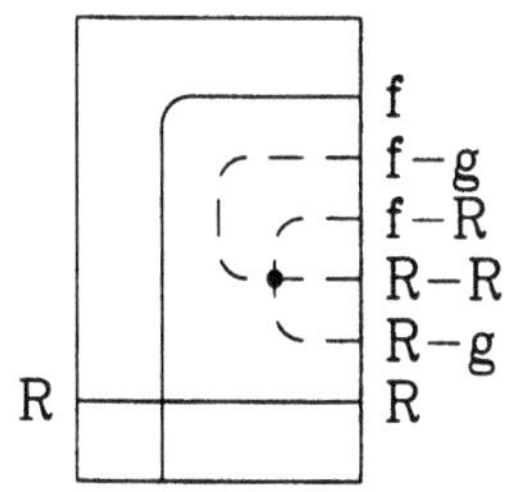

Fig. 21.5. Junctions for $n = 1$ (full lines) and $n = 2$ (dashed lines); $X_1 > 0$.

21.3.2.4 $n = 3$. There are 3 groups of 4 branches. Each group lies in one region (Fig. 21.6). The central group forms a trident. The upper group is made of asymmetric orbits only. Here we make use of the second part of Proposition 21.3.1. The symmetry Σ sends the 4 branches into the lower half-plane $X_1 < 0$; therefore it is more convenient to use the symmetry $\Sigma\Sigma'$. Under that symmetry, two of the branches (`f-g+f` and `f-R+f`) are invariant and remain in the upper group, while the two other branches (`f-g+R` and `f-R+R`) are changed into `R-g+f` and `R-R+f` and move to the lower group. Therefore the first two branches must be joined, and the last two also. A similar argument establishes the junctions in the lower group.

We remark that two joined branches always have the same signs. This is generally true and is a consequence of Broucke's principle.

Four branches are made of orbits symmetric under $\Sigma\Sigma'$: `f-g+f`, `f-R+f`, `R-R+R`, `R-g+R`. As predicted in Sect. 21.3.2.1, these branches are joined between themsleves.

21.3.2.5 $n = 4$. In Fig. 21.7, the already known junctions are represented only for $n = 3$; the junctions for $n = 2$ and $n = 1$ lie outside of them and do not contribute any further division of the plane into regions. This appears to be generally true: it is sufficient to consider the already known junctions for $n - 1$ when we try to find the junctions for n.

The new branches form 5 groups of 4 branches, which we label G_1 to G_5, and 2 groups of 8 branches, H_1 and H_2. We study these groups in turn. If we

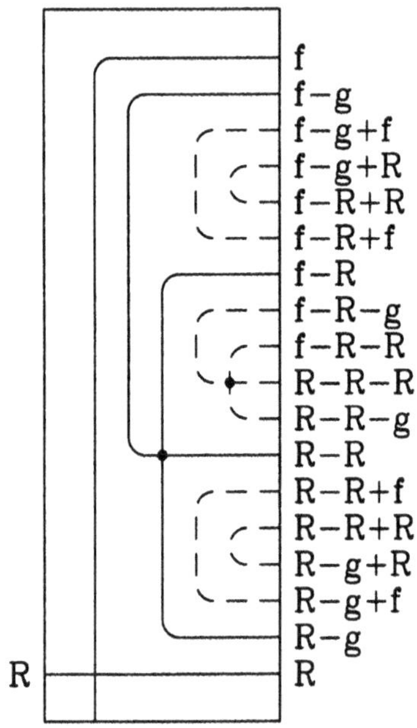

Fig. 21.6. Junctions for $n < 3$ (full lines) and $n = 3$ (dashed lines); $X_1 > 0$.

apply the symmetry $\Sigma\Sigma'$ to G_1, we find that two branches go over to G_2, while the two other branches go over to G_4; the junctions are thus established. The groups G_2, G_4, G_5 are solved in a similar way. G_3 is a trident. We turn now to the groups of 8 branches. Applying the symmetry Σ to the group H_1, we find that 4 branches remain in H_1, while the 4 other branches go over to H_2. The first 4 branches form a trident and therefore the junctions are solved: `f-g+f-g` is joined to `f-R+R-g`, and `f-g+R-g` to `f-R+f-g`. After these junctions have been drawn in Fig. 21.7, we find that the last 4 branches are now separated into 2 groups of 2 branches, so that their junctions are determined. (Their characteristics cannot intersect those of the previous trident, because the symmetrical of a common orbit would have to lie both in H_1 and in H_2.) The last group H_2 is solved in a similar way.

21.3.2.6 n = 5. For $n \geqq 5$, establishing the ordered list of branches by hand becomes tedious and error-prone, and a computer program was set up to generate this list automatically.

For $n = 5$, there are 108 branches with $X_1 > 0$. Using the same methods as before, we find again that all junctions can be established. We omit the details and give only the results on Fig. 21.8. (That figure is in three parts, represented here side by side for convenience, but which should be rearranged one above the other, in descending order, to represent the (W, X_1) quarter-plane.)

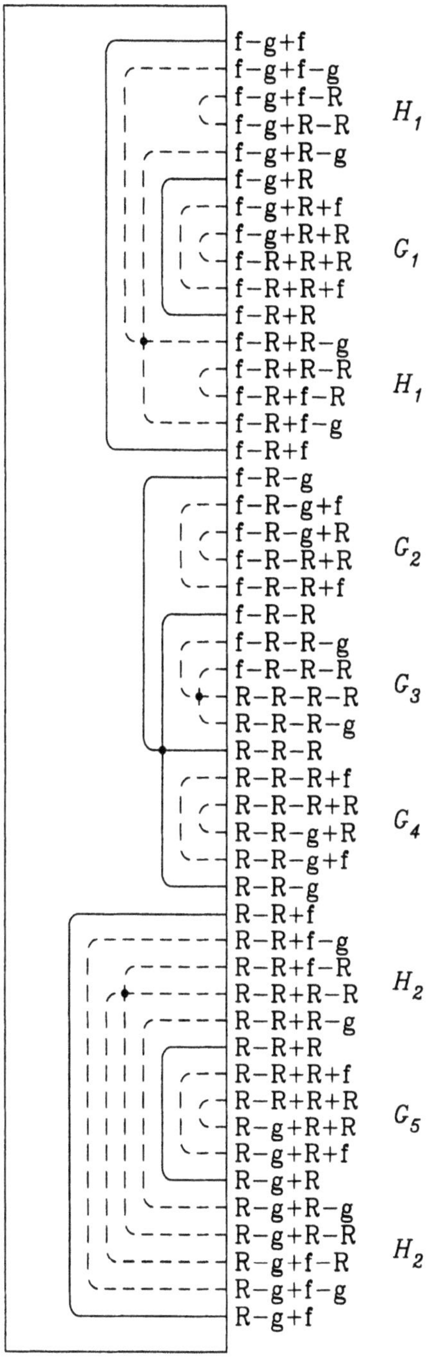

Fig. 21.7. Junctions for $n = 3$ (full lines) and $n = 4$ (dashed lines); $X_1 > 0$.

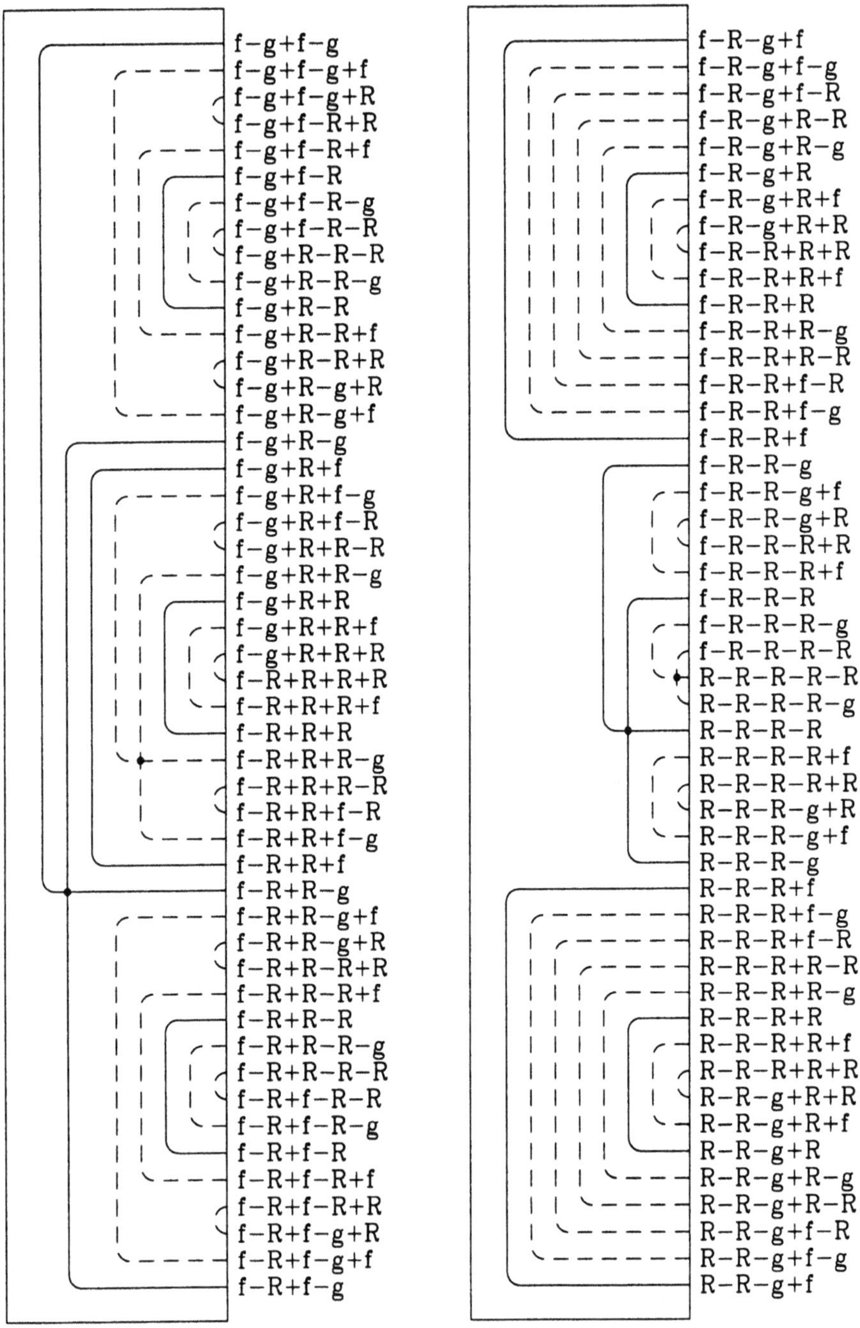

Fig. 21.8. Junctions for $n = 4$ (full lines) and $n = 5$ (dashed lines); $X_1 > 0$.

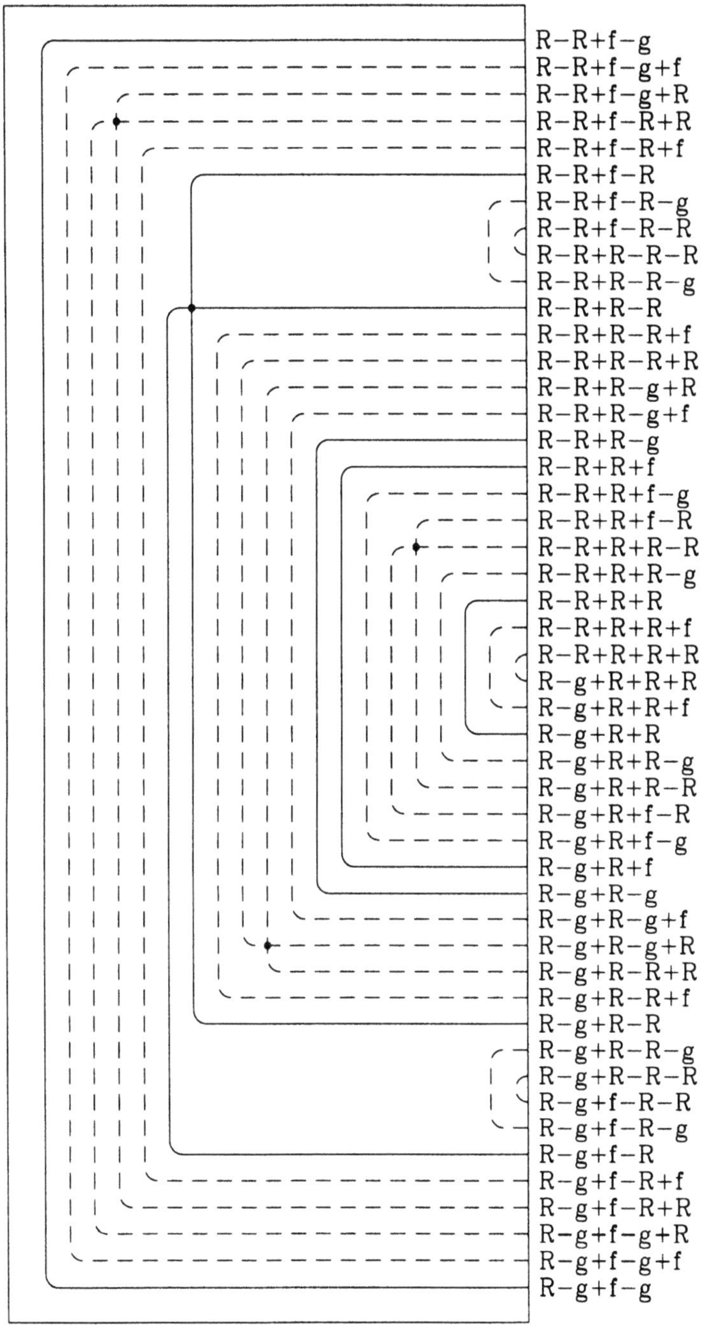

Fig. 21.8. (continuation)

A new kind of trident is observed here in two cases. The situation is exactly the same as for the standard trident (see Sect. 13.3.3.2), with the symmetry Σ replaced by $\Sigma\Sigma'$. It involves thus two branches of orbits symmetric under $\Sigma\Sigma'$ and two branches of asymmetric orbits; the latter two branches are changed into each other by $\Sigma\Sigma'$. One such trident consists of the branches R-R+f-R+R, R-g+f-g+R, R-R+f-g+R, R-g+f-R+R; the other of R-R+R-R+R, R-g+R-g+R, R-R+R-g+R, R-g+R-R+R. The characteristics of the symmetric and asymmetric orbits intersect in a common orbit, represented by a dot.

21.3.2.7 $n = 6$. There are 324 branches with $X_1 > 0$. We do not show the results here as they would take up too much space.

All junctions can be established by the same methods as above, except for two groups of 4 branches, which are represented on Fig. 21.9. The already established junctions are shown as full lines.

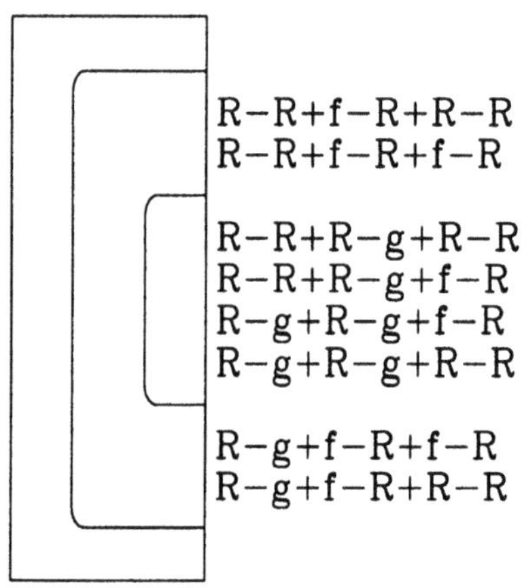

Fig. 21.9. Some hard branches for $n = 6$.

However, a new kind of argument can be used here. We remember that the figure is essentially a sketch of the (W, X_1) plane, and we notice that the region containing one of the groups is encircled by the region containing the other group. On the other hand, every characteristic has a minimal value of W. That minimal value is the same for two characteristics symmetrical of each other. Suppose now that in the outer region, the branch R-R+T-R+R-R is joined to the branch R-T+T-R+T-R. Applying Σ, we find that in the inner region, R-R+R-T+R-R is joined to R-T+R-T+T-R. But it is clear from the figure that these two characteristics cannot have the same minimal value of W, since the first characteristic encircles the region which itself contains the second characteristic. We have thus a contradiction. Similarly, we find that

R-R+T-R+R-R cannot be joined to R-T+T-R+R-R. It must therefore be joined to R-R+T-R+T-R, and all junctions are established.

21.3.2.8 $n = 7$. There are 972 branches with $X_1 > 0$. A detailed study shows that, using the above methods, all junctions can be established except for 40 branches.

A numerical computation seems necessary for these cases.

Thus, the positional method is not able to solve all cases. However, its success in the present case (transition 2.2) is spectacular: all junctions up to $n = 6$ have been established.

An alternative would be to find a set of rules prescribing the junctions for all values of n. Some regularities can be observed in the computed cases. If a junction is not involved in a trident, a single symbol changes between the two branches. Moreover, that symbol cannot be one of the two end symbols. If the junction is involved in a trident, two symbols change; they occupy symmetrical positions in the sequence, i.e. they correspond to basic arcs i and $n + 1 - i$.

21.4 Results for Bifurcations of Type 2

Figs. 21.5 to 21.8 give the junctions between branches at transition 2.2. We use now these results to come back to our original objective and deduce the corresponding junctions between branches in a partial bifurcation of type 2.

Only branches with a - sign are involved (Sect. 21.1, Property 3). For these branches, the path followed through the partial bifurcation of type 2 is as follows. When the branch approaches the bifurcation , ν increases from 0. When $\nu = 1/3$ is reached, the branch goes through a transition 2.1, with $w > 0$ (Sect. 19.1, Property 5). This transition was studied in Sect. 19.3.3.1 and was found to have a simple and systematic structure: successive S-arcs coalesce to form a single R-arc, while T-arcs are not affected; and the family continues toward higher values of ν. When $\nu = 1/2$ is reached, we have a transition 2.2, and the branch is joined to another branch as described by the above results. ν starts then to decrease and the above evolution proceeds now in reverse, going through another transition 2.1. The whole junction is shown schematically in Table 21.2.

Table 21.2. Partial bifurcation of type 2: junction between two branches with a - sign.

ν	0	↗	1/3	↗	1/2	↘	1/3	↘	0
Transition			2.1		2.2		2.1		
Arcs		S, T		R, T		R, T		S, T	
Example	- f31		f-R-R-R+R		f-R-R-g+R		- f2g1		

Each junction shown in Figs. 21.5 to 21.8 corresponds thus to one junction between two branches in a partial bifurcation of type 2. The symbolic names of these branches are easily found: each R-arc corresponds to a sequence of S-arcs; each + sign inside the R-arc corresponds to a node and thus separates two consecutive S-arcs(Sect. 18.1.2).

As an example, consider the branch - **f31** (see last line of Table 21.2). When it reaches transition 2.1, the two S-arcs represented by the symbols 3 and 1 fuse into a R-arc of length 4, and the orbit becomes `f-R-R-R+R`. At transition 2.2, that branch is joined to `f-R-R-g+R`, as shown by Fig. 21.8. This branch contains now two R-arcs of lengths 2 and 1. When transition 2.1 is again visited, these two R-arcs change into S-arcs, and we end on the branch - **f2g1**.

The junctions of the branches obtained in this way up to $n = 5$ are listed in Table 21.3. This table is in the same format as Table 8.12, which was obtained using only qualitative methods. It can be verified that there is agreement for all branches which appear in both tables (sign - and $2 \leqq n \leqq 4$). (For this comparison, the symbols `i` and `e` in Table 8.12 should be replaced by `f` and `g` respectively; see Sect. 17.1.)

Tables 19.1 and 21.3 give all junctions for partial bifurcations of type 2, for $n = 2$ to 5. It can be seen that the quantitative approach is much more powerful than the qualitative one: all the cases left undecided in Table 8.12 (the majority) are now solved; in addition all junctions are established for $n = 5$. Nothing prevents in principle the solution of higher values of n, using numerical computation. (The junctions for $n = 6$ are listed in Table 19.1 for branches with a + sign. They were also computed for for branches with a - sign (see Sect. 21.3.2.7), but not given here.) However, the amount of work grows exponentially.

Table 21.3. Partial bifurcation, type 2, n = 2 to 5: junctions between the - branches, determined by the quantitative study.

Branch	Junction
2P2+S	- 11 - gf
2P2+A	- 1f - g1
2P2-S	- 2 - fg
2P2-A	- 1g - f1
2P3++S	- 111 - g1f
2P3++A	- 11f - g11
2P3+-A	- g1g - gfg
	- g2 - gf1
	- 1fg - 11g
	- 1f1 - 12
2P3-+A	- f1f - fgf
	- f11 - fg1
	- 1gf - 2f
	- 1g1 - 21
2P3--S	- 3 - f1g
2P3--A	- 2g - f2
2P4+++S	- 1111 - g11f
2P4+++A	- 111f - g111
2P4++-A	- 11fg - 111g
	- 11f1 - 112
	- g11g - g1fg
	- g12 - g1f1
2P4+-+S	- 1fg1 - 121
	- g2f - gfgf
2P4+-+A	- 1fgf - 11gf
	- 1f11 - 11g1
	- 1f1f - 12f
	- g1gf - gf1f
	- g1g1 - g21
	- gf11 - gfg1
2P4+--A	- 1f1g - 12g
	- 1f2 - 13
	- g2g - gf1g
	- g3 - gf2
2P4-++A	- 1g1f - 21f
	- 1g11 - 211
	- f11f - fg1f
	- f111 - fg11
2P4-+-S	- 1gf1 - 22
	- f11g - fgfg
2P4-+-A	- 1gfg - 2fg
	- 1g2 - 2f1
	- 1g1g - 21g
	- f1fg - fg1g
	- f1f1 - f12
	- fg2 - fgf1
2P4--+A	- 2gf - 3f
	- 2g1 - 31
	- f2f - f1gf
	- f21 - f1g1
2P4---S	- 4 - f2g
2P4---A	- 3g - f3

Table 21.3. (continuation)

2P5++++S

- 11111
- g111f

2P5++++A

- 1111f
- g1111

2P5+++-A

- 111fg
- 1111g

- 111f1
- 1112

- g111g
- g11fg

- g112
- g11f1

2P5++-+A

- 11fgf
- 111gf

- 11fg1
- 111g1

- 11f11
- 1121

- 11f1f
- 112f

- g11gf
- g1fgf

- g11g1
- g1fg1

- g121
- g1f11

- g12f
- g1f1f

2P5++--A

- 11f1g
- 112g

- 11f2
- 113

- g12g
- g1f1g

- g13
- g1f2

2P5+-++A

- g1g1f
- g21f

- g1g11
- g211

- gf11f
- gfg1f

- gf111
- gfg11

- 1fg1f
- 1f11f

- 1fg11
- 1f111

- 121f
- 11g1f

- 1211
- 11g11

2P5+-+-A

- g1gfg
- g2fg

- g1gf1
- g1g2

- g1g1g
- g21g

- g22
- g2f1

- gf1fg
- gfgfg

- gf1f1
- gf12

- gf11g
- gfg1g

- gfg2
- gfgf1

- 1fgfg
- 11gfg

- 1fgf1
- 11g2

- 1fg2
- 11gf1

- 1fg1g
- 11g1g

- 1f11g
- 121g

- 1f12
- 12f1

- 1f1f1
- 122

- 1f1fg
- 12fg

2P5+--+S

- g3f
- gf1gf

- 1f1g1
- 131

2P5+--+A

- g2gf
- gf2f

- g2g1
- g31

- gf21
- gf1g1

- 1f1gf
- 12gf

- 1f21
- 12g1

- 1f2f
- 13f

2P5+---A

- g3g
- gf2g

- g4
- gf3

- 1f2g
- 13g

- 1f3
- 14

Table 21.3. (continuation)

2P5-+++A
- f111f - fg11f
- f1111 - fg111
- 1g11f - 211f
- 1g111 - 2111

2P5-++-S
- f111g - fg1fg
- 1g1f1 - 212

2P5-++-A
- f11fg - fg11g
- f11f1 - f112
- fg12 - fg1f1
- 1g1fg - 21fg
- 1g12 - 21f1
- 1g11g - 211g

2P5-+-+A
- f1fgf - f11gf
- f1fg1 - f1f11
- f1f1f - f12f
- f121 - f11g1
- fg1gf - fgfgf
- fg1g1 - fg21
- fg2f - fgf1f
- fgf11 - fgfg1
- 1gfgf - 2fgf
- 1gfg1 - 2f11
- 1gf11 - 2fg1
- 1gf1f - 2f1f
- 1g2f - 22f
- 1g21 - 21g1
- 1g1g1 - 221
- 1g1gf - 21gf

2P5-+--A
- f1f1g - f12g
- f1f2 - f13
- fg2g - fgf1g
- fg3 - fgf2
- 1gf1g - 1g2g
- 1gf2 - 1g3
- 22g - 2f1g
- 23 - 2f2

2P5--++A
- 2g1f - 31f
- 2g11 - 311
- f21f - f1g1f
- f211 - f1g11

2P5--+-A
- 2gfg - 3fg
- 2gf1 - 3f1
- 2g2 - 32
- 2g1g - 31g
- f2fg - f1gfg
- f2f1 - f1gf1
- f22 - f1g2
- f21g - f1g1g

2P5---+A
- 3gf - 4f
- 3g1 - 41
- f3f - f2gf
- f31 - f2g1

2P5----S
- 5 - f3g

2P5----A
- 4g - f4

22. Total Transition 2.2

The equations are

$$X_i^2 - Z_i - W = 0 ,$$
$$Y_i - Y_{i-1} - Z_i X_i = 0 ,$$
$$Y_i(X_{i+1} - X_i) - 1 = 0 , \qquad (22.1)$$

where i is to be taken modulo n and takes all values from 1 to n.

These equations form a system of $3n$ equations for the $3n + 1$ variables W, Y_0 to Y_{n-1}, X_1 to X_n, Z_1 to Z_n. As in the case of partial bifurcations (Chap. 21), we expect one-parameter families of solutions, which correspond to ordinary one-parameter families of orbits.

22.1 Properties

The properties are essentially the same as for partial transition 2.2 (Sect. 21.1). The proof of Property 2 is slightly different: we use the fact that a continually increasing or decreasing sequence of Y_i cannot be periodic.

A lower bound can again be established for W, but it is not the same as (21.7). Two successive values X must be different in view of (22.1c). Therefore there exists a X_i such that $X_{i-1} < X_i$ and $X_i > X_{i+1}$. (21.4) is still true, and we have then from (22.1c):

$$Y_{i-1} > \frac{1}{2} W^{-1/2} , \qquad Y_i < -\frac{1}{2} W^{-1/2} , \qquad (22.2)$$

from which

$$|Y_i - Y_{i-1}| > W^{-1/2} . \qquad (22.3)$$

On the other hand, (22.1b) gives

$$|Y_i - Y_{i-1}| < W^{3/2} . \qquad (22.4)$$

Combining, we obtain

$$W > 1 . \qquad (22.5)$$

In regard to property 6, the bifurcation orbit corresponds to a cycle of period n of the mapping (21.11).

22.1.1 Stability and Jacobian

We proceed as usual (see Sects. 20.1.2, 18.2.2, 14.1.2). To compute the Jacobian, we order the variables as follows:

$$Y_0, X_1, Z_1, Y_1, X_2, Z_2, Y_2, \ldots, Y_{n-1}, X_n, Z_n \,. \tag{22.6}$$

We obtain

Proposition 22.1.1. *In a total transition 2.2, the Jacobian vanishes if and only if the bifurcation orbit is a critical orbit of the first kind (the stability index is $z = 1$).*

Any pair of variables can be used to compute the stability index. It will be convenient to use X_0 and X_1. From (22.1) we find

$$\begin{aligned}
\begin{pmatrix} \mathrm{d}X_i \\ \mathrm{d}Y_{i-1} \end{pmatrix} &= \begin{pmatrix} 0 & 1 \\ Y_{i-1}^2 & -Y_{i-1}^2 \end{pmatrix} \begin{pmatrix} \mathrm{d}X_{i-1} \\ \mathrm{d}X_i \end{pmatrix} , \\
\begin{pmatrix} \mathrm{d}X_i \\ \mathrm{d}Y_i \end{pmatrix} &= \begin{pmatrix} 1 & 0 \\ 3X_i^2 - W & 1 \end{pmatrix} \begin{pmatrix} \mathrm{d}X_i \\ \mathrm{d}Y_{i-1} \end{pmatrix} , \\
\begin{pmatrix} \mathrm{d}X_i \\ \mathrm{d}X_{i+1} \end{pmatrix} &= \begin{pmatrix} 1 & 0 \\ 1 & -1/Y_i^2 \end{pmatrix} \begin{pmatrix} \mathrm{d}X_i \\ \mathrm{d}Y_i \end{pmatrix} ,
\end{aligned} \tag{22.7}$$

from which

$$\begin{pmatrix} \mathrm{d}X_i \\ \mathrm{d}X_{i+1} \end{pmatrix} = \begin{pmatrix} 0 & 1 \\ -\dfrac{Y_{i-1}^2}{Y_i^2} & \dfrac{u_i}{Y_i^2} \end{pmatrix} \begin{pmatrix} \mathrm{d}X_{i-1} \\ \mathrm{d}X_i \end{pmatrix} \tag{22.8}$$

with

$$u_i = Y_{i-1}^2 + Y_i^2 - 3X_i^2 + W \,. \tag{22.9}$$

We can then compute the matrix

$$\begin{pmatrix} \dfrac{\partial X_n}{\partial X_0} & \dfrac{\partial X_n}{\partial X_1} \\ \dfrac{\partial X_{n+1}}{\partial X_0} & \dfrac{\partial X_{n+1}}{\partial X_1} \end{pmatrix} = \begin{pmatrix} 0 & 1 \\ -\dfrac{Y_{n-1}^2}{Y_n^2} & \dfrac{u_n}{Y_n^2} \end{pmatrix} \cdots \begin{pmatrix} 0 & 1 \\ -\dfrac{Y_0^2}{Y_1^2} & \dfrac{u_1}{Y_1^2} \end{pmatrix} \tag{22.10}$$

and the stability index is

$$z = \frac{1}{2} \left(\frac{\partial X_n}{\partial X_0} + \frac{\partial X_{n+1}}{\partial X_1} \right) . \tag{22.11}$$

22.2 Small Values of n

We will represent the characteristics in the (W, Y_0) plane. Note that a point of this plane is not sufficient to define an orbit: the three values W, Y_0, X_1 are needed.

22.2.1 $n = 1$

There are no solutions in that case: (22.1c) cannot be satisfied since the first term vanishes. This corresponds to the fact that there are no R-orbits for $n = 1$ (Sect. 18.2.3), and that we cannot either have a T-arc, because we would then have successive identical T-arcs (Proposition 4.3.2).

The total bifurcation 2T1 exists (see Chap. 23). However, it does not visit the transition 2.2. It will be studied in Sect. 23.1.

22.2.2 $n = 2$

The equations are

$$\begin{aligned} &X_1^2 - Z_1 - W = 0\,, \qquad Y_1 - Y_0 - Z_1X_1 = 0\,, \\ &Y_1(X_2 - X_1) - 1 = 0\,, \\ &X_2^2 - Z_2 - W = 0\,, \qquad Y_0 - Y_1 - Z_2X_2 = 0\,, \\ &Y_0(X_1 - X_2) - 1 = 0\,. \end{aligned} \tag{22.12}$$

Combining these equations, we obtain

$$Y_1 = -Y_0 \tag{22.13}$$

and

$$(X_1 + X_2)(X_1^2 - X_1X_2 + X_2^2 - W) = 0\,. \tag{22.14}$$

Because of the symmetry Σ' (Sect. 21.1, Property 5), we can consider only solutions with $Y_0 > 0$. Equating to 0 the first factor in (22.14), we obtain a first family of solutions

$$\begin{aligned} &Y_0 = -Y_1 = \sqrt{\frac{W \pm \sqrt{W^2-4}}{8}}\,, \qquad X_1 = -X_2 = \sqrt{\frac{W \mp \sqrt{W^2-4}}{2}}\,, \\ &Z_1 = Z_2 = \frac{-W \mp \sqrt{W^2-4}}{2}\,. \end{aligned} \tag{22.15}$$

These solutions exist for $W \geqq 2$. The asymptotic branches for $W \to +\infty$ are, for the upper signs

$$Y_0 = -Y_1 \sim \frac{\sqrt{W}}{2}\,, \qquad X_1 = -X_2 \sim \frac{1}{\sqrt{W}}\,, \qquad Z_1 = Z_2 \sim -W \tag{22.16}$$

and for the lower signs

$$Y_0 = -Y_1 \sim \frac{1}{2\sqrt{W}}\,, \qquad X_1 = -X_2 \sim \sqrt{W}\,, \qquad Z_1 = Z_2 \sim -\frac{1}{W}\,. \tag{22.17}$$

Using the asymptotic expressions of Sect. 21.1.1, the numerical values of y_0 and y_1 obtained in Sect. 18.2.3 and the branch notation defined in Sect. 21.1.5, we find that these two branches correspond to `+R-R+` and `+f-g+` respectively.

The stability is computed from (22.10), which gives

$$\begin{pmatrix} \dfrac{\partial X_2}{\partial X_0} & \dfrac{\partial X_2}{\partial X_1} \\ \dfrac{\partial X_3}{\partial X_0} & \dfrac{\partial X_3}{\partial X_1} \end{pmatrix} = \begin{pmatrix} 0 & 1 \\ -1 & \dfrac{u_2}{Y_0^2} \end{pmatrix} \begin{pmatrix} 0 & 1 \\ -1 & \dfrac{u_1}{Y_0^2} \end{pmatrix}$$

$$= \begin{pmatrix} -1 & \dfrac{u_2}{Y_0^2} \\ -\dfrac{u_1}{Y_0^2} & -1 + \dfrac{u_1 u_2}{Y_0^4} \end{pmatrix} \tag{22.18}$$

with

$$u_1 = 2Y_0^2 - 3X_1^2 + W , \qquad u_2 = 2Y_0^2 - 3X_2^2 + W . \tag{22.19}$$

The stability index is

$$z = -1 + \frac{u_1 u_2}{2Y_0^4} . \tag{22.20}$$

In the present case, this gives

$$z - 1 = 8 \left[(W^2 - 4)(2W^2 - 3) \mp (2W^3 - 7W)\sqrt{W^2 - 4} \right] . \tag{22.21}$$

This vanishes for $W = 2$, which corresponds to an extremum in W, and for $W = 3/\sqrt{2}$ with the upper sign, which corresponds to the intersection with the second family (presently to be described).

A second solution is obtained by setting the second factor in (22.14) equal to 0. Eliminating variables with the help of (22.12), we obtain an equation for Z_1:

$$Z_1^4 + WZ_1^3 + 2Z_1^2 + 2WZ_1 + 4 = 0 . \tag{22.22}$$

It will be convenient here to use Z_1 as independent variable instead of W. The other variables are expressed as functions of Z_1, in the case $Y_1 > 0$, by

$$W = -\frac{Z_1^4 + 2Z_1^2 + 4}{Z_1(Z_1^2 + 2)} , \qquad Y_0 = -Y_1 = \sqrt{\frac{-Z_1}{Z_1^2 + 2}} ,$$

$$X_1 = \frac{2}{\sqrt{-Z_1(Z_1^2 + 2)}} , \qquad X_2 = Z_1 \sqrt{\frac{-Z_1}{Z_1^2 + 2}} , \qquad Z_2 = \frac{2}{Z_1} . \tag{22.23}$$

The variations as Z_1 runs from $-\infty$ to 0 are shown by Table 22.1.

There are two branches $W \to +\infty$, corresponding to $Z_1 \to -\infty$ and $Z_1 \to 0$ respectively. The asymptotic values are, for the first branch:

$$Y_0 = -Y_1 \sim W^{-1/2} , \qquad X_1 \sim 2W^{-3/2} ,$$

$$Z_1 \sim -W , \qquad X_2 \sim -W^{1/2} , \qquad Z_2 \sim -2W^{-1} , \tag{22.24}$$

and for the second branch:

Table 22.1. Variations of the variables when Z_1 increases from $-\infty$ to 0.

Z_1	$-\infty \nearrow -\sqrt{2} \nearrow 0$
W	$+\infty \searrow \frac{3}{\sqrt{2}} \nearrow +\infty$
$Y_0 = -Y_1$	$0 \nearrow 2^{-3/4} \searrow 0$
X_1	$0 \nearrow 2^{-1/4} \nearrow +\infty$
X_2	$-\infty \nearrow -2^{-1/4} \nearrow 0$
Z_2	$0 \searrow -\frac{1}{\sqrt{2}} \searrow -\infty$

$$\begin{aligned}&Y_0 = -Y_1 \sim W^{-1/2}\,, \qquad X_1 \sim W^{1/2}\,,\\ &Z_1 \sim -2W^{-1}\,, \qquad X_2 \sim -2W^{-3/2}\,, \qquad Z_2 \sim -W\,.\end{aligned} \tag{22.25}$$

Using the asymptotic expressions of Sect. 21.1.1, we find that these two branches correspond to +R-g+ and +f-R+ respectively.

The stability index is computed from (22.20), which gives here

$$z - 1 = -\frac{(Z_1^2-2)^2(Z_1^4+6Z_1^2+4)}{Z_1^4}\,. \tag{22.26}$$

This has a double root for $Z_1^2 = 2$, or $W = 3/\sqrt{2}$, corresponding to both an extremum and an intersection with the other family.

The solutions are shown on Fig. 22.1. It is sufficient to show the quarter-plane $W > 0$, $Y_0 > 0$. We have a *trident* (Sect. 13.3.3.2), formed by the two branches of symmetric orbits +R-R+ and +f-g+ and the two branches of asymmetric orbits +R-g+ and +f-R+. The latter two branches have the same projection on the (W, Y_0) plane.

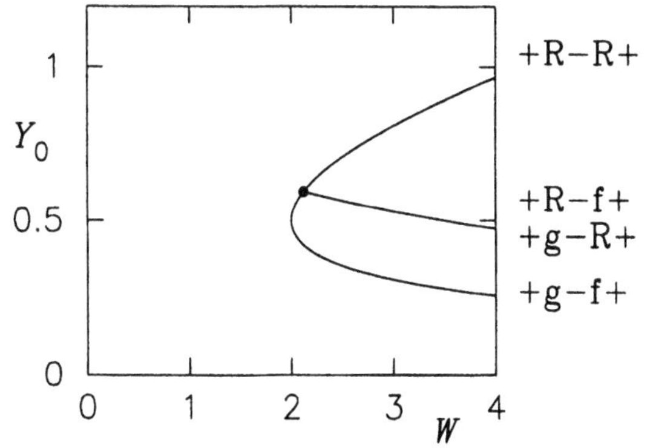

Fig. 22.1. Characteristics for 2.2T2 transitions.

22.2.3 $n = 3$

It seems that for $n = 3$ the general solution can only be obtained numerically (see below Sect. 22.2.4). The equations can be solved, however, in the particular case of symmetric orbits.

One of the crossings of the symmetry axis is at the junction of two basic arcs. We take it as origin ($h = 0$). There is then $Y_1 = Y_2$, $X_2 = 0$, $X_1 = -X_3$, and the equations are easily solved; for $Y_0 > 0$ we obtain

$$
\begin{aligned}
&Y_0 = \sqrt{\frac{W \pm \sqrt{W^2 - 6}}{12}}\,, \qquad X_1 = -X_3 = \sqrt{\frac{W \mp \sqrt{W^2 - 6}}{2}}\,,\\
&Z_1 = Z_3 = \frac{-W \mp \sqrt{W^2 - 6}}{2}\,, \qquad Y_1 = Y_2 = -\sqrt{\frac{W \pm \sqrt{W^2 - 6}}{3}}\,,\\
&X_2 = 0\,, \qquad Z_2 = -W\,.
\end{aligned}
\tag{22.27}
$$

The computation of the stability gives

$$
z - 1 = \frac{27}{2}\left(\frac{1}{24Y_0^4}\right)\left(\frac{1}{288Y_0^8} + \frac{1}{8Y_0^4} - 3\right)\,. \tag{22.28}
$$

This vanishes for $Y_0 = 24^{-1/4}$, $W = \sqrt{6}$, which corresponds to the minimum of W, and for

$$
Y_0 = \left(\frac{3 + \sqrt{33}}{144}\right)^{1/4} \tag{22.29}
$$

which should correspond to an intersection with a family of asymmetric orbits.

The characteristic is shown on Fig. 22.2. The two branches are identified from the asymptotic expressions. Two other families can be obtained by shifting the origin, i.e. rotating the indices.

22.2.4 Numerical Computation

For $n > 2$, a numerical computation seems necessary. A program similar to those described in Sects. 14.2.2 and 20.2.5.2 was used. It solves the system of $2n$ equations for the Y_i and X_i formed by (17.205a) and (17.207):

$$
Y_i(X_{i+1} - X_i) - 1 = 0\,, \qquad Y_i - Y_{i-1} - X_i({X_i}^2 - W) = 0\,. \tag{22.30}
$$

The computation of a given branch is begun at a large value of W and proceeds toward smaller values. The method of computation differs depending on whether the branch corresponds to a R-orbit or not.

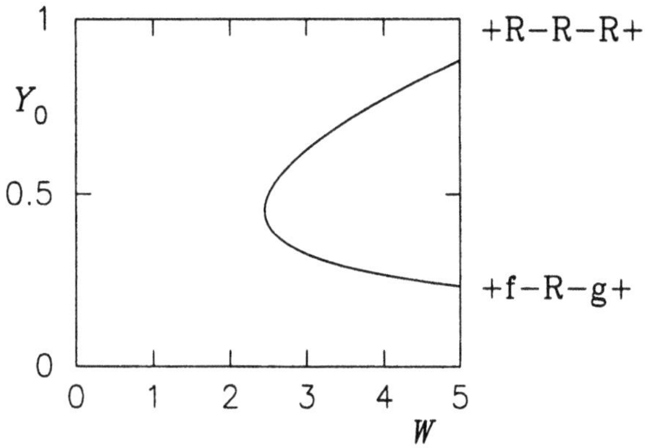

Fig. 22.2. Characteristics for 2.2T3 transitions, symmetric orbits, $h = 0$.

22.2.4.1 *R*-Orbit. In the case of a R-orbit, the asymptotic form is defined by (21.15b) and (21.15c), with numerical values for the y_i and x_i which characterize the branch; it is thus well isolated from other branches. On the other hand, the stability index tends toward a finite value (see Sect. 18.2.2). Therefore we can immediately use a shooting method; it is not necessary to begin with a relaxation method (which in any case would be inapplicable since there are no nodes). Starting from the approximate values of Y_0 and X_1 given by (21.15b) and (21.15c):

$$Y_0 \simeq W^{1/2} y_0 \,, \qquad X_1 \simeq W^{-1/2} x_1 \,, \tag{22.31}$$

with y_0 and x_1 determined in Sect. 18.2, we compute successively Y_1, X_2, $\ldots$, Y_n, X_{n+1}. We compute also the variations, as functions of $\mathrm{d}Y_0$ and $\mathrm{d}X_1$. This allows an iterative correction of the initial values, until a periodic orbit is obtained ($Y_n = Y_0$, $X_{n+1} = X_1$). It allows also a computation of the stability index z.

22.2.4.2 Orbit Containing *T*-Arcs. In that case, we start with a relaxation method as usual (Sects. 14.2.2, 20.2.5.2), based on the asymptotic decomposition of the orbit into T- and R-arcs. The asymptotic expressions (Sect. 21.1.1) show that $Y_i = \Theta(W^{-1/2})$ at a node, $Y_i = \Theta(W^{1/2})$ at an antinode (inside a R-arc). The steps are as follows:

1. We take as initial approximation $Y_i = 0$ in nodes.
2. For each arc, we compute the values of the internal Y_i and X_i (see below).
3. For each node, we recompute Y_i, using (22.30a).
4. We go back to step 2, until the solution has converged.

Some details are necessary for point 2. In the case of a T-arc, we must compute X_i from (22.30b), where Y_i and Y_{i-1} are given. For large W, this third-degree equation has 3 real roots, close to 0, $+W^{1/2}$, $-W^{1/2}$ respectively.

As indicated by (21.14), we select the root close to $+W^{1/2}$ in the case of a T^f arc, $-W^{1/2}$ in the case of a T^g arc.

In the case of a R-arc, the end values Y_i and Y_{i+j} are known, and we must compute the intermediate values $Y_{i+1}, \ldots, Y_{i+j-1}$ and $X_{i+1}, \ldots, X_{i+j}$. We start from an approximate value of X_{i+1}, given by (21.15c): $X_{i+1} \simeq W^{-1/2}x_1$. We compute then successively $Y_{i+1}, X_{i+2}, \ldots, Y_{i+j}$ with (22.30). We compute also the variations, as functions of dX_{i+1}. This allows a correction of the initial value of X_{i+1}, and we continue until the final value Y_{i+j} agrees with the given value.

The branch is followed towards decreasing W. Usually at some point the iteration does not converge anymore, because the decomposition into T- and R-arcs ceases to be a good approximation. We shift then to a shooting method, which works well for moderate values of W where the branches are well separated.

22.3 Results for Bifurcations of Type 2

All branches were computed and joined up to $n = 5$. Showing characteristics in the (W, Y_0) plane would not be useful because the number of branches is too great and the figure is crowded. Instead, detailed print-outs of the branches were made, from which the junctions were found by inspection.

As in the case of partial bifurcation (Sect. 21.4), we can use the knowledge of these junctions at transition 2.2 to deduce the corresponding junctions between branches in a total bifurcation of type 2. These junctions follow again the path illustrated by Table 21.2, going successively through a transition 2.1, a transition 2.2, and another transition 2.1. In each transition 2.1, a sequence of S-arcs is replaced by a R-arc (or a R-orbit if the whole orbit is involved), or conversely. The rules are the same as in Sect. 21.4.

We do not list here the junctions computed for the transition 2.2 since they represent only an intermediate step. Instead we list in Table 22.2 the final results of interest, i.e. the junctions between branches for the total bifurcation. This table is in the same format as Table 8.18, which was obtained using only qualitative methods. It can be verified that there is agreement for all branches which appear in both tables (sign - and $n \leqq 4$). (For this comparison, the symbols `i` and `e` in Table 8.18 should be replaced by `f` and `g` respectively; see Sect. 17.1.)

Table 22.2, together with Table 20.1, gives all junctions for total bifurcations of type 2, from $n = 2$ to $n = 5$. The cases left undecided in Table 8.18 are now solved; in addition all junctions are established for $n = 5$. Nothing prevents in principle the solution of higher values of n, using numerical computation. (The junctions for $n = 6$ are listed in Table 20.1 for branches with a + sign.) However, the amount of work grows exponentially.

Table 22.2. Total bifurcation, type 2, $n = 2$ to 5: junctions between the - branches, determined by the quantitative study.

2T2+-+0	- 2 - fg
2T2+-+A	- f1 - 1g
2T2-+-0	- -11- - gf
2T2-+-A	- 1f - g1
2T3++-+1	- 12 - 1fg
2T3++-+A	- 1f1 - 11g
2T3+-++2	- 21 - fg1
2T3+-++A	- f11 - 1g1
2T3+--+0	- 3 - f1g
2T3+--+A	- f2 - 2g
2T3-++-0	- -111- - g1f
2T3-++-A	- 11f - g11
2T3-+--2	- -12- - gf1
2T3-+--A	- -1f1- - g2
2T3--+-1	- -21- - 1gf
2T3--+-A	- 2f - -1g1-
2T4+++-+2	- 112 - 11fg
2T4+++-+A	- 11f1 - 111g
2T4++-++0	- 121 - 1fg1
2T4++-++A	- 1f11 - 11g1
2T4++--+1	- 13 - 1f1g
2T4++--+A	- 1f2 - 12g
2T4+-+++2	- 211 - fg11
2T4+-+++A	- f111 - 1g11
2T4+-+-+0	- 1gf1 - f11g
2T4+-+-+2	- fg2 - 2fg
2T4+-+-+A	- 21g - 1g2
	- 2f1 - f12
	- 1gfg - fg1g
	- fgf1 - f1fg
2T4+--++3	- 31 - f1g1
2T4+--++A	- f21 - 2g1
2T4+---+0	- 4 - f2g
2T4+---+A	- f3 - 3g
2T4-+++-0	- -1111- - g11f
2T4-+++-A	- 111f - g111
2T4-++--3	- -112- - g1f1
2T4-++--A	- -11f1- - g12
2T4-+-+-0	- -1fg1- - g2f
2T4-+-+-2	- gf11 - 11gf
2T4-+-+-A	- -11g1- - g21
	- -1f11- - 12f
	- gfg1 - g1gf
	- gf1f - 1fgf
2T4-+---2	- -13- - gf2
2T4-+---A	- -1f2- - g3
2T4--++-1	- -211- - 1g1f
2T4--++-A	- 21f - -1g11-
2T4--+--0	- -22- - -1gf1-
2T4--+--A	- -2f1- - -1g2-
2T4---+-2	- -31- - 2gf
2T4---+-A	- 3f - -2g1-

Table 22.2. (continuation)

2T5++++-+3	- 1112 - 111fg
2T5++++-+A	- 111f1 - 1111g
2T5+++-++1	- 1121 - 11fg1
2T5+++-++A	- 11f11 - 111g1
2T5+++--+2	- 113 - 11f1g
2T5+++--+A	- 11f2 - 112g
2T5++-+++4	- 1211 - 1fg11
2T5++-+++A	- 1f111 - 11g11
2T5++-+-+1	- 122 - 1f11g
	- 11gf1 - 1fgfg
2T5++-+-+A	- 11g2 - 11g1g
	- 1f12 - 121g
	- 1fg2 - 1fg1g
	- 12f1 - 1f1f1
	- 12fg - 1f1fg
	- 1fgf1 - 11gfg
2T5++--++0	- 131 - 1f1g1
2T5++--++A	- 1f21 - 12g1
2T5++---+1	- 1f2g - 14
2T5++---+A	- 1f3 - 13g
2T5+-++++2	- 2111 - fg111
2T5+-++++A	- f1111 - 1g111
2T5+-++-+0	- 212 - 1g1f1
	- f111g - fg1fg
2T5+-++-+A	- 211g - 1g11g
	- 21f1 - 1g12
	- 21fg - 1g1fg
	- f112 - f11f1
	- fg12 - fg1f1
	- f11fg - fg11g
2T5+-+-++4	- 221 - f11g1
	- 1gf11 - fgfg1
2T5+-+-++A	- 1g21 - 1g1g1
	- f121 - 21g1
	- fg21 - fg1g1
	- 2f11 - f1f11
	- 2fg1 - f1fg1
	- fgf11 - 1gfg1
2T5+-+--+2	- 23 - 2f1g
	- fg3 - fgf1g
2T5+-+--+A	- 1g3 - 1g2g
	- f13 - f1f2
	- 2f2 - 22g
	- 1gf2 - 1gf1g
	- fgf2 - fg2g
	- f12g - f1f1g
2T5+--+++3	- 311 - f1g11
2T5+--+++A	- f211 - 2g11
2T5+--+-+3	- 32 - f1g2
	- 3fg - f1gfg
2T5+--+-+A	- 31g - 2g1g
	- 3f1 - f2f1
	- f22 - 2g2
	- f21g - f1g1g
	- f2fg - 2gfg
	- 2gf1 - f1gf1
2T5+---++4	- f2g1 - 41
2T5+---++A	- f31 - 3g1
2T5+----+0	- 5 - f3g
2T5+----+A	- f4 - 4g

Table 22.2. (continuation)

2T5-++++-0
- -11111- - g111f
2T5-++++-A
- 1111f - g1111
2T5-+++--4
- -1112- - g11f1
2T5-+++--A
- -111f1- - g112
2T5-++-+-3
- -1121- - g1f11
- 111gf - g1fgf
2T5-++-+-A
- -111g1- - g11g1
- -11f11- - g121
- -11fg1- - g1fg1
- 112f - 11f1f
- g12f - g1f1f
- 11fgf - g11gf

2T5-++---3
- g1f2 - -113-
2T5-++---A
- -11f2- - g13
2T5-+-++-2
- -1211- - 11g1f
- gf111 - gfg1f
2T5-+-++-A
- g211 - g1g11
- 121f - -11g11-
- g21f - g1g1f
- -1f111- - 1f11f
- -1fg11- - 1fg1f
- gf11f - gfg11
2T5-+-+--4
- -122- - g2f1
- -1fg2- - gfgf1

2T5-+-+--A
- -11g2- - g1g2
- -1f12- - -1f1f1-
- -12f1- - g22
- -11gf1- - g1gf1
- -1fgf1- - gfg2
- gf12 - gf1f1
2T5-+--+-0
- -131- - -1f1g1-
- g3f - gf1gf
2T5-+--+-A
- g31 - g2g1
- 13f - 1f2f
- -1f21- - -12g1-
- gf21 - gf1g1

- gf2f - g2gf
- 12gf - 1f1gf
2T5-+----2
- -14- - gf3
2T5-+----A
- -1f3- - g4
2T5--+++-1
- -2111- - 1g11f
2T5--+++-A
- 211f - -1g111-
2T5--++--0
- -1g1f1- - -212-
2T5--++--A
- -21f1- - -1g12-
2T5--+-+-1
- -221- - 1g2f
- -2fg1- - 1gfgf
2T5--+-+-A
- -21g1- - -1g1g1-
- -2f11- - 2f1f

- 22f - -1g21-
- 21gf - 1g1gf
- 2fgf - -1gfg1-
- -1gf11- - 1gf1f
2T5--+---4
- -23- - -1gf2-
2T5--+---A
- -2f2- - -1g3-
2T5---++-2
- 2g1f - -311-
2T5---++-A
- 31f - -2g11-
2T5---+--1
- -32- - -2gf1-
2T5---+--A
- -3f1- - -2g2-
2T5----+-3
- -41- - 3gf
2T5----+-A
- 4f - -3g1-

23. Bifurcations 2T1 and 2P1

The two cases with $n = 1$ are very different from the general case $n > 1$ and we study them separately in the present chapter.

The junctions for these two cases have been established in Chap. 8 (Tables 8.12, 8.18), so that we could dispense with their study if we were only interested in the junctions. But we will pursue the quantitative approach for these two cases also, for homogeneity, and in order to obtain in all cases a quantitative approximation of the families of periodic orbits in the neighbourhood of the bifurcation.

23.1 Total Bifurcation of Type 2, $n = 1$ (2T1)

We have the 4 fundamental equations (17.63) to (17.66) for the variables h_0, Δa_1, u_1', u_1'', ΔC.

For $\mu = 0$, there are 4 branches arriving at the bifurcation: + E, - E, + 1, - 1 (Table 6.5). All these branches are made of symmetric generating orbits. We will assume that for $\mu > 0$, the corresponding periodic orbits are also symmetric. Then there is

$$u_1'' = -u_1' \; . \tag{23.1}$$

(17.66) is identically satisfied, and the system of fundamental equations reduces to

$$\begin{aligned} h_0 u_1' &= -\frac{1}{\nu}\mu[1 + O(h_0, u_1')] \; , \\ u_1' &= -\frac{3\pi I(a-1)}{2a^2 v_y}\Delta a_1[1 + O(\Delta C, \Delta a_1, u_1'^2, \mu)] \; , \\ 2(a-1)h_0 &= aV_Y\Delta C + \frac{v_y}{a}\Delta a_1 + a u_1'^2 + O(\Delta C^2, \Delta a_1^2, u_1'^4, \mu) \; . \end{aligned} \tag{23.2}$$

23.1.1 The Case $\nu = 0$

We consider first the case $\mu = 0$, $\Delta C \neq 0$, corresponding to $\nu = 0$. As in Sects. 12.3 and 17.3, this case is simple and it is not necessary to use the machinery described in Sect. 11.4. The solutions are known: they correspond to a displacement from the bifurcation orbit along one of the branches + E, - E, + 1, - 1. The first equation (23.2a) reduces to

$$h_0 u_1' = 0 . \tag{23.3}$$

We have two solutions. Using (23.2b) and (23.2c), we find that the first solution is

$$u_1' = 0 , \qquad \Delta a_1 = 0 , \qquad h_0 = \frac{aV_Y}{2(a-1)}\Delta C + O(\Delta C^2) , \tag{23.4}$$

and the second solution is

$$h_0 = 0 , \qquad \Delta a_1 = -\frac{a^2 V_Y}{v_y}\Delta C + O(\Delta C^2) ,$$

$$u_1' = \frac{3\pi I(a-1)V_Y}{2v_y^2}\Delta C + O(\Delta C^2) . \tag{23.5}$$

These two solutions correspond respectively to orbits of the first species (branches ± E) and second species (branches ± 1). In the latter case, we may remark that (23.5) is a particular case of the equations (17.84) and (17.86), for $m = 1$.

23.1.2 The Case $0 < \nu < 1/2$

We consider now the case $\mu > 0$, $\Delta C \neq 0$, corresponding to $\nu > 0$. We will try to extend each solution obtained in the previous section for $\nu = 0$. This corresponds to the asymptotic branches of the families of orbits for $\mu \neq 0$, which are close to the branches for $\mu = 0$ (see Fig. 1.1).

We use once more the general method described in Sect. 11.4.

Steps 1 and 2. The results for $\nu = 0$ suggest the change of variables

$$u_1' = \Delta C x'^* , \qquad h_0 = \Delta C y^* , \qquad \Delta a_1 = \Delta C z^* . \tag{23.6}$$

Step 3. It will be convenient to make a second change of variables

$$x'^* = \frac{3\pi I(a-1)V_Y}{v_y^2}x' , \qquad y^* = \frac{aV_Y}{2(a-1)}y , \qquad z^* = \frac{a^2 V_Y}{v_y}z . \tag{23.7}$$

Step 4. Substituting in the fundamental equations (23.2), collecting the dominant terms in the left-hand member, and dividing by the appropriate powers of ΔC, we obtain

$$yx' = O(\mu \Delta C^{-2}) ,$$
$$2x' + z = O(\Delta C) + O(\mu) ,$$
$$y - 1 - z = O(\Delta C, \mu \Delta C^{-1}) . \tag{23.8}$$

Step 5. We want all terms in the right-hand members to vanish for $\mu \to 0$. This is the case if

$$0 < \nu < \frac{1}{2} . \tag{23.9}$$

Note the difference with the general case $n > 1$, where the limits of the interval of validity were $0 < \nu < 1/3$ (Equ. (17.103)).

The asymptotic equations are then obtained by equating the right-hand members to zero:

$$yx' = 0\,, \qquad 2x' + z = 0\,, \qquad y - 1 - z = 0\,. \tag{23.10}$$

Step 6. We solve now these asymptotic equations. There are two solutions. The first is

$$x' = 0\,, \qquad z = 0\,, \qquad y = 1\,. \tag{23.11}$$

It is the continuation of the first species solution (23.4). The second solution is

$$y = 0\,, \qquad z = -1\,, \qquad x' = 1/2\,. \tag{23.12}$$

It is the continuation of the second species solution (23.5).

Step 7. We compute the Jacobian of the system (23.10) for the three variables y, x', z:

$$|J| = \begin{vmatrix} x' & y & 0 \\ 0 & 2 & 1 \\ 1 & 0 & -1 \end{vmatrix} = -2x' + y\,. \tag{23.13}$$

This gives $|J| = 1$ for the first solution, $|J| = -1$ for the second solution. Thus, *for $0 < \nu < 1/2$, there exist for $\mu \neq 0$ solutions of the fundamental equations (23.8) close to the solutions (23.11) and (23.12) of the asymptotic equations.* The error is of the order of the largest term in the right-hand members of (23.8), i.e. $O(\Delta C, \mu\Delta C^{-2})$.

Steps 8 and 9. We go back to the physical variables, using (23.7) and (23.6), and we refine the error estimates for some variables. For the first solution (23.11), we obtain

$$\begin{aligned} h_0 &= \frac{aV_Y}{2(a-1)}\Delta C[1 + O(\Delta C, \mu\Delta C^{-2})]\,, \\ u_1' &= -u_1'' = O(\Delta C^2, \mu\Delta C^{-1})\,, \\ \Delta a_1 &= O(\Delta C^2, \mu\Delta C^{-1})\,. \end{aligned} \tag{23.14}$$

Using (23.2a) and (23.2b), we can refine the estimates and obtain the principal term for u_1', u_1'', Δa_1:

$$\begin{aligned} u_1' &= -u_1'' = -\frac{2(a-1)}{aV_Y v}\mu\Delta C^{-1}[1 + O(\Delta C, \mu\Delta C^{-2})]\,, \\ \Delta a_1 &= \frac{4a\,\mathrm{sign}(v_y)}{3\pi I V_Y}\mu\Delta C^{-1}[1 + O(\Delta C, \mu\Delta C^{-2})]\,. \end{aligned} \tag{23.15}$$

For the second solution, we have

$$h_0 = O(\Delta C^2, \mu\Delta C^{-1}) ,$$
$$u_1' = -u_1'' = \frac{3\pi I(a-1)V_Y}{2v_y^2}\Delta C[1 + O(\Delta C, \mu\Delta C^{-2})] ,$$
$$\Delta a_1 = -\frac{a^2 V_Y}{v_y}\Delta C[1 + O(\Delta C, \mu\Delta C^{-2})] . \tag{23.16}$$

Using (23.2a), we can refine the estimate for h_0:

$$h_0 = -\frac{2v}{3\pi I(a-1)V_Y}\mu\Delta C^{-1}[1 + O(\Delta C, \mu\Delta C^{-2})] . \tag{23.17}$$

These equations are particular cases of (17.122), (17.123), (17.127), for $m = 1$.

For $\mu \to 0$, the equations (23.14), (23.15), (23.16), (23.17) reduce to the equations (23.4) and (23.5) of the previous section.

23.1.3 The Case $\nu = 1/2$

We study now what happens when we reach the value $\nu = 1/2$, corresponding to $\Delta C = \Theta(\mu^{1/2})$.

Step 1. The results of the previous section show that when the value $\nu = 1/2$ is reached, all quantities h_0, u_1', u_1'', Δa_1 become of the order of $\mu^{1/2}$, for both solutions.

Step 2. This suggests the change of variables

$$h_0 = \mu^{1/2} y^* , \qquad u_1' = \mu^{1/2} x'^* , \qquad \Delta a_1 = \mu^{1/2} z^* , \qquad \Delta C = \mu^{1/2} w^* . \tag{23.18}$$

Step 3. It will be convenient to make the second change of variables

$$y* = \frac{1}{a-1}\sqrt{\frac{av}{3\pi I}}\, y , \qquad x'^* = (a-1)\sqrt{\frac{3\pi I}{av^3}}\, x' ,$$
$$z^* = \frac{a}{v_y}\sqrt{\frac{av}{3\pi I}}\, z , \qquad w^* = -\frac{2}{V_Y}\sqrt{\frac{v}{3\pi I a}}\, w . \tag{23.19}$$

Step 4. Substituting in the fundamental equations (23.2), we obtain

$$yx' + 1 = O(\mu^{1/2}) , \qquad 2x' + z = O(\mu^{1/2}) , \qquad 2y + 2w - z = O(\mu^{1/2}) . \tag{23.20}$$

Step 5. The right-hand members tend to zero for $\mu \to 0$. The asymptotic equations are obtained by equating them to zero:

$$yx' + 1 = 0 , \qquad 2x' + z = 0 , \qquad 2y + 2w - z = 0 . \tag{23.21}$$

Step 6. The system of asymptotic equations (23.21) is easily solved. Eliminating x' and z, we obtain

$$y^2 + wy - 1 = 0 \tag{23.22}$$

and

$$y = \frac{-w \pm \sqrt{w^2+4}}{2}\,, \qquad z = w \pm \sqrt{w^2+4}\,, \qquad x' = \frac{-w \mp \sqrt{w^2+4}}{2}\,. \tag{23.23}$$

The characteristics in the (w, y) plane are shown in Fig. 23.1.

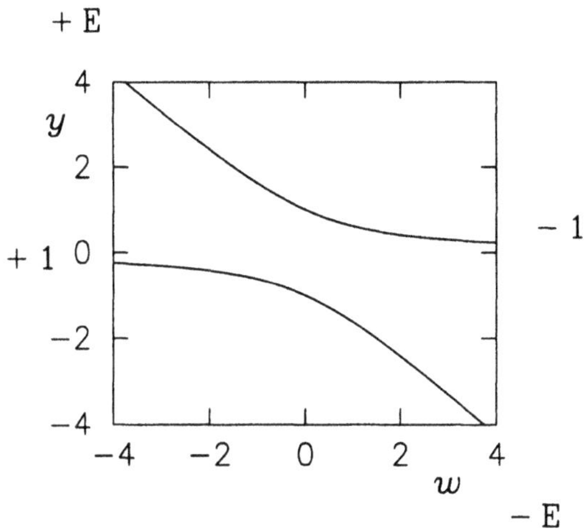

Fig. 23.1. Characteristics of the **2T1** bifurcation.

Step 7. We compute the Jacobian of the system (23.21) for the three variables y, x', z:

$$|J| = \begin{vmatrix} x' & y & 0 \\ 0 & 2 & 1 \\ 2 & 0 & -1 \end{vmatrix} = -2x' + 2y = \pm 2\sqrt{w^2+4}\,. \tag{23.24}$$

It never vanishes. Therefore, *for $\nu = 1/2$, there exist for $\mu \neq 0$ solutions of the fundamental equations (23.20) close to the solutions (23.23) of the asymptotic equations.* The error is of the order of $\mu^{1/2}$.

Step 9. We can go back to the physical variables, using (23.18) and (23.19). We obtain

$$h_0 = \frac{a}{4(a-1)}\left(V_Y\Delta C \pm \sqrt{V_Y^2\Delta C^2 + \frac{16v}{3\pi I a}\mu}\right) + O(\mu)\,,$$

$$\Delta a_1 = \frac{a^2}{2v_y}\left(-V_Y\Delta C \pm \sqrt{V_Y^2\Delta C^2 + \frac{16v}{3\pi I a}\mu}\right) + O(\mu)\,,$$

$$u_1' = \frac{3\pi I(a-1)}{4v^2}\left(V_Y\Delta C \mp \sqrt{V_Y^2\Delta C^2 + \frac{16v}{3\pi I a}\mu}\right) + O(\mu)\,. \tag{23.25}$$

If ν is decreased, so that μ becomes negligible compared with ΔC^2, these expressions reduce to those of the case $0 < \nu < 1/2$: either (23.14a) and (23.15), or (23.16b), (23.16c), and (23.17), depending on the $\pm$ sign. This allows the identification of the branches shown in Fig. 23.1. We recover the junctions deduced from Broucke's principle (Table 8.18).

23.1.4 The Case $\nu > 1/2$

Fig. 23.1 shows that w vanishes in two points. As for type 1, Sect. 12.6, the limit $w \to 0$ corresponds to moving from the case $\nu = 1/2$ to the case $\nu > 1/2$, which should in principle be studied separately. However, the solutions found in this way turn out to be a subset of the solutions of the system (23.25), corresponding to $\Delta C = 0$. It is not necessary to consider these solutions separately; they are not fundamentally different from the other solutions.

23.1.5 Recapitulation

The quantitative study of the bifurcation 2T1 is now complete. For a given small μ, consider for instance a family of periodic orbits coming along the + E branch. It corresponds at first to a small value of ν as the distance ΔC to the bifurcation is still comparatively large; the orbits are approximately described by the equations (23.14) and (23.15) of Sect. 23.1.2. When ν reaches the value 1/2, the orbits are approximately described by the equations (23.25) of Sect. 23.1.3. The branch + E joins there the branch - 1 (Fig. 23.1). As we continue along the family, ν decreases from 1/2 to small values; the orbits are approximately described by the equations (23.16) and (23.17) of Sect. 23.1.2, as we move away from the bifurcation.

Fig. 23.2 shows in log-log plots the variations in order of magnitude of the quantities h_0, Δa_1, u_1' and u_1'' as functions of ΔC.

This figure for the 2T1 bifurcation differs much from the figure for the general case $n > 1$ (Fig. 17.2). Here there is a single transition at $\mu = 1/2$, instead of the two transitions at $\mu = 1/3$ and $\mu = 1/2$. This explains why a particular treatment was necessary.

The junctions are shown in Table 23.1 in the usual format. They agree with those established in Chap. 8 with the help of Broucke's principle (Table 8.18).

The Tables 20.1, 22.2, and 23.1, taken together, give all junctions for total bifurcations of type 2, for $n = 1$ to 5.

In a sense the bifurcation 2T1 is completely described by the equations (23.25), which include the equations of Sects. 23.1.1 and 23.1.2 as limit cases.

The bifurcation 2T1 was studied by Guillaume (1971, pp. 125–126). It can be verified that his equation (IV–33) is identical with the above equations.

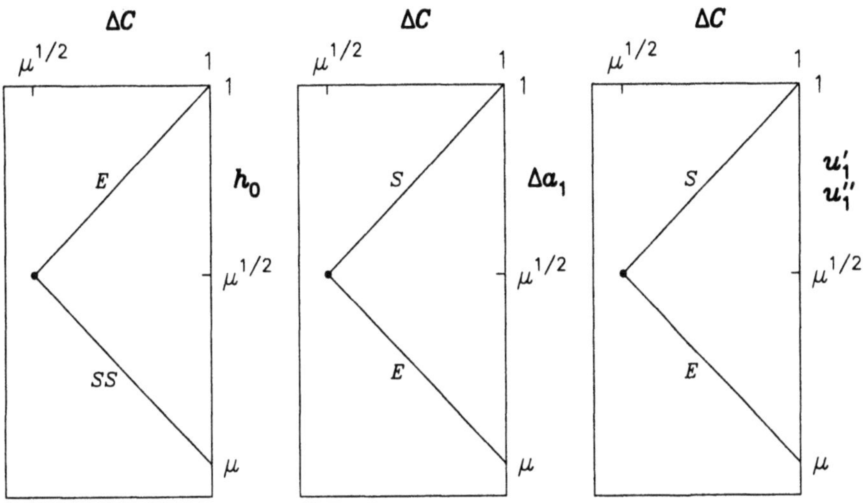

Fig. 23.2. 2T1 bifurcation: variations of h_0, Δa_1, u_1' and u_1'' as functions of ΔC. Both scales are logarithmic. In the left panel, SS represents values of h_0 at a node, while E represents values at an antinode.

Table 23.1. Total bifurcation, type 2, $n = 1$: junctions determined by the quantitative study.

2T1++0
+ E
- 1

2T1--0
- E
+ 1

23.2 Partial Bifurcation of Type 2, $n = 1$ (2P1)

We have the 5 fundamental equations (17.64) to (17.67) for the variables h_0, h_1, Δa_1, u_1', u_1'', ΔC. However, from (17.67) we have $h_0 = O(\mu)$, $h_1 = O(\mu)$. In (17.65) and (17.66), h_0 and h_1 can then be absorbed in the right-hand side. We have thus 3 equations for Δa_1, u_1', u_1'', ΔC:

$$u_1'' - u_1' = \frac{3\pi I(a-1)}{a^2 v_y}\Delta a_1[1 + O(\Delta C, \Delta a_1, u_1'^2, u_1''^2, \mu)] ,$$

$$0 = aV_Y \Delta C + \frac{v_y}{a}\Delta a_1 + a u_1'^2 + O(\Delta C^2, \Delta a_1^2, u_1'^4, \mu) ,$$

$$0 = \frac{3\pi I}{2av_y}\Delta a_1(u_1' + u_1'')[1 + O(\Delta C, \Delta a_1, u_1'^2, u_1''^2)] + O(\mu) . \tag{23.26}$$

23.2.1 The Case $\nu = 0$

We consider first the case $\mu = 0$, $\Delta C \neq 0$, corresponding to $\nu = 0$. The solutions correspond to a displacement from the bifurcation orbit along one of the branches which emanate from it. These branches are + 1, - 1, - i, - e (Table 6.9). (23.26c) reduces to

$$\Delta a_1(u_1' + u_1'') = 0 . \tag{23.27}$$

Therefore we have two solutions. The first solution is

$$\Delta a_1 = 0 , \qquad u_1' = u_1'' = \pm\sqrt{-V_Y \Delta C}[1 + O(\Delta C)] . \tag{23.28}$$

This solution corresponds to the T-arcs (branches - i and - e). The $-$ sign corresponds to a T^i arc, the + sign to a T^e arc.

The second solution is $u_1' = -u_1''$. From (23.26a) we have then

$$u_1' = \Theta(\Delta a_1) \tag{23.29}$$

and from (23.26b)

$$\Delta a_1 = -\frac{a^2 V_Y}{v_y}\Delta C[1 + O(\Delta C)] . \tag{23.30}$$

(23.26a) gives then

$$u_1' = -u_1'' = \frac{3\pi I(a-1)V_Y}{2v_y^2}\Delta C[1 + O(\Delta C)] . \tag{23.31}$$

The results (23.28), (23.30), (23.31) are particular cases of the equations (17.73), (17.76), (17.84), (17.86) (for $m = 1$).

We consider now the case $\mu > 0$, $\Delta C \neq 0$, corresponding to $\nu > 0$. We will try to extend each solution obtained in the previous section for $\nu = 0$. This corresponds to the asymptotic branches of the families of orbits for $\mu \neq 0$, which are close to the branches for $\mu = 0$ (see Fig. 1.1).

We use the general method described in Sect. 11.4. The intervals in ν and the appropriate changes of variables turn out to be different for the T-arcs and for the S-arcs. Also we know already that the two symmetric branches + 1 and - 1 (S-arcs) are joined, and the two asymmetric branches - i and - e (T-arcs) are joined (Tables 7.2, 8.12). Therefore we consider the two families separately.

23.2.2 T-Arcs: The Case $0 < \nu < 2/3$

Steps 1 and 2. The results for $\nu = 0$ suggest the change of variables

$$\Delta a_1 = \Delta C Z^* , \qquad u_1' = (-\epsilon' \Delta C)^{1/2} X'^* , \qquad u_1'' = (-\epsilon' \Delta C)^{1/2} X''^* . \tag{23.32}$$

Step 3. It will be convenient to make a second change of variables

$$Z^* = \frac{a^2 V_Y}{v_y} Z , \qquad X'^* = \pm\sqrt{\epsilon' V_Y} X' , \qquad X''^* = \pm\sqrt{\epsilon' V_Y} X'' . \tag{23.33}$$

In the last two equations, we take the sign + for a T^e arc, – for a T^i arc; in that way, X' and X'' are always positive.

Step 4. Substituting in the fundamental equations (23.26), we obtain

$$\begin{aligned} &X'' - X' = O(\Delta C^{1/2}) , \\ &X'^2 - 1 - Z = O(\Delta C, \mu \Delta C^{-1}) , \\ &Z(X' + X'') = O(\Delta C, \mu \Delta C^{-3/2}) . \end{aligned} \tag{23.34}$$

Step 5. All terms in the right-hand members vanish for $\mu \to 0$ if

$$0 < \nu < 2/3 . \tag{23.35}$$

The asymptotic equations are then obtained by equating the right-hand members to zero:

$$X'' - X' = 0 , \qquad X'^2 - 1 - Z = 0 , \qquad Z(X' + X'') = 0 . \tag{23.36}$$

Step 6. The asymptotic equations (23.36) are easily solved. We obtain 3 solutions:

$$\begin{aligned} &Z = 0 , \qquad X' = X'' = 1 ; \\ &Z = 0 , \qquad X' = X'' = -1 ; \\ &Z = -1 , \qquad X' = X'' = 0 . \end{aligned} \tag{23.37}$$

The relevant solution is determined by continuity with the previous case $\mu = 0$: it is (23.37a). (This can also be shown from the fact that, as mentioned above, X' and X'' are positive.)

Step 7. We compute the Jacobian of the system (23.36) for the three variables X', X'', Z:

$$|J| = \begin{vmatrix} -1 & 1 & 0 \\ 2X' & 0 & -1 \\ Z & Z & X'+X'' \end{vmatrix} = -2X'(X'+X'') = -4 . \tag{23.38}$$

Thus, *for $0 < \nu < 2/3$, there exists for $\mu \neq 0$ a solution of the fundamental equations (23.34) close to the solution (23.37a) of the asymptotic equations.* The error is of the order of the largest term in the right-hand members of (23.34), i.e. $O(\Delta C^{1/2}, \mu\Delta C^{-3/2})$.

Steps 8 and 9. We go back to the physical variables, using (23.32) and (23.33). We obtain

$$\begin{aligned} u_1' &= \pm\sqrt{-V_Y \Delta C}\,[1 + O(\Delta C^{1/2}, \mu\Delta C^{-3/2})] , \\ u_1'' &= \pm\sqrt{-V_Y \Delta C}\,[1 + O(\Delta C^{1/2}, \mu\Delta C^{-3/2})] , \\ \Delta a_1 &= O(\Delta C^{3/2}, \mu\Delta C^{-1/2}) . \end{aligned} \tag{23.39}$$

The error estimates can be refined. From (23.26c) and (23.39) we obtain

$$\Delta a_1 = O(\mu\Delta C^{-1/2}) . \tag{23.40}$$

Note the difference with the general case (17.127). The present bifurcation 2P1 is in fact the only case where g' and g'' both vanish.

From (23.26b) we have then

$$u_1' = \pm\sqrt{-V_Y \Delta C}\,[1 + O(\Delta C, \mu\Delta C^{-3/2})] , \tag{23.41}$$

and from (23.26a)

$$u_1' = \pm\sqrt{-V_Y \Delta C}\,[1 + O(\Delta C, \mu\Delta C^{-3/2})] . \tag{23.42}$$

It is not possible here to compute the principal term of Δa_1, because it depends on the last term $O(\mu)$ in (23.26c), which is not known in the present order of approximation.

23.2.3 *T*-Arcs: The Case $\nu \geqq 2/3$

We consider now the limit $\nu \to 2/3$, still using the change of variables (23.32). In the limit, the first term of the right-hand side of (23.26c) becomes of order $O(\mu)$, i.e. of the same order as the error term. The equation is then indeterminate. The $O(\mu)$ approximation, introduced in Sect.11.3.2, is no longer sufficient; it would be necessary to go to a higher-order approximation to obtain a quantitative description of the T-arcs in the range $\nu \geqq 2/3$.

23.2.4 *S*-Arcs: The Case $0 < \nu < 1$

For $\mu = 0$, the two branches + 1 and - 1 are made of symmetric bifurcation arcs. We assume as usual that the complement is symmetric (Restriction 7.3.2) and that for $\mu > 0$ the orbits remain symmetric. Then there is

$$u_1'' = -u_1' . \tag{23.43}$$

(23.26c) is identically satisfied, and (23.26a) reduces to

$$u_1' = -\frac{3\pi I(a-1)}{2a^2 v_y}\Delta a_1[1 + O(\Delta C, \Delta a_1, u_1'^2, u_1''^2, \mu)] . \tag{23.44}$$

Steps 1 and 2. The equations (23.30) and (23.31) for $\nu = 0$ suggest the change of variables

$$\Delta a_1 = \Delta C z^* , \qquad u_1' = \Delta C x'^* . \tag{23.45}$$

Step 3. It will be convenient to make a second change of variables

$$z^* = \frac{a^2 V_Y}{v_y} z , \qquad x'^* = \frac{3\pi I(a-1)V_Y}{v_y^2} x' . \tag{23.46}$$

Step 4. The fundamental equations (23.26) become

$$2x' + z = O(\Delta C, \mu) , \qquad z + 1 = O(\Delta C, \mu\Delta C^{-1}) . \tag{23.47}$$

Step 5. All terms in the right-hand sides vanish for $\mu \to 0$ if

$$0 < \nu < 1 . \tag{23.48}$$

The asymptotic equations are then obtained by equating the right-hand sides to zero:

$$2x' + z = 0 , \qquad z + 1 = 0 . \tag{23.49}$$

Step 6. The asymptotic equations (23.49) are immediately solved:

$$z = -1 , \qquad x' = 1/2 . \tag{23.50}$$

Step 7. The Jacobian is

$$|J| = \begin{vmatrix} 2 & 1 \\ 0 & 1 \end{vmatrix} = 2 . \tag{23.51}$$

Thus, *for $0 < \nu < 1$, there exists for $\mu \neq 0$ a solution of the fundamental equations (23.47) close to the solution (23.50) of the asymptotic equations.* The error is of the order of the largest term in the right-hand sides of (23.47), i.e. $O(\Delta C, \mu\Delta C^{-1})$.

Steps 8 and 9. We go back to the physical variables, using (23.45) and (23.46). We obtain

$$\begin{aligned} \Delta a_1 &= -\frac{a^2 V_Y}{v_y}\Delta C[1 + O(\Delta C, \mu\Delta C^{-1})] , \\ u_1' &= \frac{3\pi I(a-1)V_Y}{2v_y^2}\Delta C[1 + O(\Delta C, \mu\Delta C^{-1})] . \end{aligned} \tag{23.52}$$

23.2.5 S-Arcs: The Case $\nu \geqq 1$

We consider now the limit $\nu \to 1$, still using the change of variables (23.45). In the limit, the main terms in (23.26b) become of the same order as the error term $O(\mu)$. Thus, here again, our $O(\mu)$ approximation fails; it would be necessary to go to a higher approximation to obtain a quantitative description of the S-arcs in the range $\nu \geqq 1$.

23.2.6 Recapitulation

The quantitative method, as developed in the present volume, does not entirely describe the 4 branches of the bifurcation 2P1, and is thus unable to establish their junctions.

These junctions, however, have been determined in Chap. 7 on the basis of symmetry considerations (Table 7.2). We reproduce them here in Table 23.2 (using the new notations f and g; see Sect. 17.1), so as to make the present volume self-contained.

Table 23.2. Partial bifurcation, type 2, $n = 1$: junctions determined by symmetry from Table 7.2.

2P1S	2P1A
+ 1	- f
- 1	- g

The Tables 19.1, 21.3, and 23.2, taken together, give all junctions for partial bifurcations of type 2, for $n = 1$ to 5.

23.3 Conclusions for Type 2

The study of bifurcations of type 2, begun in Chapter 17, is now complete. Results for $n = 1$ are given earlier in the present chapter, in Tables 23.1 and 23.2. For $n > 1$, branches with a + sign are joined at transition 2.1 (see Sects. 19.3.3.2 and 20.2.5.2); results are given by Table 19.1 for partial bifurcations and by Table 20.1 for total bifurcations. Branches with a - sign are joined at transition 2.2 (see Sects. 21.4 and 22.3); results are given by Table 21.3 for partial bifurcations and by Table 22.2 for total bifurcations.

23.3.1 The Newton Approach

The Newton approach, which was developed in Chap. 15 for type 1, could equally well be applied to type 2. In the case of a partial bifurcation of type 2, we have a system with $n_B = 4n + 1$ variables μ, ΔC, h_i $(i = 1, \ldots, n-1)$, Δa_i, u'_i, u''_i $(i = 1, \ldots, n)$, and $m_B = 4n - 1$ nonlinear equations (17.63) to (17.66). Thus here again, $m_B = n_B - 2$, and the solutions lie on two-dimensional manifolds.

In the case of a total bifurcation of type 2, there are $n_B = 4n+2$ variables, $m_B = 4n$ equations, and again $m_B = n_B - 2$.

Here again, the Newton polyhedra are simple: the fundamental equations (17.63) and (17.64) have 3 significant terms, while (17.65) and (17.66) have 4 significant terms. The polyhedra are simplexes, either triangles or tetrahedra. Thus, a detailed analysis of the polyhedra, faces, truncated equations, and normal cones can easily be made as in Sect. 15.3, and used to find the valid combinations in some simple cases, as in Sects. 15.5.4 and 15.8.1. We will not make this analysis here, as it would take up too much space, and also because the Newton approach has the same fundamental shortcomings as in the case of type 1 (Sect. 15.9): (i) it can only be applied for a specific value of n; (ii) the amount of work grows exponentially with n (even faster than for type 1).

23.3.2 Proving General Results

It might be possible to develop for type 2 an approach similar to that presented in Chap. 16, i.e. to prove general properties of the solutions directly from the fundamental equations. We will not attempt this here. Such an approach would be likely to be more involved than for type 1, since the system of fundamental equations is more complicated.

23.4 Type 3

Having completed the quantitative analysis of bifurcations of types 1 and 2, we should now logically tackle the last type, i.e. type 3.

However, the analysis of type 3 proves to be much more complex than that of types 1 and 2. In fact, this analysis is still unfinished at the time of this writing. Transitions have been identified for the following values of ν:

$$\frac{1}{5}, \frac{2}{9}, \frac{1}{4}, \frac{n-1}{4n-5}, \frac{1}{3}, \frac{n-1}{2n-1}, \frac{1}{2}, \frac{2}{3}, \tag{23.53}$$

and it is quite possible that other transitions remain to be discovered. This should be contrasted with the single transition at $\nu = 1/2$ for type 1, and the two transitions at $\nu = 1/3$ and $\nu = 1/2$ for type 2. Note also that in two cases, the value of ν depends on n, the number of basic arcs in the bifurcation.

In the case of type 2, in addition to the S-arcs and T-arcs, we had to introduce one new kind of arc: the R-arc. For type 3, at least 12 new kinds of arcs have to be introduced.

In any case, the study of type 3 could not fit into the present volume II. Ideally this study should be completed and then published in a third volume. But it seems now unlikely that I will be able to fulfill this program.

I have accumulated about 400 pages of manuscript notes, in french, on type 3; any interested colleague is welcome to a copy of these notes, for what they are worth.

Index of Definitions

This index refers to the section in the present Volume II where a word or expression is first encountered and defined. (See also the Index of Definitions in Volume I for terms already defined in that volume.)

Index of Notations

This index refers to the section in the present Volume II where a notation is first introduced and defined. (See also the Index of Notations in Volume I for notations already defined in that volume.)

Notation	Section	Definition
a_i	12.1.1	value for intermediate orbit
A_i	12.1.1	value for intermediate orbit
$\mathcal{A}$	11.3.1	set of basic arcs
b_i	15.2	coefficient of asymptotic form
B	15	subscript for Bruno's variables
$\mathbf{B}$, B^j	15.5	maximal linear subspace
C_i	12.1.1	value of C for intermediate orbit
$\mathbf{C}$	15.5.1	polyhedral cone
$\mathcal{C}$	11.3.1	set of internal collisions
d	15.6	dimension of the truncated system
E_i	15.3.1	unit vector
f_j	15.1	fundamental equations in Newton approach
g', g''	12.4.2, 17.4	switching variable
G_1, G_2, G_3	12.1.4	constants
i_α	13.1.2	node position
$\|J\|$	11.4	Jacobian
K	12.1.4	quantity characterizing a bifurcation
$\mathbf{K}$	15.2	cone of the problem
L_1	17.5	auxiliary constant
L_2	17.6	auxiliary constant
L_3	17.7	auxiliary constant
m_a, m_b	12.4.2	values of m before and after collision
m_B	15.1	number of equations
m_α	13.1.2	value of m for arc α
n_B	15.1	number of variables
$\tilde{n}$	12.2	number of T-arcs in a T-sequence
$\tilde{n}$	17.5	number of basic arcs in a R-region
N	13.1.2	number of S or T arcs
$\mathbf{N}$, N^i	15.5.1	fundamental system of solutions
$O(x)$	11.2.1	of the order of x
p_i	15.2	exponent of asymptotic form

$Q = (q_1, \ldots, q_{n_B})$		
	15.2	exponents in a monomial
R	17.5	R-region, formed by fusion of S-arcs
s_i	12.1.1, 12.1.4	indicates whether collision i is in P or Q
$\mathbf{S}$	15.2	set of points in Newton approach
t_i	11.3.1	time of passage near collision i
t_{0i}	11.3.2	time far from collisions on basic arc i
t'_i, t''_i	12.1.1	time of intersection of unit circle by intermediate orbit
T^f, T^g	17.1	new notations for T-arcs
u'_i, u''_i	17.2	radial velocity
U_α	13.1.2	unspecified arc
$\mathbf{U}_{ik}^{(d)}$	15.3	normal cone
v_x, v_y	17.2.2.1	velocity at collision, in rotating axes
V_X, V_Y	17.2.2.1	velocity at collision, in fixed axes
w_i	15.7	new variables in Newton approach
$\bar{X}_i$, $\bar{Y}_i$	12.4	scaled variables for type 1, $0 < nu < 1/2$
W, X_i, Y_i	12.5	scaled variables for type 1, $nu = 1/2$
x'_i, x''_i, X'_i, X''_i, y_i, z_i		
	17.4	scaled variables for type 2, $0 < nu < 1/3$
w, x'_i, x''_i, X'_i, X''_i, y_i, z_i, Z_i		
	17.5	scaled variables for type 2, $nu = 1/3$
x'_i, x''_i, X'_i, X''_i, y_i, z_i, Z_i		
	17.6	scaled variables for type 2, $1/3 < nu < 1/2$
W, X'_i, X''_i, Y_i, Z_i		
	17.7	scaled variables for type 2, $nu = 1/2$
$X = (x_1, \ldots, x_{n_B})$		
	15.1	variables in Newton approach
Z_i	12.1.1	value for intermediate orbit
α	13.1.2	arc number
$\mathbf{\Gamma}$	15.2	Newton polyhedron
Δx_i	11.3.2, 12.1.1	variation of a quantity x for intermediate orbit
λ_i	15.5.1	coefficient
ν	11.4	exponent defining the relative orders of magnitude of ΔC and μ
$\mathbf{\Pi}$	15.5	cone of truncation
$\boldsymbol{\rho}$	11.3.1	vector from M_2 to M_3
$\boldsymbol{\rho}_i$	11.3.2	$\boldsymbol{\rho}$ for intermediate orbit
σ'	17.1	relative side of passage
Σ'	13.1	symmetry in reduced variables Y, X, W
τ	15.2	independent variable in Newton approach
Ω	11.3.1	second species bifurcation orbit

References

Bruno, A. D. (1976): Periodic solutions of the second kind in the restricted three-body problem (in russian). Institute of Applied Mathematics, Moscow, preprint No. 95. English translation in Bruno 1994, chap. 7.
Bruno, A. D. (1994): The restricted 3-body problem: Plane periodic orbits (english translation of Bruno 1990). De Gruyter Expositions in Mathematics 17, Walter de Gruyter, Berlin, New York.
Bruno, A. D. (1998): Power Geometry in Algebraic and Differential Equations (in russian), Fizmatlit, Moscow. English translation: Bruno 2000.
Bruno, A. D. (2000): Power Geometry in Algebraic and Differential Equations (english translation of Bruno 1998). North-Holland Mathematical Library, Elsevier, Amsterdam.
Devaney, Robert L. (1981): The baker transformation and a mapping associated to the restricted three body problem. Commun. Math. Phys. **80**, 465–476.
Graham, R.L., Knuth, D. E., Patashnik, O. (1989): Concrete mathematics. Addison-Wesley, Reading, Massachusetts.
Guillaume, P. (1971): Solutions périodiques symétriques du problème restreint des trois corps pour de faibles valeurs du rapport des masses. Ph. D. Thesis, Liège University.
Guillaume, P. (1973b): Periodic symmetric solutions of the restricted problem. Celest. Mech. **8**, 199–206.
Guillaume, P. (1975b): The restricted problem: an extension of Breakwell-Perko's matching theory. Celest. Mech. **11**, 449–467.
Hénon, M. (1997): Generating Families in the Restricted Three-Body Problem, Lecture Notes in Physics **m 52**, Springer, Berlin.
Hénon, M., Guyot, M. (1970): Stability of periodic orbits in the restricted problem. In: G. E. O. Giacaglia (ed.), Periodic Orbits, Stability and Resonances, Reidel, Dordrecht-Holland, 349–374.
Hitzl, D. L., Hénon, M. (1977b): The stability of second species periodic orbits in the restricted problem ($\mu = 0$). Acta Astronautica **4**, 1019–1039.
Mihalas, D., Routly, P. M. (1968): Galactic Astronomy. Freeman, San Francisco.
Perko, L. M. (1965): Asymptotic matching in the restricted three-body problem, Ph. D. Thesis, University Microfilms, Ann Arbor, Michigan.
Perko, L. M. (1976b): Second species periodic solutions with an $O(\mu)$ near-Moon passage. Celest. Mech. **14**, 395–427.
Perko, L. M. (1977a): Second species solutions with an $O(\mu^\nu)$, $\frac{1}{3} < \nu < 1$, near-Moon passage. Celest. Mech. **16**, 275–290.
Perko, L. M. (1981a): Periodic orbits in the restricted problem: an analysis in the neighbourhood of a 1st species-2nd species bifurcation. SIAM J. Appl. Math. **41**, 181–202.
Perko, L. M. (1981b): Second species solutions with an $O(\mu^\nu)$, $0 < \nu < 1$, near-Moon passage. Celest. Mech. **24**, 155–171.

Lecture Notes in Physics

For information about Vols. 1–533
please contact your bookseller or Springer-Verlag

Vol. 534: F. Moreno, F. González (Eds.), Light Scattering from Microstructures. Proceedings, 1998. XII, 300 pages. 2000

Vol. 535: H. Dreyssé (Ed.), Electronic Structure and Physical Properties of Solids: The Uses of the LMTO Method. Proceedings, 1998. XIV, 458 pages. 2000.

Vol. 536: T. Passot, P.-L. Sulem (Eds.), Nonlinear MHD Waves and Turbulence. Proceedings, 1998. X, 385 pages. 1999.

Vol. 537: S. Cotsakis, G. W. Gibbons (Eds.), Mathematical and Quantum Aspects of Relativity and Cosmology. Proceedings, 1998. XII, 251 pages. 1999.

Vol. 538: Ph. Blanchard, D. Giulini, E. Joos, C. Kiefer, I.-O. Stamatescu (Eds.), Decoherence: Theoretical, Experimental, and Conceptual Problems. Proceedings, 1998. XII, 345 pages. 2000.

Vol. 539: A. Borowiec, W. Cegła, B. Jancewicz, W. Karwowski (Eds.), Theoretical Physics. Fin de Siècle. Proceedings, 1998. XX, 319 pages. 2000.

Vol. 540: B. G. Schmidt (Ed.), Einstein's Field Equations and Their Physical Implications. Selected Essays. 1999. XIII, 429 pages. 2000.

Vol. 541: J. Kowalski-Glikman (Ed.), Towards Quantum Gravity. Proceedings, 1999. XII, 376 pages. 2000.

Vol. 542: P. L. Christiansen, M. P. Sørensen, A. C. Scott (Eds.), Nonlinear Science at the Dawn of the 21st Century. Proceedings, 1998. XXVI, 458 pages. 2000.

Vol. 543: H. Gausterer, H. Grosse, L. Pittner (Eds.), Geometry and Quantum Physics. Proceedings, 1999. VIII, 408 pages. 2000.

Vol. 544: T. Brandes (Ed.), Low-Dimensional Systems. Interactions and Transport Properties. Proceedings, 1999. VIII, 219 pages. 2000

Vol. 545: J. Klamut, B. W. Veal, B. M. Dabrowski, P. W. Klamut, M. Kazimierski (Eds.), New Developments in High-Temperature Superconductivity. Proceedings, 1998. VIII, 275 pages. 2000.

Vol. 546: G. Grindhammer, B. A. Kniehl, G. Kramer (Eds.), New Trends in HERA Physics 1999. Proceedings, 1999. XIV, 460 pages. 2000.

Vol. 547: D. Reguera, G. Platero, L.L. Bonilla, J.M. Rubí(Eds.), Statistical and Dynamical Aspects of Mesoscopic Systems. Proceedings, 1999. XII, 357 pages. 2000.

Vol. 548: D. Lemke, M. Stickel, K. Wilke (Eds.), ISO Surveys of a Dusty Universe. Proceedings, 1999. XIV, 432 pages. 2000.

Vol. 549: C. Egbers, G. Pfister (Eds.), Physics of Rotating Fluids. Selected Topics, 1999. XVIII, 437 pages. 2000.

Vol. 550: M. Planat (Ed.), Noise, Oscillators and Algebraic Randomness. Proceedings, 1999. VIII, 417 pages. 2000.

Vol. 551: B. Brogliato (Ed.), Impacts in Mechanical Systems. Analysis and Modelling. Lectures, 1999. IX, 273 pages. 2000.

Vol. 552: Z. Chen, R. E. Ewing, Z.-C. Shi (Eds.), Numerical Treatment of Multiphase Flows in Porous Media. Proceedings, 1999. XXI, 445 pages. 2000.

Vol. 553: J.-P. Rozelot, L. Klein, J.-C. Vial Eds.), Transport of Energy Conversion in the Heliosphere. Proceedings, 1998. IX, 214 pages. 2000.

Vol. 554: K. R. Mecke, D. Stoyan (Eds.), Statistical Physics and Spatial Statistics. The Art of Analyzing and Modeling Spatial Structures and Pattern Formation. Proceedings, 1999. XII, 415 pages. 2000.

Vol. 555: A. Maurel, P. Petitjeans (Eds.), Vortex Structure and Dynamics. Proceedings, 1999. XII, 319 pages. 2000.

Vol. 556: D. Page, J. G. Hirsch (Eds.), From the Sun to the Great Attractor. X, 330 pages. 2000.

Vol. 557: J. A. Freund, T. Pöschel (Eds.), Stochastic Processes in Physics, Chemistry, and Biology. X, 330 pages. 2000.

Vol. 558: P. Breitenlohner, D. Maison (Eds.), Quantum Field Theory. Proceedings, 1998. VIII, 323 pages. 2000

Vol. 559: H.-P. Breuer, F. Petruccione (Eds.), Relativistic Quantum Measurement and Decoherence. Proceedings, 1999. X, 140 pages. 2000.

Vol. 560: S. Abe, Y. Okamoto (Eds.), Nonextensive Statistical Mechanics and Its Applications. IX, 272 pages. 2001.

Vol. 561: H. J. Carmichael, R. J. Glauber, M. O. Scully (Eds.), Directions in Quantum Optics. XVII, 369 pages. 2001.

Vol. 562: C. Lämmerzahl, C. W. F. Everitt, F. W. Hehl (Eds.), Gyros, Clocks, Interferometers...: Testing Relativistic Gravity in Space. XVII,507 pages. 2001.

Vol. 564: T. Pöschel, S. Luding (Eds.), Granular Gases. VIII, 457 pages. 2001.

Vol. 566: J. L. Lumley (Ed.), Fluid Mechanics and the Environment: Dynamical Approaches. VIII, 412 pages. 2001.

Vol. 568: P. A. Vermeer, S. Diebels, W. Ehlers, H. J. Herrmann, S. Luding, E. Ramm (Eds.), Continuous and Discontinuous Modelling of Cohesive-Frictional Materials. XIV, 307 pages. 2001.

Vol. 569: M. Ziese, M. J. Thornton (Eds.), Spin Electronics. XVIII, 493 pages. 2001.

Vol. 570: S. G. Karshenboim, F. S. Pavone, F. Bassani, M. Inguscio, T. W. Hänsch (Eds.), The Hydrogen Atom: Precision Physics of Simple Atomic Systems. XXIV, 291 pages. 2001.

Monographs

For information about Vols. 1–23
please contact your bookseller or Springer-Verlag

Vol. m 24: A. Ern, V. Giovangigli, Multicomponent Transport Algorithms. XIV, 427 pages. 1994.

Vol. m 25: A. V. Bogdanov, G. V. Dubrovskiy, M. P. Krutikov, D. V. Kulginov, V. M. Strelchenya, Interaction of Gases with Surfaces. XIV, 132 pages. 1995.

Vol. m 26: M. Dineykhan, G. V. Efimov, G. Ganbold, S. N. Nedelko, Oscillator Representation in Quantum Physics. IX, 279 pages. 1995.

Vol. m 27: J. T. Ottesen, Infinite Dimensional Groups and Algebras in Quantum Physics. IX, 218 pages. 1995.

Vol. m 28: O. Piguet, S. P. Sorella, Algebraic Renormalization. IX, 134 pages. 1995.

Vol. m 29: C. Bendjaballah, Introduction to Photon Communication. VII, 193 pages. 1995.

Vol. m 30: A. J. Greer, W. J. Kossler, Low Magnetic Fields in Anisotropic Superconductors. VII, 161 pages. 1995.

Vol. m 31 (Corr. Second Printing): P. Busch, M. Grabowski, P.J. Lahti, Operational Quantum Physics. XII, 230 pages. 1997.

Vol. m 32: L. de Broglie, Diverses questions de mécanique et de thermodynamique classiques et relativistes. XII, 198 pages. 1995.

Vol. m 33: R. Alkofer, H. Reinhardt, Chiral Quark Dynamics. VIII, 115 pages. 1995.

Vol. m 34: R. Jost, Das Märchen vom Elfenbeinernen Turm. VIII, 286 pages. 1995.

Vol. m 35: E. Elizalde, Ten Physical Applications of Spectral Zeta Functions. XIV, 224 pages. 1995.

Vol. m 36: G. Dunne, Self-Dual Chern-Simons Theories. X, 217 pages. 1995.

Vol. m 37: S. Childress, A.D. Gilbert, Stretch, Twist, Fold: The Fast Dynamo. XI, 406 pages. 1995.

Vol. m 38: J. González, M. A. Martín-Delgado, G. Sierra, A. H. Vozmediano, Quantum Electron Liquids and High-Tc Superconductivity. X, 299 pages. 1995.

Vol. m 39: L. Pittner, Algebraic Foundations of Non-Com-mutative Differential Geometry and Quantum Groups. XII, 469 pages. 1996.

Vol. m 40: H.-J. Borchers, Translation Group and Particle Representations in Quantum Field Theory. VII, 131 pages. 1996.

Vol. m 41: B. K. Chakrabarti, A. Dutta, P. Sen, Quantum Ising Phases and Transitions in Transverse Ising Models. X, 204 pages. 1996.

Vol. m 42: P. Bouwknegt, J. McCarthy, K. Pilch, The W3 Algebra. Modules, Semi-infinite Cohomology and BV Algebras. XI, 204 pages. 1996.

Vol. m 43: M. Schottenloher, A Mathematical Introduction to Conformal Field Theory. VIII, 142 pages. 1997.

Vol. m 44: A. Bach, Indistinguishable Classical Particles. VIII, 157 pages. 1997.

Vol. m 45: M. Ferrari, V. T. Granik, A. Imam, J. C. Nadeau (Eds.), Advances in Doublet Mechanics. XVI, 214 pages. 1997.

Vol. m 46: M. Camenzind, Les noyaux actifs de galaxies. XVIII, 218 pages. 1997.

Vol. m 47: L. M. Zubov, Nonlinear Theory of Dislocations and Disclinations in Elastic Body. VI, 205 pages. 1997.

Vol. m 48: P. Kopietz, Bosonization of Interacting Fermions in Arbitrary Dimensions. XII, 259 pages. 1997.

Vol. m 49: M. Zak, J. B. Zbilut, R. E. Meyers, From Instability to Intelligence. Complexity and Predictability in Nonlinear Dynamics. XIV, 552 pages. 1997.

Vol. m 50: J. Ambjørn, M. Carfora, A. Marzuoli, The Geometry of Dynamical Triangulations. VI, 197 pages. 1997.

Vol. m 51: G. Landi, An Introduction to Noncommutative Spaces and Their Geometries. XI, 200 pages. 1997.

Vol. m 52: M. Hénon, Generating Families in the Restricted Three-Body Problem. XI, 278 pages. 1997.

Vol. m 53: M. Gad-el-Hak, A. Pollard, J.-P. Bonnet (Eds.), Flow Control. Fundamentals and Practices. XII, 527 pages. 1998.

Vol. m 54: Y. Suzuki, K. Varga, Stochastic Variational Approach to Quantum-Mechanical Few-Body Problems. XIV, 324 pages. 1998.

Vol. m 55: F. Busse, S. C. Müller, Evolution of Spontaneous Structures in Dissipative Continuous Systems. X, 559 pages. 1998.

Vol. m 56: R. Haussmann, Self-consistent Quantum Field Theory and Bosonization for Strongly Correlated Electron Systems. VIII, 173 pages. 1999.

Vol. m 57: G. Cicogna, G. Gaeta, Symmetry and Perturbation Theory in Nonlinear Dynamics. XI, 208 pages. 1999.

Vol. m 58: J. Daillant, A. Gibaud (Eds.), X-Ray and Neutron Reflectivity: Principles and Applications. XVIII, 331 pages. 1999.

Vol. m 59: M. Kriele, Spacetime. Foundations of General Relativity and Differential Geometry. XV, 432 pages. 1999.

Vol. m 60: J. T. Londergan, J. P. Carini, D. P. Murdock, Binding and Scattering in Two-Dimensional Systems. Applications to Quantum Wires, Waveguides and Photonic Crystals. X, 222 pages. 1999.

Vol. m 61: V. Perlick, Ray Optics, Fermat's Principle, and Applications to General Relativity. X, 220 pages. 2000.

Vol. m 62: J. Berger, J. Rubinstein, Connectivity and Superconductivity. XI, 246 pages. 2000.

Vol. m 63: R. J. Szabo, Ray Optics, Equivariant Cohomology and Localization of Path Integrals. XII, 315 pages. 2000.

Vol. m 64: I. G. Avramidi, Heat Kernel and Quantum Gravity. X, 143 pages. 2000.

Vol. m 65: M. Hénon, Generating Families in the Restricted Three-Body Problem. Quantitative Study of Bifurcations. XII, 301 pages. 2001.

Vol. m 66: F. Calogero, Classical Many-Body Problems Amenable to Exact Treatments. XIX, 749 pages. 2001.

GPSR Compliance

The European Union's (EU) General Product Safety Regulation (GPSR) is a set of rules that requires consumer products to be safe and our obligations to ensure this.

If you have any concerns about our products, you can contact us on ProductSafety@springernature.com

In case Publisher is established outside the EU, the EU authorized representative is:

Springer Nature Customer Service Center GmbH
Europaplatz 3
69115 Heidelberg, Germany

Batch number: 10584501

Printed by Printforce, the Netherlands